Design of
reinforced and prestressed masonry

Design of
reinforced and prestressed masonry

W. G. CURTIN, CEng, FICE, FIStructE, MConsE, PhD, MEng

G. SHAW, CEng, FIStructE, MConsE

J. K. BECK, CEng, MIStructE

Curtins Consulting Engineers

Thomas Telford, London

Published by Thomas Telford Ltd, Thomas Telford House, 1 Heron Quay, London E14 9XF

First published 1988

British Library Cataloguing in Publication Data
Curtin, W. G. (William George)
    Design of reinforced and prestressed masonry.
    1. Construction materials: Reinforced masonry & prestressed masonry
    I. Title   II. Shaw, G.   III. Beck, J. K.
    624.1'83

ISBN: 0 7277 1314 0

Typeset in Great Britain by Santype International Limited, Salisbury, Wiltshire
Printed and bound in Great Britain by Redwood Burn Ltd, Trowbridge, Wiltshire

Preface

Our introduction to reinforced masonry was not happy. At the end of the Second World War there was a severe shortage of structural steelwork and many local contractors in the North West were unskilled in reinforced concrete construction. They tried to solve the problem by using reinforced brickwork for factories and similar buildings—but gave no cover to the reinforcement apart from the brickwork. By the late 1950s many of the structures were showing signs of distress, such as corrosion of reinforcement, splitting and spalling brickwork and near structural collapse. At the same time there was no famine of structural steelwork or shortage of contractors skilled in concrete construction and so reinforced brickwork practically died out.

In the early 1960s, we became interested in structural brickwork (there was little blockwork available and what there was lacked quality and reliability) and subsequently reinforced brickwork.

One of the earliest projects was a church in North Wales where the architect requested tall, slender, brick columns which had to be reinforced to resist the bending stresses caused by wind loading. The success of the project was encouraging and stimulated us to other applications. Then by luck we solved a difficult problem in structural masonry by prestressing it. This too was successful and we began to develop and exploit the potential of both reinforced and prestressed brickwork through the 1960s and early 1970s. By this time concrete blockwork was improving rapidly in strength and reliability and it was not long before we began reinforcing hollow blockwork.

Our increasing experience and applications led, naturally, to further developments and we quickly appreciated that reinforced and prestressed masonry had high potential, low cost and good buildability. The potential for brickwork and for blockwork has now advanced with economic benefits to construction.

There was no severe difficulty in applying the principles of reinforced and prestressed concrete to masonry (brick and blockwork). However, there was some difficulty, in the early days, in convincing new graduates that masonry was an engineering material and not an old-fashioned building material fit only for housing or cladding. That misconception has long gone from our practice and to help others we decided to write this book.

The book is a joint effort and we hope it passes on our experience in design, detailing and construction. We are, as ever, grateful for the help and encouragement from our enthusiastic staff, and thank particularly Paul Bolton, Geoff Othick, Steve Hunt and Karl Nesheim for their many suggestions, criticism and advice. We appreciate Marion McNally's and Pamela Jackson's patience, care and interest (and tolerance) in typing the manuscript.

We realise that it is likely that some engineers will disagree with our interpretation of the codes, our methods, priorities and so on, just as we argued, debated and disagreed among ourselves and with our staff. This is all to the good for it makes for progress—the conformist innovates nothing! The manual is, however, a reasonable consensus of our opinion and our staff have found it useful.

We have worked whenever possible with BS 5628 as it has become a mandatory document. Where, in our opinion, there are inadequacies in or omissions from the code we have suggested methods which have been acceptable to building control authorities. Also, BS 5628: Part 2 is the first European code dealing with this technique and could possibly form a basis for a Eurocode. American codes on reinforced masonry suffer from the disadvantage of using Imperial measures and working stress, not SI units and limit state philosophy. We have tried hard to keep the manual straightforward. We would welcome criticisms, corrections, suggestions from engineers (and other professions), and readers and hope that they, too, will find it useful.

W. G. Curtin
G. Shaw
J. K. Beck

Acknowledgments

Over the past twenty-five years we have been very fortunate in working with patient architects, tolerant contractors and trusting clients, particularly the public authority and large private corporation engineer clients. From discussions at design team meetings, out on the site and in the drawing office, we learnt much and received much freely given advice, suggestions and help. We appreciate the friendly encouragement of many engineers in both the private and public sector. We are particularly grateful to those contractors who assisted in our site tests at either no cost or minimal charge and also for their interest in the techniques. To list all those who helped would be impossible but we are very grateful to them all.

We are happy to give credit to those organisations for permission to quote from their publications or for the supply of information. A number of individuals were exceptionally helpful in the writing of this book. Some of the organisations and individuals are listed alphabetically below—we apologise if there are any omissions from the list.

> British Standards Institution
> Building Research Establishment
> Cement and Concrete Association
> Institution of Civil Engineers
> Institution of Structural Engineers
> Adrian Bell, University of Manchester Institute of Science and Technology, for computer work on the column design graphs
> Rodney Bradshaw, Bradshaw, Buckton & Tonge, for comments on composite action
> Bill Sharp, County Structural Engineer, Lancashire County Council, for help on the manuscript.

Material from BS 5628 is included by permission of the British Standards Institution, from whom complete copies can be obtained.

Any shortcomings, of any sort, in this book are entirely our responsibility and in no way the responsibility of those listed.

Contents

Notation

A_{m}	cross-sectional area of primary reinforcing steel (in mm^2)
A_{s1}	area of compression reinforcement in the most compressed face (in mm^2)
A_{s2}	area of reinforcement in the least compressed face (in mm^2)
A_{sv}	cross-sectional area of reinforcing steel resisting shear forces (in mm^2)
a	shear span (in mm)
a_{v}	distance from face of support to the nearest edge of a principal load (in mm)
B_{r}	diaphragm wall cross-rib spacing (in mm)
b	width of section (in mm)
b_{c}	width of compression face midway between restraints (in mm)
C	compressive force (in N)
C_{s}	average stress concentration factor
c	elastic contraction of masonry (in mm)
D	overall depth of diaphragm wall (in mm)
d	effective depth (in mm)
d_{c}	depth of masonry in compression (in mm)
d_1	depth from the surface to the reinforcement in the more highly compressed face (in mm)
d_2	depth of the centroid of the reinforcement from the least compressed face (in mm)
E_{c}	modulus of elasticity of concrete (in kN/mm^2)
E_{m}	modulus of elasticity of masonry (in kN/mm^2)
E_{n}	nominal earth or water load (in N)
E_{s}	modulus of elasticity of steel (in kN/mm^2)
e	eccentricity of post-tensioning force (in mm)
e_{a}	additional eccentricity due to deflection (in mm)
e_{x}	resultant eccentricity in plane of bending (in mm)
F_{Bc}	compressive force in masonry (in N)
F_{bst}	tensile bursting force (in N)
F_{pu}	characteristic tensile strength of prestressing steel (in N)
F_{s}	force in reinforcement in opposite face to F_{sc} (in N)
F_{sc}	compressive force in reinforcement (in N)
F_{vs}	composite action—variable stress concentration factor
f_{a}	composite action—basic allowable stress (in N/mm^2)
f_{b}	characteristic anchorage bond strength between mortar or concrete infill and steel (in N/mm^2)
f_{bs}	characteristic local bond strength (in N/mm^2)
f_{c}	composite action—maximum allowable stress
f_{cl}	strength of concrete at transfer (in N/mm^2)
f_{f}	increased value of f_{k} (in N/mm^2)
f_{k}	characteristic compressive strength of masonry (in N/mm^2)
f_{ki}	flexural compressive stress before losses (in N/mm^2)
f_{kx}	characteristic flexural strength (tension) of masonry (in N/mm^2)

f_s	stress in the reinforcement (in N/mm²)
f_{s1}	stress in the reinforcement in the least compressed face (in N/mm²)
f_{s2}	stress in the reinforcement in the most compressed face (in N/mm²)
f_{sa}	composite action—allowable steel stress
f_t	principal tensile stress (in N/mm²)
f_{t1}	compressive stress in face 1 (in N/mm²)
f_{t2}	compressive stress in face 2 (in N/mm²)
f_{uac}	compressive stress due to direct compression (in N/mm²)
f_{ubc}	compressive stress due to flexure (in N/mm²)
f_{udl}	composite action—uniformly distributed compressive stress (in N/mm²)
f_v	characteristic shear strength of masonry (in N/mm²)
f_w	composite action—average wall stress (in N/mm²)
f_y	characteristic tensile strength of reinforcing steel (in N/mm²)
f_{yr}	characteristic strength of shear reinforcement (in N/mm²)
G_k	characteristic dead load (in N)
g_B	design load per unit area due to loads acting at right angles to the bed joints (in N/mm²)
H	composite action—horizontal thrust on beam
h_{ef}	effective height of wall or column (in mm)
I	second moment of area (in mm⁴)
K_t	coefficient to allow for type of prestressing tendon
L	length of wall (in mm) and span of beam
L_{ef}	effective length of wall (in mm)
l_t	transmission length (in mm)
M	bending moment due to design load (in N mm)
M_a	increase in moment due to slenderness (in N mm)
M_b	bending moment at base of wall (in N mm)
M_{BC}	moment of resistance of reinforced masonry (in N mm)
M_d	design moment of resistance (in N mm)
M_w	bending moment in height of wall (in N mm)
MR_s	moment of resistance (cracked section) (in N mm)
N	design axial load (in N)
N_d	design axial load resistance (in N)
N_{dz}	design axial load resistance of column, ignoring all bending (in N)
n_w	design load per unit length (in N/m)
P	prestressing force (in N)
P_k	characteristic post-tensioning force (in N)
P_t	post-tensioning force (in N)
p_{ubc}	ultimate compressive flexural stress (in M/mm²)
Q	moment of resistance factor (in N/mm²)
Q_{ca}	composite action—bending moment factor
Q_k	characteristic imposed load (in N)
q	overall section dimension in direction perpendicular to y axis (in mm)
R	composite action—reduction factor
r	A_s/bt
S_v	actual spacing of links (in mm)
SR	slenderness ratio
T	flange thickness of diaphragm wall (in mm)
s_v	spacing of shear reinforcement along member (in mm)
t	overall thickness of wall or column (in mm)
t_{ef}	effective thickness of wall or column (in mm)
t_r	cross-rib thickness (in mm)
V	shear force due to design loads (in N)
v_h	shear stress (in N/mm²)

W_k	characteristic wind load (in N)
W_s	width of stress block (in mm)
$\bar{y}$	distance from neutral axis to centroid of area (in mm)
Z	section modulus (in mm^3)
z	lever arm (in mm)
β	capacity reduction factor allowing for effects of slenderness and eccentricity
γ_f	partial safety factor for load
γ_m	partial safety factor for material
γ_{mb}	partial safety factor for bond strength between mortar or concrete infill and steel
γ_{mm}	partial safety factor for compressive strength of masonry
γ_{ms}	partial safety factor for strength of steel
γ_{mv}	partial safety factor for shear strength of masonry
ε	strain in reinforcement
Σ_u	sum of perimeters of reinforcing bars
σ	stress (principal)
τ	shear stress principal

1

Introduction and general considerations

1.1. Introduction

Brickwork and blockwork, like concrete, are strong in compression but very weak in tension, i.e. their bending tensile strength is often less than 5% of their compressive strength. However, again like concrete, they can be reinforced to carry the tensile stresses or prestressed to eliminate them.

Normal plain structural masonry is an economic, fast and simple technique for producing durable, attractive structures; reinforcing and prestressing masonry widens the application, increases the potential and improves the cost-competitiveness. Brickwork should not be just a mere cladding or housing material, nor are blocks just a cheap substitute for bricks—both materials have high structural potential.

Engineers have been and are still conditioned to think in terms of framed construction for buildings. The frame is then enclosed with weather-resistant walls, other non-loadbearing walls are added to enclose staircases and lifts, provide corridors, subdivide the space, etc. In many cases if the walls are designed first, as loadbearing structural elements, then the need for the frame is eliminated, with obvious savings in cost, time and complications. Frequently retaining walls, tanks and silos are automatically designed in reinforced concrete and the alternatives are not considered. Many other examples could be cited where engineers have not considered or even appreciated the possibilities of these so-called 'new' materials. This book, it is hoped, will redress the balance.

The bulk of masonry construction, and therefore research, has been hidebound by the solid square and rectangular shapes of the past—and this is the only section type the codes deal with. However, competent and innovative designers will apply the modern developments in other structural materials to masonry. Typical examples are steel hollow-box girders and tubular sections, plywood box-beams, concrete T-beams and folded slabs, timber diaphragm plates and shells and similar advances. The main drawbacks to such geometric sections are the high cost of shuttering in reinforced concrete and the fabrication costs in steelwork. In many cases the cost of geometric sections is higher in masonry than the cost of solid rectangular sections but the extra-over costs do not tend to be nearly so high as in reinforced concrete and structural steel. The vastly improved structural efficiency usually compensates for the increased cost.

1.2. Changes in material behaviour

Reinforcing and prestressing masonry not only improves its bending resistance and increases both its vertical and lateral load-bearing capacity: it also changes the material behaviour from brittle to ductile. Thus it can withstand increased strain, can bend more without cracking and has a higher principal tensile strength. Plain masonry, once cracked under load, usually remains cracked after removal of the load. In reinforced and particularly prestressed masonry, cracks tend to close on removal of the load, thus helping to maintain its serviceability in respect of weather resistance and so on.

1.3. Development

Reinforced and prestressed brickwork are not new and daring techniques: in fact they predate reinforced concrete. For example, Brunel used the reinforcing technique on the shafts of the Rotherhithe Tunnel in the 1820s. Heated tie-rods were often fixed through Victorian brick structures requiring strengthening, which compressed them as the ties cooled. However, the advent of cast iron and then structural steelwork, and finally reinforced concrete saw the practical demise of traditional massive masonry structures.

Countries such as Japan and India, however, with limited resources of structural steel and concrete, made extensive use of reinforced brickwork between the First and Second World Wars and there has been some use of reinforced brickwork and blockwork since the Second World War in North America. Recently in Europe there has been increasing awareness of and interest in the technique. Modern advances in masonry design make it highly cost competitive and the resulting structures bear no resemblance to the traditional massive construction. Designers have, in effect, a new structural material.

1.4. Applications

The applications fall into two main groups

- to improve bending resistance
- to improve vertical loadbearing capacity.

There are many other applications such as

- improved shear and principal tensile stress resistance
- resistance to in-plane tensile stresses in walls subject to differential settlement
- improved resistance, if required, to racking shear in shear walls and torsion in such elements as columns supporting overhead cranes
- provision of resistance to accidental damage and impact
- considerable increase in ductility and stiffness.

1.5 Increased bending resistance: improvement of lateral load-bearing capacity

1.5.1. Reinforced and prestressed concrete block retaining walls

The hollow block forms, in effect, permanent precast shutters for reinforcement, or prestressing tendons, and grout (concrete infill) to make efficient retaining walls. There is little doubt that this technique will become widespread. In the following examples blocks can often be used instead of bricks.

1.5.2. Reinforced brick retaining walls

There are a number of types of reinforced brick retaining wall. The more common are now briefly described.

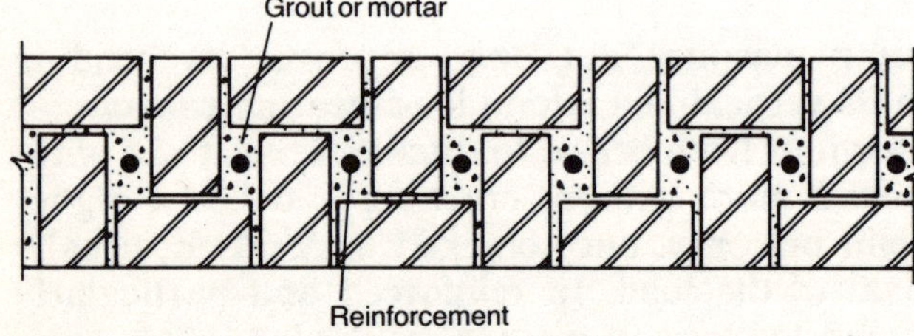

Fig. 1.1. Plan of Quetta bond wall

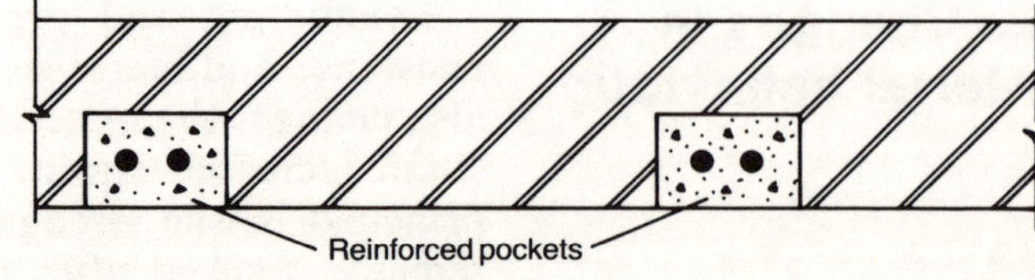

Fig. 1.2. Plan of pocket wall

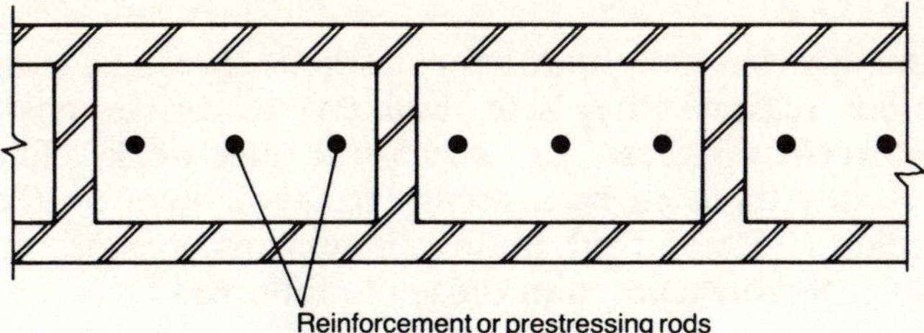

Fig. 1.3. *Plan of diaphragm wall*   Fig. 1.4. *Plan of fin wall*

1.5.2.1. Grouted cavity

This is the normal brick cavity wall with the cavity reinforced and grouted.

1.5.2.2. Quetta bond

Voids are formed by the brickwork bonding which is built around reinforcement. They are grouted or mortared up as the brickwork proceeds (see Fig. 1.1).

1.5.2.3. Pocket retaining walls

These are solid walls with closely spaced pockets which are later reinforced and concreted (see Fig. 1.2).

1.5.2.4. Diaphragm walls

Such walls are used for taller walls requiring increased lever arm. They are, basically, wide cavity walls, with bonded cross-ribs (see Fig. 1.3), and are built in blockwork as well as brickwork.

1.5.2.5. Fin walls

Fin walls are normal solid or cavity walls with deep narrow piers known as fins. They are also used for tall walls and act as vertical cantilever T-beams (see Fig. 1.4).

1.5.3. Post-tensioned retaining walls

The diaphragm, fin and other geometric sections, with their high Z/A ratio and radius of gyration, are ideal shapes for prestressing.

1.5.4. Wind-resistant walls

Walls of tall, single-storey structures for factories, warehouses, churches and similar buildings are rarely subject to appreciable direct compressive stresses from lightly loaded lightweight roofs. Their design is usually governed by the bending (flexural) tensile stresses caused by lateral wind pressure. Also, when the masonry is used as mere cladding the structural design of large panels is affected mainly by the flexural tensile stresses. Any excess tensile stresses can be catered for by reinforcing or prestressing. When diaphragm and fin walls are used structurally for tall single-storey structures their lateral resistance can be increased even further by reinforcing or prestressing.

1.5.5. Blast-resistant walls

As a result of the Flixborough explosion disaster it is now mandatory to provide blast-resistant walls in structures which may be subject to the effect of explosion. This provision is often met by reinforced masonry and could also be met by prestressing.

1.5.6. Beams

Although masonry beams are generally made for aesthetic, prestige, convenience and other similar reasons they are unlikely to prove cost-competitive against reinforced concrete or structural steelwork. The increased cost, however, usually tends to be a small extra-over item on the cost of the completed building. Reinforced hollow blockwork beams are generally cheaper and simpler to construct than those of reinforced brick.

1.5.7. Wall beams

Reinforcing the lower courses of walls has long been a method of converting the wall into a deep beam which is more resistant to differential settlement and also allows the wall to span openings.

1.5.8. Composite action

Walls can be built to act integrally with reinforced concrete strip footings to form the compression flange of a composite beam. Openings and damp-proof courses which could destroy the integral action must be carefully considered.

1.5.9. Accidental damage

Reinforcing parts of walls can increase their resistance to accidental damage.

1.6. Improved vertical loadbearing capacity

1.6.1. Columns: reinforced

In the same way that concrete columns can have increased loadbearing capacity by the addition of reinforcement, so too can masonry.

1.6.2. Columns: solid square and rectangular

These are the most common form of column (at the present) and the only type that the code deals with. They can be made in either brickwork or blockwork. For small square columns it has been found from experience that hollow blockwork is simpler to reinforce than brickwork, since it forms in effect a precast permanent shutter. For larger columns, of size greater than the largest hollow block, the designer should check the best application of either material.

1.6.3. Geometric columns

In nature the commonest strut is the hollow tube (stalks of grain, legs of seagulls and so on), which is highly structurally efficient because of its very high r/A ratio. Tubular columns can be made in masonry. They tend to be used mainly for aesthetic reasons because higher labour construction costs are likely. A compromise between structural efficiency and costs is to use the hollow-box column, where the void can be used to house services. Other geometric forms are discussed in sections 7.2 and 7.3.

1.6.4. Walls: rectangular

The cavity, Quetta bond and pocket wall can all be reinforced to improve their vertical loadbearing capacity.

1.6.5. Geometric walls

As well as diaphragms, fins and other geometric sections, the folded slab technique can also be applied in chevron walls (see Fig. 1.5). Such walls have a higher resistance, not only to bending but also to vertical loading, because of their increased section modulus and radius of gyration.

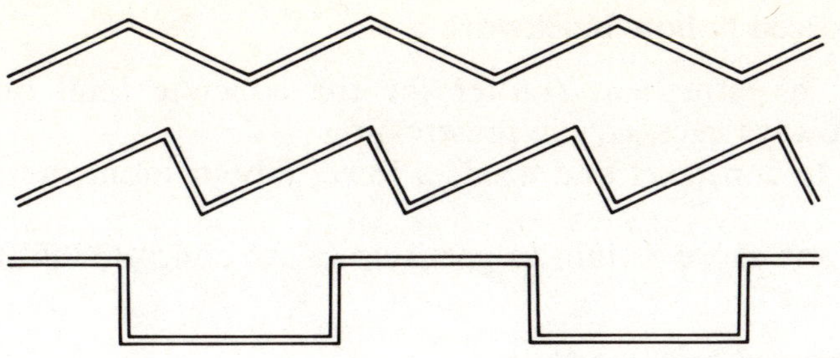

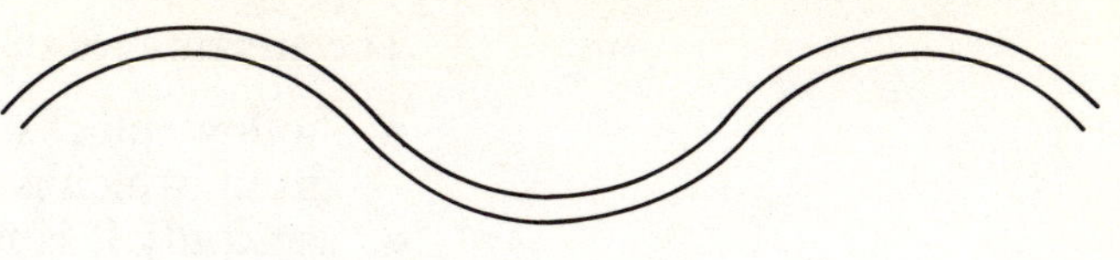

Fig. 1.5 (left). Chevron-type walls

Fig. 1.6 (above). Serpentine wall

Another example is the serpentine wall (see Fig. 1.6), which acts similarly to the chevron and is partially an application of shell and arch technology. The higher cost of the serpentine wall may be compensated for by its improved lateral load resistance and possible aesthetic appeal.

1.7. Improved vertical loadbearing capacity and bending resistance

1.7.1. Columns

Columns in modern masonry structures are more subject to both vertical load and bending than those in traditional massive construction. The bending action can be created either by high eccentricity of the vertical load or the application of lateral loading. The bending can be about one axis or both axes (i.e. biaxial).

The obvious solution is to use solid square or rectangular columns and, where applicable and with modifications, the approach used in reinforced concrete columns. However, consideration should be given to geometric sections, such as the cruciform column for biaxial bending.

1.7.2. Columns subject to vertical tension and bending

Columns in such situations as supports for tall single-storey structures with low pitch or flat roofs are subject to tension in strapping (tailing) down the roof to resist wind suction forces. The direct tensile strength of masonry should rarely be relied on since not only is it low but also it can be unreliable. The roof can, of course, be strapped down to the column but there are many situations where post-tensioned rods can anchor the roof to the column and down to the foundation if necessary (and thus provide improved bending resistance to lateral wind forces on the column).

1.7.3. Columns subject to vertical compressive load and torsion

Crane gantry columns are a typical example of combined vertical loading and torsion (caused not only by the crane load but also by surge and sway forces). This can be considered as a special case of biaxial bending and such columns can be prestressed or, more commonly, reinforced to withstand such action.

1.7.4. Walls subject to vertical load and bending

The cavity, Quetta bond and pocket wall can also be reinforced to improve their resistance to combined vertical loading and bending. However, consideration should be given to the application of geometric sections.

1.8. Whether to reinforce or to prestress

The decision as to whether to reinforce or prestress brickwork or blockwork will depend mainly on cost, whether bricks or blocks are the predominant material in the structure and occasionally for aesthetic reasons. A simple guide would be to reinforce hollow blockwork and prestress brickwork.

The advantages of reinforced hollow blockwork are

- hollow blocks form a permanent shutter for the concrete infill or grout, which is not always necessary in prestressing
- generally it is faster to construct and needs a lower labour input than brickwork
- it is usually simpler and more certain to grout up (place concrete infill) effectively.

The advantages of prestressed brickwork are

- bricks generally have much higher compressive strength than blocks
- it is simpler to form sections with a high Z/A ratio and radius of gyration in brickwork
- loss of prestress due to shrinkage is less with brickwork than with blockwork.

1.9. Economics

It is not possible to lay down hard and fast facts on costs because every job differs, but it is possible to state that reinforced or prestressed masonry is often more economical than are other structural alternatives. The best advice is to prepare a preliminary design for reinforced and prestressed masonry and cost it against the alternatives.

For example, although a reinforced concrete retaining wall is often cheaper than reinforced masonry, if the concrete has to have a ribbed or other expensive shutter treatment to improve its appearance (or is clad in brickwork to hide it) then reinforced brickwork may be cheaper. Another example is that of a tall single-storey shed structure subject to mining subsidence which can have its roof economically supported on a post-tensioned masonry diaphragm wall.

The techniques are simple and experience has shown that they are well within the capacity of small and relatively unsophisticated contractors. They make for rapid and economic construction and the resulting structures can be attractive and durable. In fact they are not just a substitute for reinforced or prestressed concrete, but a new structural alternative to structural steelwork, reinforced concrete and structural timber.

The economy results from simplicity, a reduction in the number of different trades, site operations and materials, the elimination, or reduction in number, of subcontractors and the employment of the main contractor's labour using a relatively cheap material.

1.10. Durability

There is understandable concern among engineers that, because masonry is porous, reinforcement or prestressing rods will be vulnerable to corrosion. However, reliance should never be placed on the masonry as cover to the reinforcement or prestressing. Reinforcing bars and prestressing rods, or tendons, are placed in cavities, voids or other openings which are then grouted or concreted up (or provided with some other form of corrosion protection).

The amount of cover provided by the grout, concrete or other means, depends on the degree of exposure of the structural element. A reinforced brick column placed internally in a centrally-heated sheltered building will not need the same amount of protection as an external column on a gale-swept mountain top. Alternatively, the use of stainless or other austenitic steel may be considered.

This topic and other matters related to durability are discussed in more detail in Appendix B.

1.11. Methods of reinforcing

Brickwork and blockwork is built first to form, in effect, a permanent shutter. Reinforcement is then placed in voids, pockets, cavities or other openings and grouted or concreted up so as to act integrally with the masonry. Sometimes, as in grouted cavity walls and similar constructions, the reinforcement is fixed first and then the masonry is built up around the reinforcement.

During construction, consideration should be given to the following points.

- The cavity, or other opening, should be kept clear of mortar droppings or other obstructions.
- The cavity, or other opening, should be fully grouted or concreted up. To ensure this, it may be necessary to leave vent openings to prevent air locks. The vent holes and cleaning out holes may need to be filled in as grout rises to them. The vent openings can also serve as an indicator that the grout has filled the cavity or other opening.
- The grout or concrete should be thoroughly compacted (by vibrators if necessary) to avoid honeycombing or other voids in the filling.
- A check on the quantity of concrete or grout filling should be made to assist in ensuring full filling.
- No grout or concrete should be allowed to leak down the face of fair-faced work.
- Small lifts may be necessary when filling narrow cavities.

The position and spacing of reinforcement, lapping, curtailment, provision for shear and so on is similar to reinforced concrete design and is dealt with more fully in chapter 3.

1.12. Methods of prestressing

As in prestressed concrete, there are two methods of prestressing: pretensioning and post-tensioning. In pretensioning the wires or strands are tensioned first, and then the masonry is built round the wires or strands. This is mainly a factory operation for manufacturing such members as prestressed masonry lintels or beams.

Most prestressed masonry is currently post-tensioned, i.e. the rods or strands are tensioned after the mortar has set and reached its designed strength.

In pretensioned masonry one end of the high tensile steel wire is attached to a fixed anchorage and the other to a movable anchor (see Fig. 1.7(a)). The movable anchor is jacked out to stretch the wires or strands. The masonry beams, lintels and so on are built around the stretched wire and when the masonry has reached its designed strength (i.e. the mortar and concrete infill strength required), the movable anchor is released and the tensioned wires

Fig. 1.7. Pre-tensioned masonry

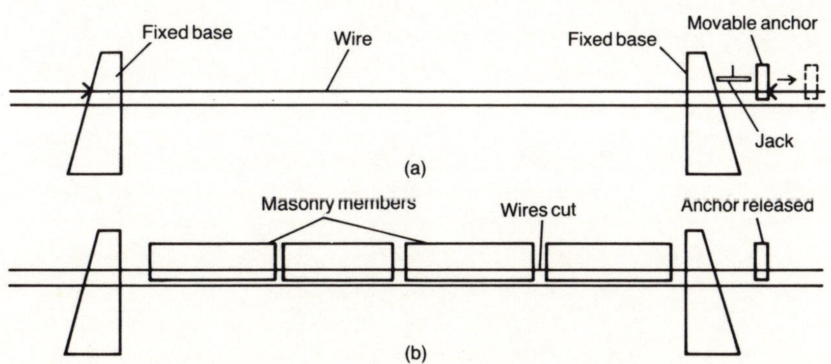

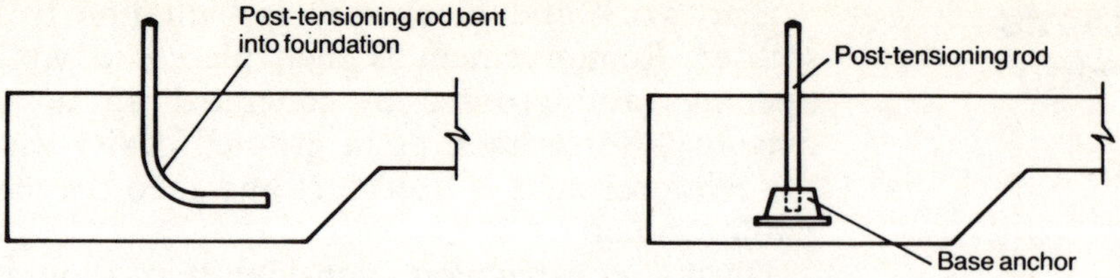

Fig. 1.8. Anchoring of base of post-tensioned wall

are cut between the ends of the separated masonry units (see Fig. 1.7(b)).

In post-tensioning, which is mainly used for walls, rods are more commonly used than wires or strands because of the site construction difficulties of keeping relatively flimsy wires or strands in position. The bottom end of the high tensile steel rod is either bent into the foundation or floor slab, or is screwed into a steel base plate anchor (see Fig. 1.8).

The masonry is then built up around the rods to the required height. An anchor plate is placed over the rod and bears on to the top of the masonry. The top of the rod is threaded to receive a nut. The rod is tensioned either by a torque spanner (for low levels of prestress) or hydraulic jack (for high levels of prestress). When the required extension has been provided the nut is firmly tightened down and locked when the torque spanner or jack is removed.

The subject is dealt with more fully in chapter 3.

1.13. General design theory

The design theory is similar to that of reinforced and prestressed concrete. The theory is dealt with in detail in chapters 3 and 6 and an outline is given here.

Reinforcement is added to carry the tensile stresses in the masonry or it is prestressed to eliminate them. The low tensile strength of masonry is ignored, as is the concrete tensile strength in reinforced concrete design.

The limit state design method is used to provide for an adequate margin of safety against the ultimate limit state being reached (collapse, overturning or buckling). Serviceability limit states such as deflection and cracking should be considered.

Fine hair cracking at the joints of reinforced masonry can occur and such cracking should be limited so as not to affect the durability, serviceability or appearance of the structure.

The designer must also consider the effects of accidental damage.

Basic design data

2.1. Limit state design

The well-known principles of limit state design philosophy, as applied to most structural materials including plain masonry, reinforced concrete and so on, are applied to reinforced and prestressed masonry. There are some necessary adjustments and additions (to plain masonry design) and these variations are discussed.

2.1.1. Ultimate limit state

The basic principle of ultimate limit state design philosophy is that under the most onerous design loading conditions the various design strengths of the materials comprising a structural element are greater than, or at least equal to, the design loadings. This is known as the ultimate limit state.

The design strength of an element is a function of the characteristic strength of the materials (say f_k) divided by the relevant partial factors of safety on the materials γ_{mm}.

The design load to be resisted is a function of the characteristic load (say F_k) multiplied by the partial factors of safety on the load γ_f.

Thus the aim of ultimate limit state design may be expressed by

$$f(f_k/\gamma_{mm}) \geqslant f(F_k \gamma_f) \tag{2.1}$$

where f is a mathematical function involving the symbols in parentheses.

The overall strength of a structure must be such that not only is its design strength adequate to resist the design loads but also it must be sufficient to ensure that the structure does not buckle, overturn, become unduly damaged by accidental forces or become unstable during or after construction.

2.1.2. Serviceability limit state

In addition the structure must remain serviceable, i.e. it must not suffer excessive deflection or cracking, it must be fatigue-resistant where necessary and be able to withstand the effect of fire. These are known as serviceability limit states. In design it is common for sections to be analysed for ultimate limit and checked for serviceability limit.

2.2. Serviceability limit states: deflection and cracking

2.2.1. Deflection

The deflection of a structure, or any part of it, should not adversely affect the performance of the structure or the applied finishes. It is practically impossible to eliminate all deflection on a loaded structure, but the deflection should not be excessive and suggested limits to deflection are given by

- span/250 for all elements (except cantilevers)
- span/125 for cantilevers
- span/500 or 20 mm, whichever is the lesser, when considering the effect on partitions or applied finishes
- span/300 upward deflection (camber) of prestressed masonry elements before application of finishes.

These limits include not only the effect of load but also creep, shrinkage and temperature effects.

The calculations for deflection are discussed in section 6.22.

2.2.2. Cracking

In the same way that a loaded reinforced concrete beam can develop fine cracks in the tensile zone, so too can reinforced masonry, particularly at the masonry unit/mortar interface. The cracking must not be of such a magnitude as to affect the durability or appearance of the structural element. The design procedures given in this book should prevent excessive cracking.

Cracking in masonry can occur due to various movements of the material (e.g. thermal, moisture, shrinkage, creep). These movements cannot be prevented and it is prudent to insert controlled cracks, known as movement joints. BS 5628: Part 3[1] gives some advice on the positioning and provision of such joints. Detailed advice is given in references 2 and 3.

Temporary cracking in cracked section analysis in prestressed masonry design (see chapter 7) is permissible but this is a special condition.

2.3. Stability

It is possible, particularly in hybrid structures (i.e. structural elements of differing structural materials), for all the elements to be structurally sound in themselves but to result in an unstable structure when connected. This can occur in structures where inadequate restraints and connections are provided, i.e. cross-wall structures lacking longitudinal bracing, single-storey gable walls without end returns or restraint at the top from roof fixings (e.g. see Fig. 2.1), and multi-storey columns inadequately restrained at floor levels by beams or slabs.

To ensure that the overall structure is both stable and robust it is vital to consider the layout of the structure on plan and in vertical sections. From this the interaction of the structural masonry elements and their interaction with other parts of the structure such as roofs, floors and foundations can be evaluated.

It is recommended that, in addition, designers should ensure the following points.

- The structure is at any level capable of resisting a uniformly distributed horizontal load equal to 1·5% of the total characteristic dead

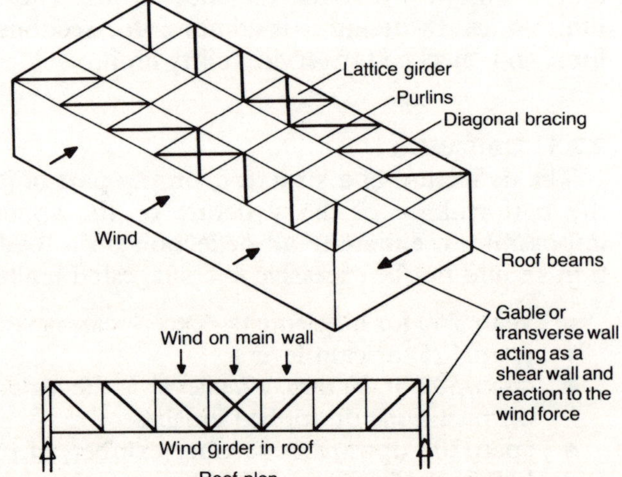

Fig. 2.1. Racking shear walls: single-storey structure

load above that level. This—the stability design load—may be distributed among the structural elements according to their stiffness, based on the element's second moment of area together with consideration of its slenderness ratio.

- There are robust connections between elements of the structure, particularly between masonry walls and columns and the connecting floors and roofs. Guidance on this is given in chapter 7 of reference 2 and in appendix C of BS 5628: Part 1.[1]

- Differences in behaviour (deflection, contraction and so on) in interconnecting structural elements made of different materials (steel, reinforced concrete and masonry) must be considered and accommodated.

- The insertion of damp-proof courses and movement joints shall not reduce the shear and bending strength of the structure below the level assumed in design. The shear and bending strengths, if any, of the materials used at these discontinuities should be determined.

2.3.1. Retaining structures—including earth, water and grain

In addition to resisting the bending, shear and other stresses and considerations discussed in section 2.4, in retaining and similar structures the element must have adequate connections, fully designed and detailed, to prevent overturning, bursting and sliding.

2.3.2. Accidental forces

In addition to design to resist the forces from normal design loading, the structure should be checked for its resistance to accidental damage, i.e. excessive loading causing damage or even destruction of an element. (Guidance on this is given in chapter 8 of reference 2 and in clause 37 of BS 5628: Part 1.[1])

2.3.3. Construction forces

It is the builder's contractural responsibility to maintain the safety of the work and to design and provide all necessary temporary propping, shoring, bracing and so on, but where unusual and special precautions are required it would be prudent for the designer to discuss these with the builder. No part of the structure should be over-loaded by temporary storage of building materials.

2.4. Loads

Characteristic loads should be based on collected data but although study is progressing on this topic there is not yet sufficient data to express loads in statistical terms. In normal design practice the values given in appropriate British Standards (which are similar to those used in other countries) are used as follows.

2.4.1. Characteristic dead load

The characteristic dead load G_k is the weight of the structure complete with finishes, fixtures and partitions. This is taken as being equal to the dead load as defined and calculated in accordance with BS 6399: Part 1[4] or other appropriate codes of practice.

2.4.2. Characteristic imposed load

The characteristic imposed load Q_k is imposed load as defined and calculated in BS 6399: Part 1[4] or other appropriate codes of practice.

2.4.3. Characteristic wind load

The characteristic wind load W_k is wind load as defined and calculated in chapter 5 of CP 3: Part 2.[5]

2.4.4. Nominal earth loads

The nominal earth loads E_n should be in acccordance with BS 8004[6] or other appropriate codes or practice.

2.4.5. Prestress force

The prestress force P_k is that force applied at the time being considered in design, i.e. before and after losses.

2.5. Structural properties of materials: masonry

2.5.1. Factors affecting characteristic compressive strength of masonry

The characteristic compressive strength of masonry f_k depends on

- the characteristic strength of the masonry unit (see section 2.5.1.1)
- the mortar designation (see section 2.5.1.2)
- the shape of the unit (see section 2.5.1.3)
- whether the work is bonded or unbonded (see section 2.5.1.4)
- the thickness of the mortar joints (see section 2.5.1.5)
- the direction of the compressive force in the member (see section 2.5.1.6)
- the standard of workmanship (see section 2.5.1.7).

2.5.1.1. Units

Typical characteristic strengths of masonry units are given in Table 2.1.

2.5.1.2. Mortar

The strength of the mortar affects the characteristic strength of the masonry. If all other factors are equal, the stronger the mortar the higher is the characteristic strength of the masonry.

The designated grades of mortar recommended for reinforced and prestressed masonry are given in Table 2.2.

The mortar strength should not be significantly stronger than the masonry unit.

2.5.1.3. Unit shape

Mortar joints are generally weaker than the masonry units so that the lower the number of mortar joints the higher is the strength of the constructed masonry. The higher the ratio of height to least horizontal dimension, the fewer is the number of mortar bed joints of the constructed masonry and thus the higher is its strength. This is the reason why large concrete blocks of relatively low strength can form masonry of the same

Table 2.1. Compressive strength of masonry units in N/mm^2

Bricks	Blocks
5·0	2·8*
10·0	3·5*
15·0	5·0*
20·0†	7·0*
27·5†	10·0*
35·0†	15·0*
50·0†	20·0*
70·0	35·0‡
100·0	50·0‡
	70·0‡

* Typical structural unit.
† Typical available value.
‡ Blocks of this strength may not be readily available.

Table 2.2. Mortar designations (table 1 of BS 5628[1])

Grade	Cement	Lime	Sand	Masonry cement	Sand	Sand with plasticiser
(i)	1	$0-\frac{1}{4}$	3	—	—	—
(ii)	1	$\frac{1}{2}$	$4-4\frac{1}{2}$	1	$2\frac{1}{2}-3\frac{1}{2}$	—
(ii)	—	—	—	1	$2\frac{1}{2}-3\frac{1}{2}$	—
(ii)	1	—	—	—	—	3–4

characteristic compressive strength as the smaller clay bricks of higher strength.

2.5.1.4. Bonding patterns

The characteristic strength values quoted in Tables 2.3–2.6 are for normally bonded masonry. Variations in established bond patterns (e.g. English and Flemish) have little comparative effect on the compressive strength of masonry. However, unbonded masonry (such as stack bonding where the perpends line up in a continuous vertical joint) should not be used in walls because it can reduce the compressive strength of the element.

Fig. 2.2. Characteristic compressive strength f_k of masonry: (a) masonry constructed with bricks or other units with a ratio of height to least horizontal dimension of 0·6; (b) block masonry constructed from solid blocks with a ratio of height to least horizontal dimension of 1·0; (c) block masonry constructed from solid blocks with a ratio of height to least horizontal dimension of 2·0–4·0; (d) block masonry constructed from structural units other than solid concrete blocks with a ratio of height to least horizontal dimension of 2·0–4·0

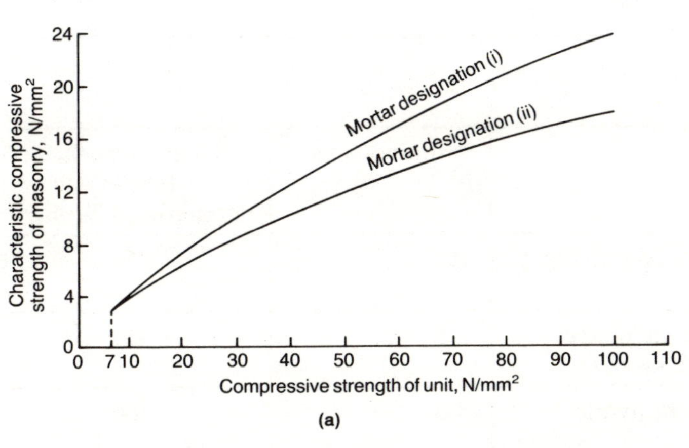

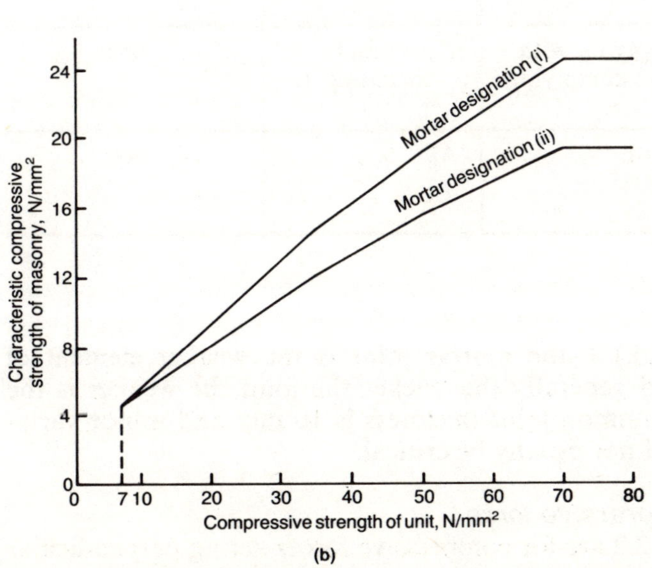

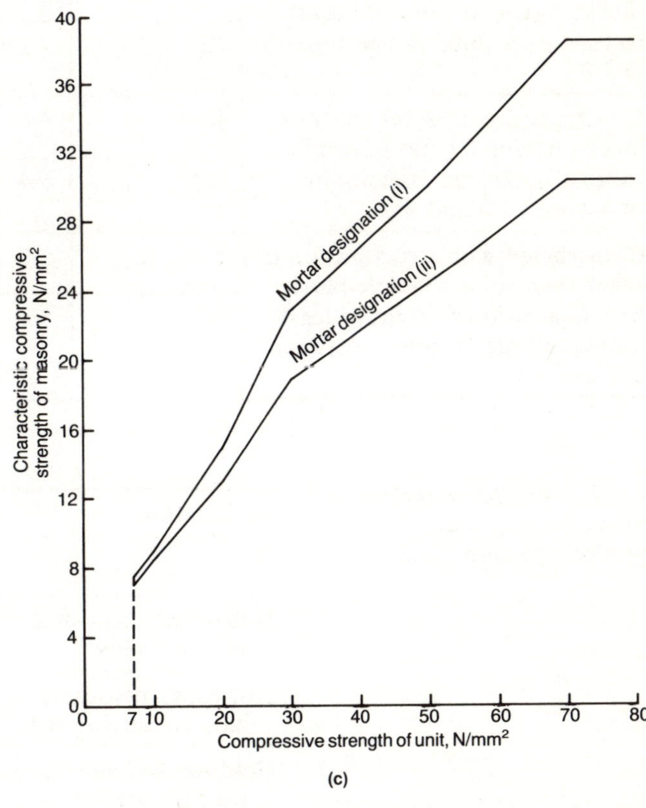

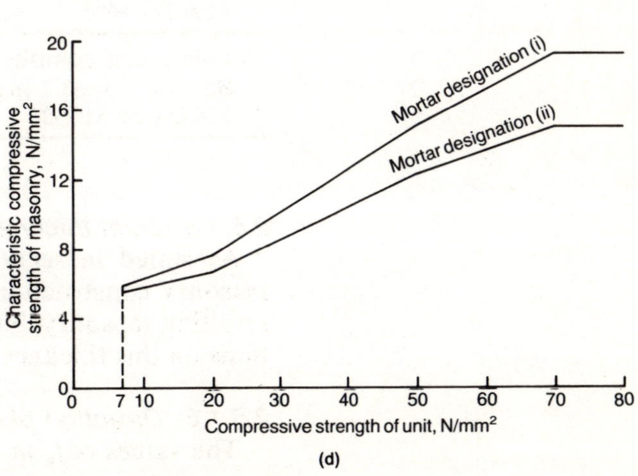

Table 2.3. Characteristic compressive strength f_k of masonry, N/mm²

Masonry type	Mortar designation	Compressive strength of unit, N/mm²								
		7	10	15	20	27·5	35	50	70	100
Constructed with bricks or other units having a ratio of height to least horizontal dimension of 0·6	(i)	3·4	4·4	6·0	7·4	9·2	11·4	15·0	19·2	24·0
	(ii)	3·2	4·2	5·3	6·4	7·9	9·4	12·2	15·1	18·2

Masonry type	Mortar designation	Compressive strength of unit, N/mm²						
		7	10	15	20	35	50	70 or greater
Constructed with solid concrete blocks having a ratio of height to least horizontal dimension of 1·0	(i)	4·4	5·7	7·7	9·5	14·7	19·3	24·7
	(ii)	4·1	5·4	6·8	8·2	12·1	15·7	19·4
Constructed with solid concrete blocks having a ratio of height to least horizontal dimension of between 2·0 and 4·0	(i)	6·8	8·8	12·0	14·8	22·8	30·0	38·4
	(ii)	6·4	8·4	10·6	12·8	18·8	24·4	30·2
Constructed with structural units other than solid concrete blocks having a ratio of height to least horizontal dimension of between 2·0 and 4·0	(i)	5·7	6·1	6·8	7·5	11·4	15·0	19·2
	(ii)	5·5	5·7	6·1	6·5	9·4	12·2	15·1

Table 2.4. Characteristic tensile strength of reinforcing steel f_y

Designation	Nominal size	Characteristic tensile strength f_y, N/mm²
Hot-rolled plain steel bars complying with BS 4449[10]	All	250
Hot-rolled deformed high yield steel bars complying with BS 4449[10]	All	460
Cold-worked steel bars complying with BS 4461[11]	All	460
Hard-drawn steel wire complying with BS 4482[28] and steel fabric complying with BS 4483[29]	Up to and including 12	485
Stainless steel complying with BS 970:[30] Part 1 grade 304S15, 316S31 or 316S33	All	460

2.5.1.5. Joint thickness

As stated in section 2.5.1.3, the mortar joint is the weaker element of masonry construction and generally the thicker the joint the weaker is the resulting masonry. The common joint thickness is 10 mm and minor variations on this thickness will not usually be critical.

2.5.1.6. Direction of compressive force

The values of f_k in Fig. 2.2 are for compressive forces acting perpendicular

Table 2.5. Dimensions and properties of hot-rolled and hot-rolled and processed high tensile alloy steel bars

Type of bar	Nominal size,* mm	Nominal tensile strength,* N/mm²	Surface	Nominal 0·1% proof stress,* N/mm²	Nominal cross-sectional area†		Nominal mass†		Tolerance		Specified properties			Maximum relaxation at 1000 h	
					Smooth bar, mm²	Ribbed bar, mm²	Smooth bar, kg/m	Ribbed bar, kg/m	On smooth bar dia.‡	On section and mass, %	Characteristic breaking load,§ kN	Characteristic 0·1% proof load, kN	Minimum elongation at fracture,** %	Initial load as % of breaking load	Value, %
Hot-rolled	20 25 32 40	1030	Smooth or ribbed	835	314 491 804 1257	349 538 874 1348	2·47 4·04 6·31 9·86	2·74 4·22 6·86 10·58	Range 0·6 mm for all sizes	Batch +4% −2% Individual bar +6% −2%	325 505 830 1300	260 410 670 1050	6	For all bars 60 70 80	For all bars 1·5 3·5 6·0
Hot-rolled and processed	20 25 32	1230	Smooth or ribbed	1080	314 491 804	349 538 874	2·47 4·04 6·31	2·74 4·22 6·86		+6% −2%	385 600 990	340 530 870	4		

* The nominal size of bar and nominal tensile strength data are given for designation purposes only.
† The nominal cross-section and nominal mass data are given for information only.
‡ The diameter tolerance does not apply to ribbed bars.
§ The load carried by the threaded portion of smooth bars is not to be less than the specified characteristic load.
¶ The modulus of elasticity may be taken as 206 ± 10 kN/mm² for as rolled and as rolled stretched and tempered bars; and 165 ± 12 kN/mm² for as rolled and stretched bars.
** The minimum elongation at fracture is measured on gauge length of $5·65\sqrt{S_0}$ where S_0 is the original cross-sectional area of the gauge length. If measurement of extension under load is possible, the minimum elongation at maximum load is to be 3·5%. The actual elongation at maximum load need not be measured.
†† The maximum standard deviations expressed as equivalent stress values of load for the various bar size are 55 N/mm² for nominal tensile strength and 60 N/mm² for nominal 0·1% proof stress.

Table 2.6. Dimensions of indented and crimped bars (see Fig. 2.7)

Nominal dimensions of indentations

Nominal wire diameter d, mm	Depth a, mm	Length, mm	Pitch, mm
5·0 and below	0·12 ± 0·05	3·5	5·5
Over 5·0	0·15 ± 0·05	5·0	8·0

Crimp dimensions

Type of crimp	Pitch	Total wave height (excluding the wire diameter	
		Helical	Uniplanar
Short pitch	$5d$ to $10d$	5%d to 10%d	10%d to 20%d
Long pitch	$8d$ to $12d$	6%d to 12%d	12%d to 25%d

Nominal internal diameter of coil

Nominal wire diameter d, mm	Coil diameter, m
4·0 and 4·5	1·25
5·0	1·5
6·0 and 7·0	2·0

to the bed face of the units. Typical of such action is an axially loaded column. In a beam where the compressive force may act parallel to the bed face there may be a need to adjust these values as discussed in section 2.5.5.

2.5.1.7. Standard of workmanship

Control of the unit manufacture, under factory conditions, is increasingly sophisticated, but site control of workmanship is more difficult. Account of this is taken in the partial factors of safety discussed in section 2.9.

2.5.2. Characteristic compressive strength of masonry

Details of the characteristic strengths of masonry of units of differing materials, strengths and shapes bedded in differing mortars are given in sections 2.5.3–2.5.7.

2.5.3. Characteristic strength of brickwork

Table 2.2 and Fig. 2.2(a) give the characteristic strength of brickwork f_k (loaded perpendicular to the bed face) of normally bonded brickwork constructed with standard format bricks or other structural masonry units with a ratio of height to least horizontal dimension of 0·6. This is known as the aspect ratio, i.e.

$$\frac{\text{height of unit}}{\text{least horizontal dimension of unit}} = \text{aspect ratio}$$

(Normal size fired-clay, calcium silicate and concrete bricks usually have an aspect ratio of 0·63. Fig. 2.2(a) is intended to apply to such units.) The values quoted are for masonry elements (walls) constructed with units laid in the normal manner, under laboratory conditions and tested at an age of 28 days under axial compression in such a manner that the effects of slenderness (see section 2.12) may be ignored.

2.5.4. Characteristic strength of blockwork f_k

When a wall, or other structural element, is constructed in blockwork the increased size of the units results in a fewer number of joints than in brickwork, and thus a higher compressive strength of the element for a given strength of unit and mortar. The compressive strength also depends on whether the units are solid or hollow.

2.5.4.1. Solid concrete blocks, aspect ratio of 1·0

Figure 2.2(b) applies to masonry built with solid concrete blocks with a ratio of height to least horizontal dimension of 1·0.

2.5.4.2. Solid concrete blocks, aspect ratio of 2·0–4·0

Figure 2.2(c) applies to masonry built with solid concrete blocks with a ratio of height to least horizontal dimension of between 2·0 and 4·0.

2.5.4.3. Hollow block masonry, aspect ratio of 2·0–4·0

Figure 2.2(d) applies to masonry built with structural units, other than solid concrete blocks (e.g. hollow blocks), with a ratio of height to least horizontal dimension of between 2·0 and 4·0.

2.5.4.4. Hollow block masonry, aspect ratio of 0·6–2·0

When masonry is built of hollow blocks having a ratio of height to least horizontal dimension of between 0·6 and 2·0, the value of f_k may be obtained by interpolation between the values given in Figs 2.2(b) and 2.2(d).

2.5.4.5. Solid block masonry, aspect ratio of 0·6–2·0

When masonry is built of solid concrete blocks having a ratio of height to least horizontal dimension of between 0·6 and 2·0, the value of f_k may be obtained by interpolation between the values given in Figs 2.2(b) and 2.2(c).

2.5.4.6. Hollow concrete block masonry filled with in situ concrete

When hollow concrete blocks are completely filled with in situ concrete, the value of f_k may be taken as for solid blocks, provided that

(*a*) the characteristic cube strength of the concrete infill is not less than minimum grade 25 for reinforced and post-tensioned masonry, or grade 40 for pretensioned masonry, or less than the compressive strength of blocks where block strength is based on their net area (see appendix C of BS 6073: Part 2[7])

(*b*) when the infill concrete strength is less than the block concrete strength the compressive strength of the masonry shall be based on solid blockwork of compressive strength equal to the cube strength of the infill concrete.

2.5.4.7. Natural stone, square dressed

Square dressed natural stone masonry is treated in the same way as solid concrete blockwork with the same value of f_k as the stonework. Natural stonework is rarely used structurally in developed countries but is used for aesthetic purposes. In developing countries with ample supplies of stone its structural use should be considered.

2.5.5. Characteristic compressive strength f_k with compressive force parallel to bed face

A compressive force parallel to the bed face of the unit can occur in the compression zone of loaded beams and can affect the value of f_k. The masonry is in effect stack bonded and is therefore likely to be weaker. However, the stresses are those due to bending and not direct compression and this tends to counteract this weakness.

The following values should be used

- for solid masonry, filled hollow blocks, frogged bricks where the frogs are fully filled and so on, the values in Figs 2.2(a)–(c)
- for cellular or perforated bricks one third of the values in Fig. 2.2(a)
- for unfilled hollow and cellular blocks the values obtained from tests or the value given in Table 2.3, using the strength of the block determined in the direction parallel to the bed face of the unit.

2.5.6. Characteristic compressive strength of masonry in bending

BS 5628: Part 2[1] recommends that the bending compressive strength be taken as equal to that in direct compression, and also recommends a reduced partial safety factor for materials when the element is bending. (Bending compressive strength is usually greater than direct compression, but the material remains unchanged, and so the Authors are of the opinion that it may be more logical to increase the characteristic compressive bending strength and use the same partial safety factor for masonry. This is discussed in detail in sections 2.9 and 3.2. It is suggested that designers should exercise their own judgement.)

2.5.6.1. Characteristic flexural (tensile) strength of masonry

The characteristic flexural strength is ignored in reinforced or prestressed masonry design because it is relatively insignificant.

2.5.7. Characteristic shear strength of masonry

2.5.7.1. Shear in bending (reinforced masonry)

Where the reinforcement is fully surrounded by mortar in bed and vertical joints, cavities, ducts and similar places the masonry characteristic shear strength f_v may be taken as 0·35 N/mm².

Where the reinforcement is fully surrounded by infill concrete, of minimum grade 25, in cavities, cores, pockets and similar places, f_v may be taken as

$$f_v = 0·35 + 17·5 A_s/bd \text{ N/mm}^2$$

but not greater than 0·7 N/mm², where A_s is the cross-sectional area of main reinforcing steel, b is the width of combined section and d is effective depth.

For simply supported reinforced beams and cantilevers where the ratio of the shear span a_v (the distance from a concentrated load to the support) to the effective depth is less than 2, f_v may be increased by a factor $2d/a_v$ (i.e. $f_v = 0·35(2d/a_v)$) but no greater than 0·7 N/mm².

For simply supported reinforced beams, cantilevers and cantilever retaining walls, when the ratio of the shear span a_v to the effective depth d lies between 2 and 6, f_v may be increased by the factor $2·5 - 0·25a_v/d$ (i.e. $f_v = 0·35(2·5 - 0·25a_v/d)$) but no greater than 1·75 N/mm².

2.5.7.2. Shear in bending (prestressed masonry)

For prestressed sections f_v may be taken as $0·35 + 0·6g_B$ N/mm² but no greater than 1·75 N/mm², where g_B is the design load per unit area due to the loads acting at right angles to the bed joints, including prestressing loads, in N/mm².

In prestressed structural elements where the prestressing is parallel to the bed joints g_B should be ignored and f_v taken as 0·35 N/mm² until more site testing and laboratory research have been completed.

For prestressed, simply supported beams or cantilevers when the ratio of the shear span to the effective depth lies between 2 and 6, f_v may be increased to $f_v = (0·35 + 0·6g_B)(2·5 - 0·25a/d)$ but no greater than 1·75 N/mm².

2.5.7.3. Racking shear in reinforced masonry shear walls

Frequently walls parallel to the wind force act as shear walls, because of the wind force being transferred (by plate action of floors, roofs and beams) to them from walls perpendicular to the wind force (see Fig. 2.1 for a typical example of a single-storey structure).

Figure 2.3 depicts the action in a multi-storey structure.

Fig. 2.3. Transfer of wind loads: multi-storey structure

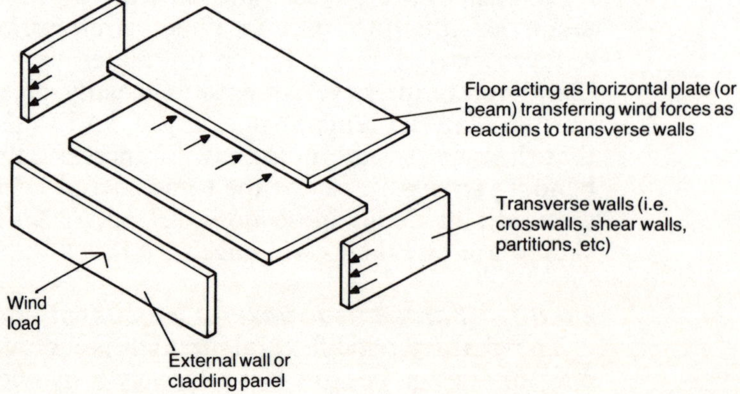

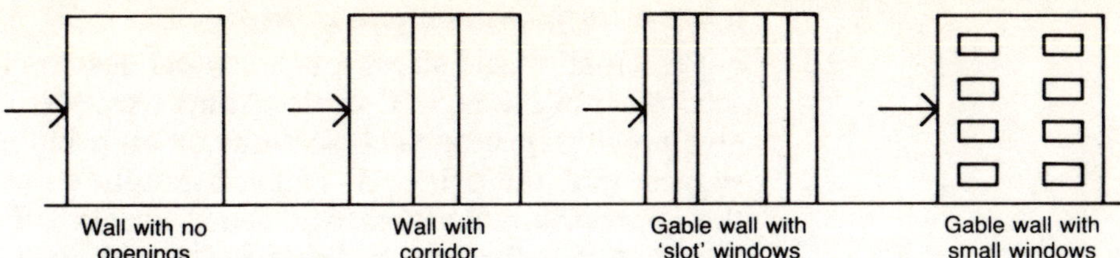

Fig. 2.4. Gable shear walls

The gable, or other transverse, walls carrying the wind force are termed shear walls. Some common types of shear wall are shown in Fig. 2.4.

In many cases there is sufficient precompression due to dead load (or the walls have high second moment of areas) to reduce the shear stresses to acceptable limits. Where there is inadequate precompression (or the second moment of area of the wall is low), the wall will develop tensile stresses because of racking (see Fig. 2.5). The stress analysis is similar to that for reinforced concrete shear walls.

Reinforcing or prestressing the wall will increase the racking shear resistance. Racking shear may be conservatively considered as being equal to the characteristic shear strength (i.e. $f_v = 0.35 + 0.6g_B$) but no greater than 1.75 N/mm^2.

Where g_B is insignificant but the reinforcement is placed in concrete infill in cavities, pockets, cores or similar places, f_v may be taken as 0.7 N/mm^2 where the ratio of height to length of the shear wall does not exceed 1.5. It has been found in practice that racking shear is rarely a critical design consideration, particularly for such low height/depth ratios.

2.5.7.4. Effect of damp-proof courses

Designers are advised to check the effect on the shear strength of the element due to the insertion of damp-proof courses and movement joints.

There is little reliable information on the shear resistance of damp-proof courses under various load conditions. In practice, so far, this has not presented a problem because of the frictional resistance generated by the relatively higher dead loads of the shear wall acting vertically on the damp-proof course (compared with the wind force on the shear wall).

Where the shearing force is high then there is danger of slip or slide (see Fig. 2.6) along a normal damp-proof course, and this must be checked. If the shear resistance along the damp-proof course is too low then the use of an engineering brick damp-proof course should be considered. Engineering bricks of very low porosity, laid three courses high, make an effective damp-proof course and have a shear resistance at least equal to that of the masonry. Three courses of slate make an effective damp-proof course for either brickwork or blockwork but this arrangement is rarely used because of its relatively high cost.

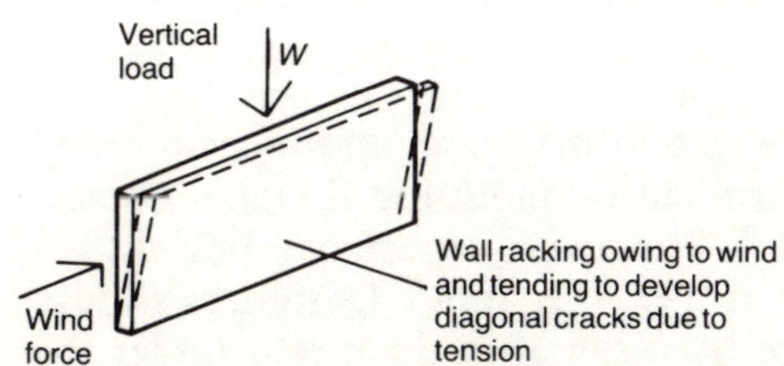

Fig. 2.5. Racking deformation

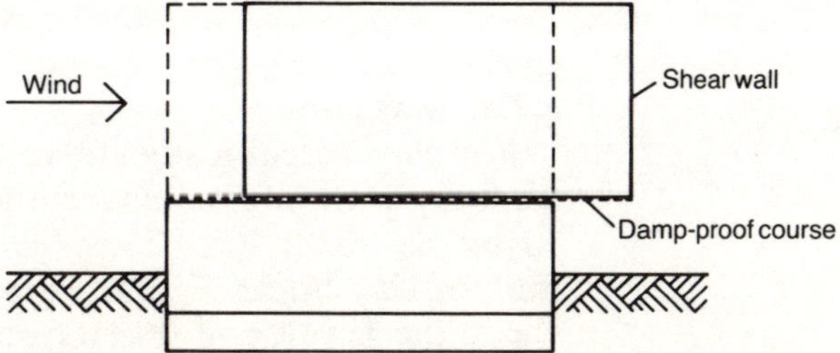

Fig. 2.6. Failure due to sliding shear along damp-proof course

2.5.7.5. Principal tensile stress

Principal tensile stress f_t is a critical design factor in prestressed geometric sections (hollow box, T sections and I sections) but BS 5628: Parts 1 and 2[1] give no guidance on such sections or on principal tensile stress. The Authors suggest that the principal tensile strength should be assumed as one half of the appropriate characteristic shear strength. This is a rather crude assumption but it is based on test results, although these tests are limited. The principal tensile strength is a function of tensile strength and masonry is anisotropic in tension, as in the following three examples.

(a) In direct tension perpendicular to the bed joints the tension strength of masonry is low, uncertain and unreliable.
(b) Tension strength parallel to the bed joints is unknown.
(c) The bending tensile strength resisting lateral loading differs horizontally and vertically. The ratio

$$\frac{\text{bending tensile strength parallel to bed joints}}{\text{bending tensile strength perpendicular to bed joints}}$$

is known as the orthogonal ratio and is commonly about 1/3. There is a need for further research in this area.

2.6. Structural properties of materials: steel

2.6.1. Characteristic tensile strength of reinforcing steel

The characteristic tensile strength of steels is given in Table 2.4. Lower values than those quoted may sometimes be necessary to reduce deflection or control cracking.

2.6.2. Characteristic compressive strength of reinforcing steel

The characteristic compressive strength of reinforcement may be taken as 0·83 times the characteristic tensile strength values given in Table 2.4. These values may need to be reduced in slender columns or walls by the capacity reduction factor β (see section 2.12.12).

2.6.3. Characteristic breaking load of prestressing steel

The characteristic strengths given in BS 4486[8] and BS 5896[9] are given in Tables 2.5–2.9. Characteristic tensile strength is denoted by F_{pu}.

2.6.4. Characteristic anchorage bond strength

The anchorage bond strength f_b for steel, in tension or compression, bedded in mortar designations (i) and (ii) may be taken as 1·5 N/mm^2 for plain bars and 2·0 N/mm^2 for deformed bars (types 1 and 2 which comply with BS 4449[10] and BS 4461[11]) (see Figs 2.7(a) and 2.7(b) and Table 2.6).

The anchorage bond strength f_b for steel, in tension or compression, bedded in concrete, minimum grade 25 infill may be taken as 1·8 N/mm^2 for plain bars and 2·5 N/mm^2 for deformed bars.

2.7. Elastic moduli E

2.7.1. Masonry

For clay, calcium silicate, concrete blocks and bricks (including reinforced masonry with infill concrete) the short-term elastic modulus E_m recommended by BS 5628: Part 2[1] is $0.9f_k$ kN/mm^2. However, experience has shown that for clay bricks E_m can vary from 0·6 to $1·0f_k$ kN/mm^2. Designers should check the E value of the masonry on site by measuring the strain under the design stress where it is critical in design calculations.

Table 2.7. Dimensions and properties of cold-drawn wire

Nominal diameter, mm	Nominal tensile strength,*† N/mm²	Nominal 0·1% proof stress,†‡ N/mm²	Nominal cross-section,‡ mm²	Nominal mass,‡ g/m	Tolerance on Diameter, mm	Tolerance on Cross-sectional area, mm²	Tolerance on Mass g/m	Specified characteristic breaking load,§¶ kN	Specified characteristic 0·1% proof load,§¶** kN	Load at 1% elongation,§¶** kN	Minimum elongation at max. load $L_0 = 200$ mm	Constriction at break	Reverse bends Minimum number	Reverse bends Bend radius, mm	Initial load (% of actual breaking load)	Max. relaxation after 1000 h Relax class 1	Max. relaxation after 1000 h Relax class 2
7	1570	1300	38·5	302	±0·05*	±0·55	±4·3	60·4	50·1	51·3	For all wires 3·5%	For all wires: a ductile break visible to the naked eye	4 for smooth wires, 3 for indented wires	20	For all wires 60%	For all wires 4·5%	For all wires 1·0%
7	1670	1390						64·3	53·4	54·7					70%	8%	2·5%
6	1670	1390	28·3	222	±0·05	±0·47	±3·7	47·3	39·3	40·2				15	80%	12%	4·5%
6	1770	1470						50·1	41·6	42·6							
5	1670	1390	19·6	154	±0·05	±0·39	±3·1	32·7	27·2	21·8				15			
5	1770	1470						34·7	28·8	29·5							
4·5	1620	1350	15·9	125	±0·05	±0·35	±2·7	25·8	21·4	21·9				15			
4	1670	1390	12·6	98·9	±0·04	±0·25	±2·0	21·0	17·5	17·9				10			
4	1770	1470						22·3	18·5	19·0							

* The nominal diameter and nominal tensile strength are for designation purposes only.
† The nominal tensile strength is calculated from the nominal cross-section and the specified characteristic breaking load.¶
‡ The nominal 0·1% proof stress, nominal cross-section and nominal mass are for information only.
§ The maximum standard deviations are 55 N/mm², or the equivalent value expressed as a load, for tensile strength, and 60 N/mm², or the equivalent value expressed as a load, for 0·1% proof stress.
¶ In view of the close tolerance on diameters and areas, specified characteristic loads have been specified rather than stresses.
** The specified characteristic 0·1% proof load is approximately 83% of the specified characteristic breaking load. By agreement, alternatively the characteristic load at 1% elongation may be specified, which is approximately 85% of the specified characteristic breaking load.
†† The modulus of elasticity is to be taken as 205 ± 10 kN/mm², unless otherwise indicated by the manufacturer.

Table 2.8. Dimensions and properties of cold-drawn wire in mill coil

Nominal diameter, mm	Nominal tensile strength,*† N/mm²	Nominal cross-section,‡ mm²	Nominal mass,‡ g/m	Tolerance on Diameter, mm	Tolerance on Cross-sectional area, mm²	Tolerance on Mass, g/m	Specified characteristic breaking load,§¶ kN	Specified characteristic load at 1% elongation,§¶** kN	Constriction at break	Reverse bends Minimum number	Reverse bends Bend radius, mm	Initial load (% of actual breaking load)	Max. relaxation after 1000 h
5	1570	19·6	154	±0·05	±0·39	±3·1	30·8	24·6	For all wires: a ductile break visible to the naked eye	4 for smooth wire, 3 for indented wire	15	For all wires 60%	For all wires 8%
5	1670						32·7	26·2				70%	10%
5	1770						34·7	27·8					
4·5	1620	15·9	125	±0·05	±0·35	±2·7	25·8	20·6			15		
4	1670	12·6	98·9	±0·04	±0·25	±2·0	21·0	16·8			10		
4	1720						21·7	17·4					
4	1770						22·3	17·8					
3	1770	7·07	55·5	±0·04	±0·19	±1·5	12·5	10·0			7·5		
3	1360						13·1	10·5					

* The nominal diameter and nominal tensile strength are for designation purposes only.
† The nominal tensile strength is calculated from the nominal cross-section and the specified characteristic breaking load.¶
‡ The nominal cross-section and nominal mass are for information only.
§ The maximum standard deviations are: 55 N/mm², or the equivalent values expressed as a load, for tensile strength and 60 N/mm², or the equivalent value expressed as a load, for stress at 1% elongation.
¶ In view of the close tolerance on diameters and areas, specified characteristic loads have been specified rather than stresses.
** In order to prove the suitability of this material, which is only used in certain applications (e.g. railway sleepers) there is a requirement for a load at 1% total elongation to be at least 80% of the specified characteristic breaking load.

Table 2.9. Dimensions and properties of strands

Type of strand*	Nominal diameter,* mm	Nominal tensile strength,* N/mm²	Nominal steel area,† mm²	Nominal mass,† g/m	Tolerance on			Specified characteristic breaking load,‡§ kN	Specified characteristic 0·1% proof load,‡§¶ kN	Load at 1% elongation,‡§¶ kN	Minimum elongation at max. load $L_0 \geqslant 500$ mm	Constriction at break	Relaxation		
					Diameter, mm	Cross-sectional area, mm²	Mass, g/m						Initial load (% of actual breaking load)	Max. relaxation after 1000 h	
														Relax class 1	Relax class 2
7-wire standard	15·2	1670	139	1090	+0·4 −0·2	For all strands +4% −2%	For all strands +4% −2%	232	197	204	For all strands 3·5%	For all strands ductile wire breaks visible to the naked eye	For all strands 60% 70% 80%	For all strands 4·5% 8% 12%	For all strands 1·0% 2·5% 4·5%
	12·5	1770	93	730				164	139	144					
	11·0	1770	71	557	+0·3 −0·15			125	106	110					
	9.3	1770	52	408				92	78	81					
7-wire super	15·7	1770	150	1180	+0·4 −0·2			265	225	233					
	12·9	1860	100	785				186	158	163					
	11·3	1860	75	590	+0·3 −0·15			139	118	122					
	9·6	1860	55	432				102	87	90					
	8·0	1860	38	298				70	59	61					
7-wire drawn	18·0	1700	223	1750	+0·4 −0·2			380	323	334					
	15·2	1820	165	1295				300	255	264					
	12·7	1860	112	890				209	178	184					

* The type of strand, nominal diameter and nominal tensile strength data are given for designation purposes only.

† The nominal steel area and nominal mass data are given for information only.

‡ The maximum standard deviations are: 55 N/mm², or the equivalent value expressed as a load, for tensile strength and 60 N/mm², or the equivalent value expressed as a load, for 0·1% proof stress.

§ In view of the close tolerance on mass and areas, specified characteristic loads have been specified rather than stresses.

¶ The specified characteristic 0·1% proof load is approximately 85% of the specified characteristic breaking load. Alternatively, by agreement between the manufacturer and the purchaser, the characteristic load at 1% elongation may be specified, which is approximately 88% of the specified characteristic breaking load.

**The nominal tensile strength is calculated from the nominal steel area and the specified characteristic breaking load.§

†† The modulus of elasticity is to be taken as 195 ± 10 kN/mm², unless otherwise indicated by the manufacturer.

Fig. 2.7. (a) Indented bars; (b) crimped bars (see Table 2.6)

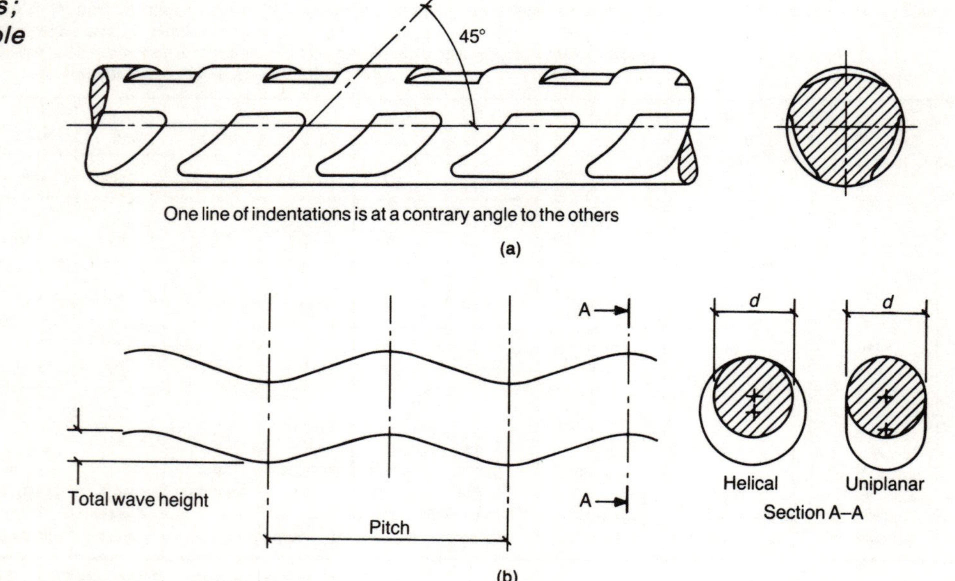

Table 2.10. Elastic modulus for concrete infill E_c

28 day cube strength, N/mm^2	E_c, kN/mm^2
20	24
25	25
30	26
40	28
50	30
60	32

2.8. Partial safety factors: loads

2.7.2. Concrete infill

The elastic modulus for infill concrete E_c used in prestressed masonry is given in Table 2.10.

2.7.3. Steel reinforcement

For all steel reinforcement (and all types of loading) the elastic modulus of steel E_s may be taken as 200 kN/mm^2.

2.7.4. Prestressing steel

The elastic modulus of prestressing steel E_s should be taken from BS 4486[8] and BS 5896.[9]

2.8.1. Introduction

Design inaccuracies can arise from minor errors in assumptions and calculations, the effect of construction tolerances, possible load increases and unusual stress redistribution. Major errors in design assumptions and calculations or faulty construction cannot be compensated for by safety factors but the normal, minor errors can be taken into account by applying a partial safety factor for loads γ_f to the characteristic loads so that

design load = characteristic load × partial factor of safety

$$= (G_k, Q_k, W_k, P_k \text{ or } E_n)\gamma_f$$

2.8.2. Ultimate limit state

The partial safety factors for loads given in BS 5628,[1] which are common with other codes, are shown in Table 2.11. It is advisable to consider all possible load combinations in design. The ultimate limit design load is the product of the characteristic load multiplied by the appropriate partial safety factor already given. Where alternative values are given for dead load, the case producing the more severe conditions should be used.

2.8.3. Serviceability limit state

The design loads for a serviceability limit state are given in Table 2.12.

Table 2.11. Partial safety factors for loads: ultimate limit state

Load combination	Ultimate limit state			
	Dead	Imposed	Wind	Earth and water
Dead plus imposed	$1 \cdot 4\, G_k$ or $0 \cdot 9\, G_k$	$1 \cdot 6\, Q_k$	—	$1 \cdot 4\, E_n$
Dead plus wind	$1 \cdot 4\, G_k$ or $0 \cdot 9\, G_k$	—	$1 \cdot 4\, W_k$	$1 \cdot 4\, E_n$
Dead plus imposed plus wind	$1 \cdot 2\, G_k$	$1 \cdot 2\, Q_k$	$1 \cdot 2\, W_k$	$1 \cdot 2\, E_n$
Dead plus wind—freestanding walls plus laterally loaded panels whose removal would in no way affect the stability of the remaining structures	$1 \cdot 4\, G_k$ or $0 \cdot 9\, G_k$	—	$1 \cdot 2\, W_k$	
Prestress force Considered as dead load	$1 \cdot 4\, P_k$ $0 \cdot 9\, P_k$			

Table 2.12. Partial safety factors for loads: serviceability limit state

Load combination	Serviceability limit state			
	Dead	Imposed	Wind	Earth and water
Dead plus imposed	$1 \cdot 0\ G_k$	$1 \cdot 0\ Q_k$	—	$1 \cdot 0\ E_n$
Dead plus wind	$1 \cdot 0\ G_k$	—	$1 \cdot 0\ W_k$	$1 \cdot 0\ E_n$
Dead plus imposed plus wind	$1 \cdot 0\ G_k$	$0 \cdot 8\ Q_k$	$0 \cdot 8\ W_k$	$1 \cdot 0\ E_n$
Prestress	$1 \cdot 0\ P_k$			

When calculating short-term deflections the load combinations (shown in Table 2.12) giving the more severe condition should be selected. It is also necessary sometimes to check additional time-dependent deflections due to creep, movement due to temperature and moisture and so on, and their effect on cracking, other local damage and the structure as a whole.

2.8.4. Accidental forces

The partial safety factors for design load γ_f in accidental damage checks are reduced as follows.

(a) Design dead load and/or prestress is taken as $0 \cdot 95 G_k$ or $1 \cdot 05 G_k$.
(b) Design imposed load is taken as $0 \cdot 35 Q_k$ (or $1 \cdot 05 Q_k$ in buildings used predominantly for storage or where the imposed load is of a permanent nature).
(c) Design wind load is taken as $0 \cdot 35 W_k$.

2.9. Partial safety factors: materials

Minor faults in the occasional material or unit can be taken into account by applying a partial safety factor.

The design strength of the materials is their characteristic strength divided by the appropriate partial safety factor γ_m. The factors are direct compression and bending γ_{mm}, shear strength of masonry γ_{mv}, bond strength between infill concrete or mortar and steel γ_{mb} and strength of steel γ_{ms}.

The partial safety factor depends both on the degree of quality control of site construction and on manufacture. The quality control on construction should be similar to the high standards exercised in construction of reinforced and prestressed concrete. Designers are referred to BS 5628: Parts 1 and 3[1] for detailed information. BS 5628 recognises two categories of manufacturing control: normal and special.

Normal control is used when the materials meet the requirements for compressive strength in the appropriate British Standard.

Special control is used when the manufacturer agrees to supply consignments of the units to a specified strength (known as the acceptance limit for compressive strength) such that the average compressive strength of a sample of units, taken from any consignment and tested in accordance with the appropriate British Standard, has a probability of not more than $2\frac{1}{2}\%$ (i.e. 1 in 40) of the sample being below the acceptance level. The manufacturer also operates quality control systems (the results of which can be made available to demonstrate to the satisfaction of the purchaser that the acceptance limit is met consistently) with the probability of failing to meet the limit never being greater than the $2\frac{1}{2}\%$.

The effect of manufacturing control on γ_{mm} is $2 \cdot 0$ for special manufacturing and $2 \cdot 3$ for normal control of structural units.

<table>
<tr><td rowspan="2">

Table 2.13. Partial factors of safety for materials

</td><td>Property</td><td>Symbol</td><td>Ultimate limit</td><td>Accidental loads</td><td>Deflections and assessing stresses or crack widths</td></tr>
<tr>
<td>Direct compression plus bending</td><td>γ_{mm}</td><td>2·0 or 2·3</td><td>1·0 or 1·15</td><td>1·5</td></tr>
</table>

Property	Symbol	Ultimate limit	Accidental loads	Deflections and assessing stresses or crack widths
Direct compression plus bending	γ_{mm}	2·0 or 2·3	1·0 or 1·15	1·5
Shear strength of masonry	γ_{mv}	2·0	1·0	—
Bond strength between concrete infill, or mortar, and steel	γ_{mb}	1·5	1·0	—
Strength of steel	γ_{ms}	1·15	1·0	1·0

When the site quality control of construction is not up to the standards recommended in BS 5628, the designer is advised to improve the quality control or, if this is impossible, to increase the value of γ_{mm}.

Other partial safety factors for different loading or design conditions are given in Table 2.13.

The values of γ_{mm} have been reduced from 2·5 and 3·0, used in plain masonry, to 2·0 and 2·3 for reinforced and prestressed masonry. The Authors have reservations about changing these factors for, basically, the material has remained unaltered. However, the compressive stress has altered in character (see section 2.5.6) from direct compression to either bending compression or a combination of bending plus direct compression, and an increase in compressive strength is reasonable. This increase is not recognised in BS 5628.[1]

The Authors recommend an increased compressive strength and an unaltered partial safety factor for material. This is in contradiction to the code. (In bending compression the compressive strength may be increased by 50%, i.e. $1·5f_k$. In direct compression the compressive strength is the basic compressive strength, i.e. $1·0f_k$. Depending on the ratio of bending and direct compression strength, it would be reasonable to use a sliding scale from 1·5 to $1·0f_k$. In practice the Authors have found it convenient to adopt $1·2f_k$ for combined stress.) However, the design strength resulting from the differing philosophies is approximately the same. Despite declarations that codes are advisory and not mandatory documents, in practice they tend to become statutory and mandatory. Designers are advised therefore that when submitting calculations to control officers they may find earlier approval when following code recommendations. The two philosophies give approximately similar values and so there is little cause for practical concern.

- Code recommendations: $f_k/\gamma_{mm} = 1/2·0 = 0·5f_k$ (special) and $1/2·3 = 0·435f_k$ (normal).
- Authors' suggestion: $f_k/\gamma_{mm} = 1·2/2·5 = 0·48f_k$ (special) and $1·2/3·0 = 0·40f_k$ (normal).

The Authors' approach is discussed in section 3.2.

2.10. Structural analysis

The design of a structural element within the structure should be based on the design strengths of the materials appropriate to the limit state being considered. When, as in the case of reinforced and prestressed masonry or concrete, materials of different structural properties are used in combination, allowances for these differences should be made.

When the element forms part of an indeterminate structure (i.e. it is not an isolated simply supported beam or separate pinned end column), then the usual engineering design and judgment must be exercised in assessing the forces in the element.

Elastic analysis is usually used to determine the force distribution in the total structure. The relative stiffness of the members may be based on any of the following

(a)   the entire masonry section, ignoring the reinforcement (this method is common in preliminary design at least)
(b)   the entire masonry section including the reinforcement on the basis of the elastic modular ratio (this method is common when it is necessary to refine the preliminary design)
(c)   the compression area of the masonry cross-section alone together with the reinforcement—on the basis of the elastic modular ratio.

It will be appreciated by the experienced designer that structural design in reinforced masonry is (as with other materials) not completely accurate. For example

(a)   the loadings are not based on accurate statistical data
(b)   the material strengths are based on characteristic strength and the strength in the structure may well be higher than this
(c)   masonry is not a truly elastic material and the values of E can vary considerably with magnitude of stress, size of element, passage of time and so on
(d)   the second moment of area calculations of an element vary depending on the assumptions used in calculation
(e)   the effective length of columns is based on judgment of the end fixity
(f)   the span of beams is also an assumption (a true simple support or fully fixed support is practically impossible to obtain in the real structure)
(g)   simple bending theory calculations tend to ignore minor stress effects due to small torsions, shear lag, alternative path load distribution to connecting elements and similar effects.

Experienced practical designers could quote further examples. In structural masonry design, as in all other structural materials, the designer must exercise his judgment in establishing his simplified mathematical model for the structure. The safety factors generally more than compensate for any minor inaccuracies.

2.11. Moments and forces in continuous beams

The common load arrangements for continuous beams should also be adopted in the design of reinforced masonry

- for alternative spans loaded with the design load $1.4G_k + 1.60Q_k$ and for other spans loaded with the minimum dead load $0.9G_k$
- for all spans loaded with the design load $1.4G_k + 1.6Q_k$.

2.12. Slenderness ratio

2.12.1. Introduction
In structural design in other materials the design strength of a column is also affected by its slenderness ratio, and beams can need lateral restraints to prevent their compression flange from buckling. The design of reinforced and prestressed masonry structural elements is influenced by the same criteria. There are minor differences because of the nature of the material, but the basic design factors are common and the design strength of a given

masonry member (in addition to the factors already discussed) is governed by

(a) its slenderness ratio which is effective height/radius of gyration (or, in BS 5628[1], effective height/effective thickness)

(b) limitations on the distance between lateral restraints to beam compression flanges

(c) the dimensions of the element and the characteristic strength of material

(d) limitations on dimensions for columns and beams

(e) eccentricity of loading.

When a column, wall or other strut is short (short is defined in BS 5628 as having a slenderness ratio of less than 8, based on effective thickness) there is no reduction in the basic design strength. When the column is long (i.e. has a slenderness ratio of greater than 8) the basic design strength not only of the masonry but also of the reinforcement or prestressing force is reduced. As the column gets longer there is a further reduction in design strength until, at the slenderness ratio limit of 27, the design strength is considered as zero. For cantilever columns and walls the slenderness ratio should not exceed 18. It can be necessary to check the deflection of free cantilevers, reinforced to carry extra load, particularly when the amount of reinforcement exceeds 0·5% of the cross-sectional area of the wall or column.

2.12.2. Slenderness ratio: definitions

The slenderness ratio is a measure of the tendency of a member to fail by buckling under compression loading before failure by crushing occurs.

The slenderness ratio of columns (or walls) is given in most codes as effective length/radius of gyration. However, masonry codes still use

$$\frac{\text{effective height (or length)}}{\text{effective thickness}} = \frac{h_{ef}}{t_{ef}} \text{ or } \frac{l_{ef}}{t_{ef}}$$

where the effective height or effective length is used, whichever gives the lower slenderness ratio.

In masonry, the term 'radius of gyration' has been replaced by the term 'effective thickness'. This is because the bulk of construction and research in masonry, in the past traditional construction and much of the present, has been done on solid rectangular or square columns and walls, and the concept of effective thickness was simpler and more appropriate. A solid wall has an effective thickness equal to its actual thickness; the addition of piers to the wall effectively increases its thickness and the insertion of a cavity (which reduces the compressive loadbearing capacity) effectively decreases it. Although this concept of effective thickness is simple and practical, it is inappropriate for hollow box sections, T beams and other geometric sections. Such sections are far more structurally efficient than solid rectangular sections and guidance on dealing with them is given in section 2.12.11.

2.12.3. Effective height

The effective height (or length) of a strut (column, wall, or similar element) depends not only on the height (or length) of the strut but also on the degree of fixity of the ends of the strut, i.e. free, pinned, fixed or partially fixed.

A fixed ended strut of the same height, section and other properties as a free cantilever will carry a much greater compressive force than the free cantilever before it buckles. In masonry, as in other materials, the end fixity (pinned, fixed or other) depends on the lateral support to the strut.

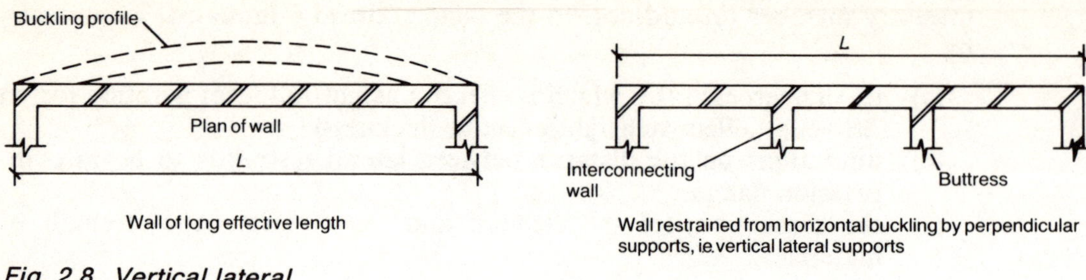

Fig. 2.8. *Vertical lateral supports*

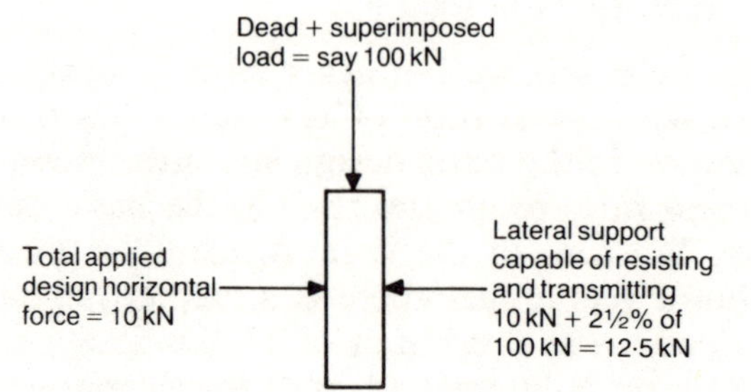

Fig. 2.9. *Lateral restraint*

2.12.4. Horizontal and vertical lateral supports

The support (the strut end condition) to the strut by adjacent members will generally be at right angles to the strut and is termed 'lateral' support. Lateral supports to a wall or column such as floors, roofs and beams acting in the horizontal plane are termed 'horizontal lateral' supports.

The lateral supports to a wall (such as an interconnecting wall) acting in the vertical plane are termed 'vertical lateral' supports (see Fig. 2.8).

To fulfil their restraining function, the horizontal and vertical supports must be able to transmit to the supporting structure (the walls, columns, beams and so forth providing lateral stability to the whole structure) the sum of the following forces

Fig. 2.10. *Simple restraints: horizontal*

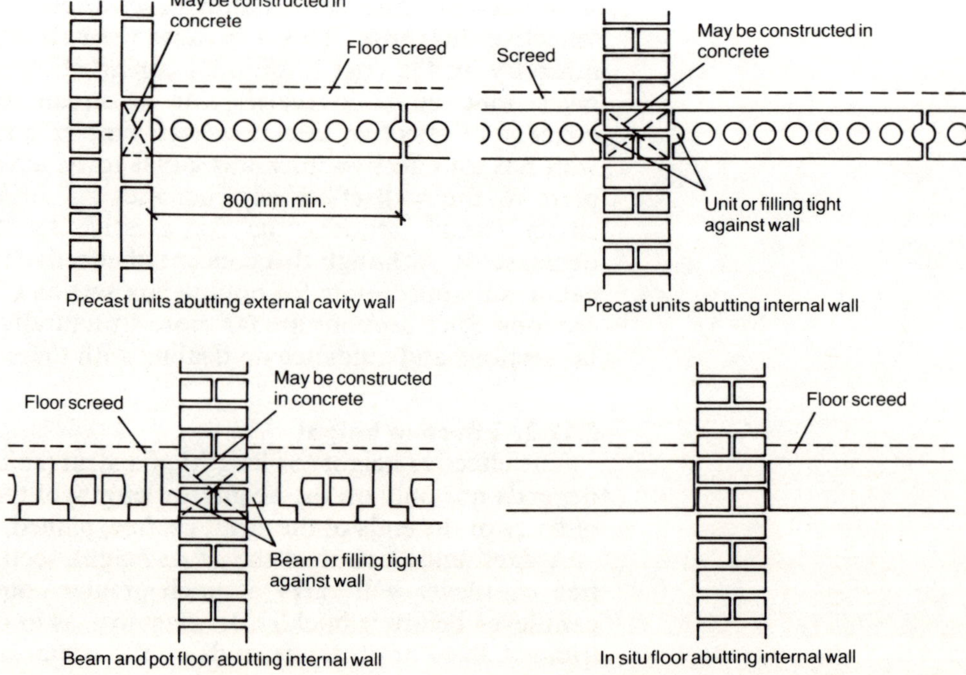

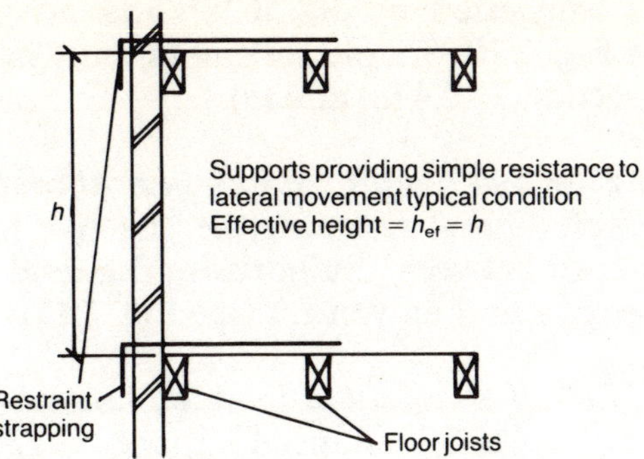

Fig. 2.11. Effective height with simple lateral restraint

(*a*) the simple static reactions to the total applied design horizontal forces at the line of lateral support

(*b*) 2·5% of the total design vertical load (i.e. the compressive force from dead and superimposed load) that the wall or other element is designed to carry at the line of lateral support (see Fig. 2.9).

When the applied horizontal force is capable of reversal (e.g. wind pressure and wind suction) the lateral restraint must be able both to prop and to tie the wall or other restrained element. The load on the prop or tie must be transferred by floors, walls and other structural elements (providing stability by acting as horizontal girders or shear walls) to the foundation.

2.12.5. Types of horizontal lateral support

A perfectly pinned joint or perfect fixed joint are convenient concepts in theory but can be difficult (and practically impossible) to achieve in practice. In buildings, there are various lateral supports such as in situ and precast concrete floors, steel beams and interconnecting masonry walls. These restraints, when providing a pinned action, are termed 'simple' restraints and, when providing some fixity, 'enhanced' restraints. Some typical simple

Fig. 2.12. Enhanced restraints: horizontal

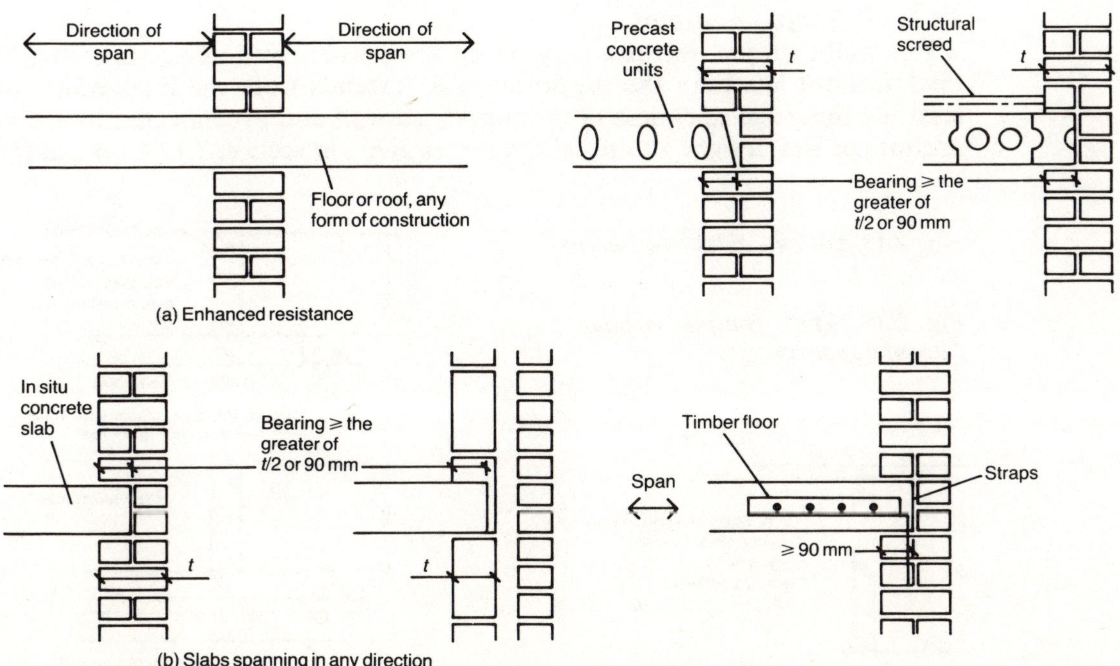

restraints (suggested by BS 5628[1] and acceptable in other countries) are shown in Fig. 2.10. Simple restraints must be capable of transmitting forces given in section 2.12.4 (*a*) and (*b*).

2.12.5.1. *Effective height, simple lateral restraint*

For simple lateral restraint the effective height h_{ef} may be taken as the clear distance between the horizontal lateral supports providing the simple resistance to lateral movement (see Fig. 2.11).

2.12.5.2. *Effective height, enhanced lateral restraint*

Some examples of enhanced support suggested by the code are given in Fig. 2.12.

Generally, enhanced resistance to lateral movement of walls and columns may be assumed when

- floors and roofs span on to the wall from both sides at the same level
- in situ concrete slabs (or precast) provide equivalent restraint irrespective of their span have a bearing of at least half of the thickness of the wall (and not less than 90 mm) on to which it spans.

For enhanced lateral restraint the effective height h_{ef} may be taken as 0·75 times the clear distance between the horizontal lateral supports providing the enhanced resistance to lateral movement (see Fig. 2.13).

When a wall or column is reinforced (increasing its capacity to resist bending moments) and is connected to reinforced restraining structural elements, then the effective height should be determined by structural analysis based on assessment of fixity of the connected reinforcement and in accordance with recommendations given in BS 8110[12].

2.12.6. Types of vertical lateral support

Vertical lateral supports, as horizontal supports, are classified as simple and enhanced and must be capable of transmitting the forces defined in section 2.12.4 (*a*) and (*b*). BS 5628[1] suggests rule of thumb details for vertical lateral restraints as shown in Fig. 2.14.

2.12.6.1. *Simple restraint*

For walls, simple restraint may be assumed where an intersecting wall, of thickness not less than the supported wall, extends from the intersection at least ten times the thickness of the supported wall and is connected by metal anchors or ties designed to resist the forces given in section 2.12.4 (*a*) and (*b*).

Fig. 2.13 (below). Effective height

Fig. 2.14 (right). Simple vertical lateral supports

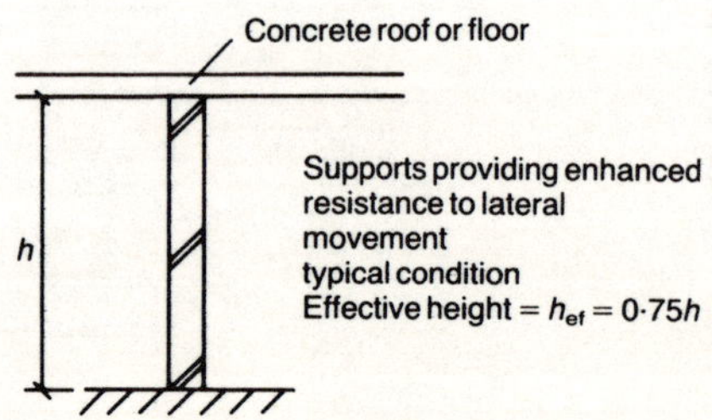

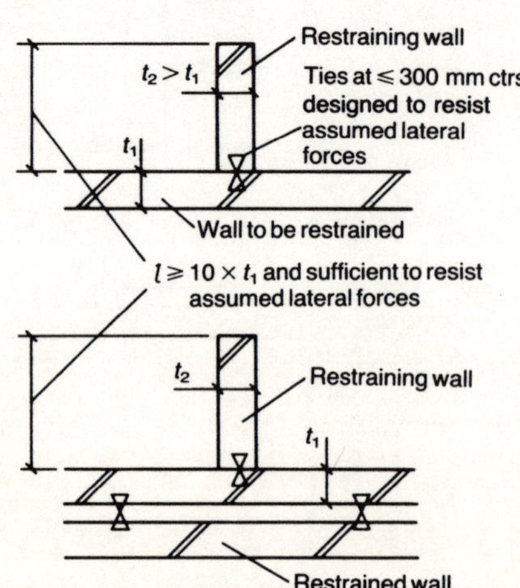

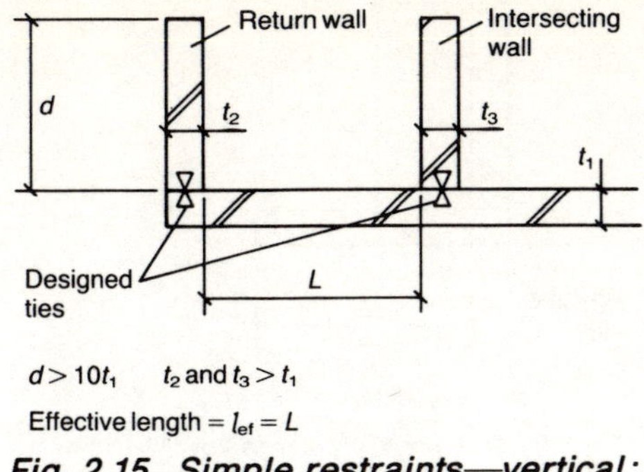

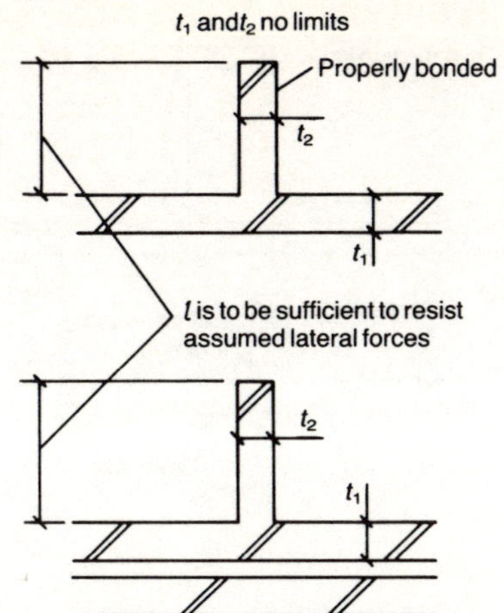

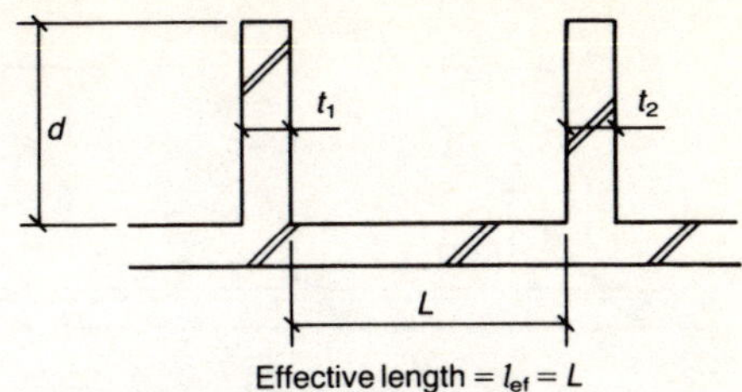

$d > 10t_1$ t_2 and $t_3 > t_1$

Effective length $= l_{ef} = L$

Fig. 2.15. Simple restraints—vertical: typical condition supports providing simple resistance to lateral movement

Fig. 2.16 (left). Enhanced restraints—vertical lateral supports providing enhanced resistance

Fig. 2.17 (above). Enhanced restraints—vertical: typical condition (a) supports providing enhanced resistance to lateral movement

BS 5628[1] states that when ties are used they should be evenly distributed and at not more than 300 mm centres. Such ties give little stiffness to the connection, have little capacity to distribute moments and provide low fixity. For design purposes they can be assumed as acting as pinned joints.

For simple lateral restraint the effective length l_{ef} may be taken as the clear distance between the vertical lateral supports providing the simple resistance to lateral movement (see Fig. 2.15).

2.12.6.2. Enhanced restraint

The code's recommendations for enhanced lateral support are shown in Fig. 2.16.

The difference between enhanced and simple restraints, shown in Figs 2.15 and 2.16, is that for enhanced restraints the interconnecting element must be properly and fully bonded to the supported wall. Such connections provide a degree of fixity, resistance to rotation and stiffness. They can be assumed to provide partial fixity.

Enhanced resistance may also be adopted when calculations verify that the restraint provides the equivalent rotational resistance or continuity as would properly bonded construction.

For enhanced vertical restraint the effective length l_{ef} may be taken as 0·75 times the clear distance between the vertical lateral supports providing enhanced resistance to lateral movement (see Fig. 2.17).

2.12.7. Free end condition

2.12.7.1. Effective height

A free-standing wall is a wall with no horizontal lateral support at the top; it thus acts as a free cantilever. The effective height of such a wall should be taken as twice the actual height from the horizontal lateral support at the base to the unrestrained top of the wall.

2.12.7.2. Effective length

An external face wall may project, along its length, beyond the gable or return wall. (In structural elements such as fin walls the fin itself is a projection from the face wall, see Fig. 2.18.)

There are two conditions for the effective length of such projections, depending on the type of vertical lateral support provided at the connected end of the projection.

(*a*)   Where the connected end has a vertical lateral support which pro-

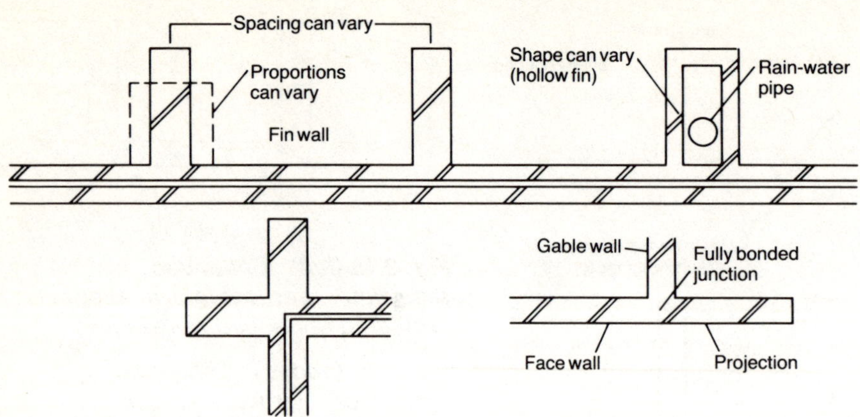

Fig. 2.18. Fin walls: typical details

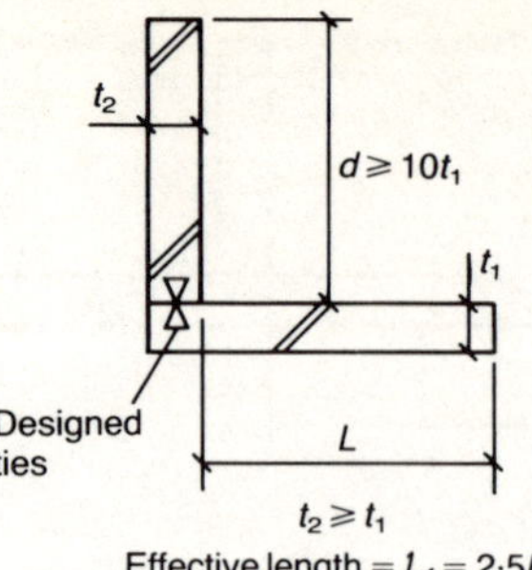

$$t_2 \geqslant t_1$$
Effective length $= l_{ef} = 2{\cdot}5L$

Fig. 2.19. Simple restraint—vertical: typical condition (d) support providing simple resistance to lateral movement

vides only simple lateral restraint to lateral movement, the effective length may be taken as 2·5 times the distance between the vertical lateral support and the free end (Fig. 2.19).

(b) When the connected end has a lateral support which provides enhanced lateral restraint to lateral movement, the effective length may be taken as twice the distance between the vertical lateral support and the free end.

2.12.8. Summary

The information given in section 2.12.7 is summarised in table 2.14 of the code.

2.12.9. Effective thickness t_{ef} of walls

As mentioned in section 2.12.2, most masonry codes employ the concept of effective thickness and not radius of gyration. The advice in BS 5628[1] (which is similar to that in other codes) for normal traditional walls is as follows. The Authors' advice on geometric sections is given in section 2.12.11.

2.12.9.1. Solid walls

For a solid wall, unstiffened by intersecting or return walls, the effective thickness is the actual thickness.

2.12.9.2. Piered walls

A solid wall stiffened by piers has an effective thickness greater than its

Table 2.14

Criterion	Simple restraint	Enhanced restraint	Free end
Effective height h_{ef}	Actual height h	0·75 × actual height	2 × actual height
—	Actual length l	0·75 × actual length	2 × actual length (enhanced restraint)
Effective length l_{ef}	—	—	2·5 × lateral length (simple restraint)

Table 2.15. Pier stiffening coefficients (see Fig. 2.20)*

Ratio of pier spacing (centre to centre) to pier width	Pier thickness		
	Wall thickness		
	1	2	3
6	1·0	1·4	2·0
10	1·0	1·2	1·4
20	1·0	1·0	1·0

* Linear interpolation is permissible, but not extrapolation.

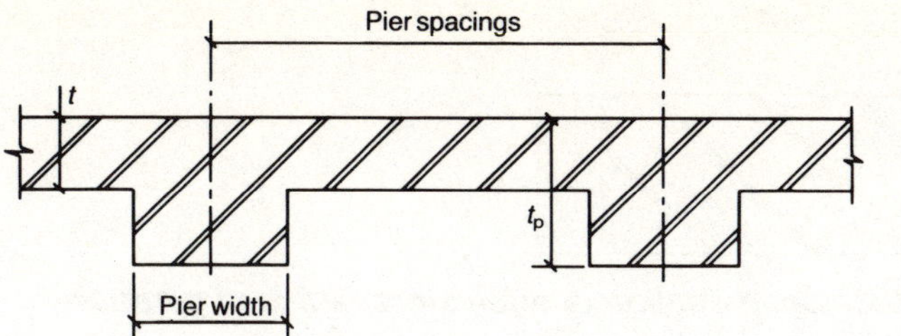
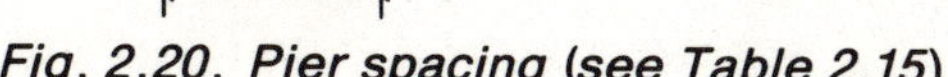

Fig. 2.20. Pier spacing (see Table 2.15)

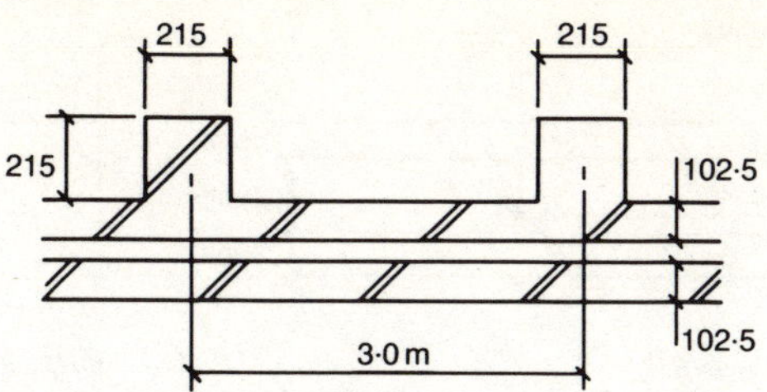

Fig. 2.21. Pier stiffened cavity wall; effective thickness of leaf with piers is $t_{ef} = 115$ mm, and so effective thickness of cavity wall is $2(115 + 102·5)/3 = 145$ mm

actual thickness, the increase being affected by the stiffening effect of the piers—the smaller the spacing and the greater the increase in the thickness of the piers, the greater is the increase in the effective thickness t of the wall. The effective thickness t_{ef} of the wall is equal to its actual thickness times a stiffening factor K, i.e. $t_{ef} = tK$ (see Fig. 2.20). Values for the appropriate stiffening factor K given in BS 5628[1] are shown in Table 2.15.

Intersecting walls are considered as piers when the effective thickness of the wall is being determined. To allow for the effect of the increased length of an intersecting wall, compared with that of a pier, the equivalent pier thickness of the intersecting wall is assumed to be three times its actual thickness.

2.12.9.3. Cavity walls

Adding another leaf of masonry, joined by adequate ties, to a solid wall will reduce its tendency to buckle whether or not only one of the leaves is loadbearing or both are. When the two leaves are separated by a cavity, however, the wall will not be as stiff as a solid bonded wall. The effective thickness of a cavity wall is considered to be only two-thirds of the sum of the thickness of the two leaves. Cases occur when one of the leaves is thicker than the other, and the effective thickness may be less than the thickness of the thicker leaf. BS 5628[1] suggests that the effective thickness of such a wall should be taken as the actual thickness of the thicker leaf, and the contribution of the thinner leaf ignored.

2.12.9.4. Piered cavity walls

When one of the leaves is stiffened by piers—a not uncommon condition—the effective thickness of the stiffened leaf should be determined as described in section 2.12.9.2. The effective thickness of the whole cavity wall is then calculated as described in section 2.12.9.3 but using the increased value of the effective thickness of the leaf stiffened by piers. An example is shown in Fig. 2.21.

2.12.9.5. Grouted cavity walls

The effective thickness of a fully grouted cavity wall may be taken as the overall thickness of the wall, provided the width of the cavity does not exceed 100 mm and the leaves are adequately tied. Should the cavity exceed 100 mm width then the effective thickness may be assumed to be the thickness of the leaves plus only 100 mm of the grouted cavity width.

2.12.10. Effective thickness of columns

The effective thickness of a solid square column is taken as the actual thickness of the column.

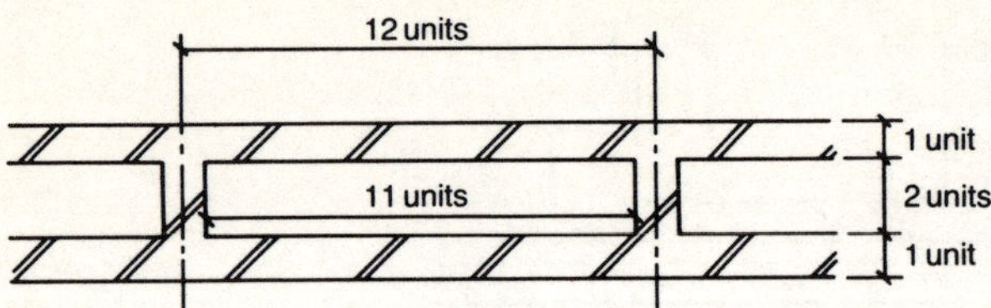

Fig. 2.22. Diaphragm wall

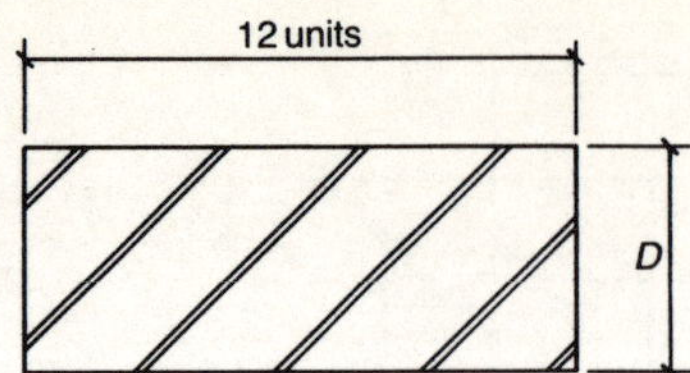

Fig. 2.23. Solid section of equivalent radius of gyration

The effective thickness of a solid rectangular column, equally restrained on both axes, is taken as the least thickness of the column.

Columns are dealt with in more detail in section 2.12.15.

2.12.11. Effective thickness of geometric sections

As discussed in section 2.12.2, most codes deal only with normal walls and columns and slenderness ratio is determined by a section's effective thickness and not, as is common in other structural materials, by a section's radius of gyration. The concept of effective thickness is, in the Authors' opinion, inappropriate for geometric sections and difficult to evaluate and define in accordance with the code.

Consider, for example, the diaphragm wall shown in Fig. 2.22 (diaphragm walls are discussed in section 7.3). The radius of gyration r of a section is equal to $\sqrt{(I/A)}$.

$$I = \frac{12 \times 4^3 - 11 \times 2^3}{12} = \frac{768 - 88}{12} = 56 \cdot 67 \text{ units}^4$$

$$A = 12 \times 4 - 11 \times 2 = 26 \text{ units}$$

Therefore

$$r = \sqrt{(I/A)} = \sqrt{\{(56 \cdot 67)/26\}} = 1 \cdot 48 \text{ units}$$

For the solid section shown in Fig. 2.23, the equivalent radius of gyration is

$$r = \sqrt{\frac{I}{A}} = \sqrt{\left(\frac{12D^3/12}{12D}\right)} = \sqrt{\frac{D^2}{12}} = 0 \cdot 29D = 1 \cdot 48 \text{ units}$$

and therefore

$$D = 1 \cdot 48/0 \cdot 29 = 5 \cdot 1 \text{ units}$$

which is greater than 4 units for the section shown in Fig. 2.22.

Since the diaphragm section is 4 units deep and, to give the same radius of gyration, a solid section would have to be 5·1 units deep, the effective depth of a diaphragm wall is greater than its overall depth! Geometric sections always have an equivalent effective depth greater than their overall depth.

This does not mean that a diaphragm wall can carry a greater load than a solid wall of the same overall thickness, but it does mean that the diaphragm can carry a higher design stress owing to its greater resistance to buckling. This is also true for other geometric sections such as fins, cruciforms and triangular sections.

At their discretion, designers may determine the r value of such sections, calculate the thickness of a solid rectangular section to give the same r value and use the thickness of the equivalent solid section as the effective thickness of the geometric section. This juggling with figures will be necessary until the code recognises such sections.

Table 2.16. Capacity reduction factor β

Slenderness ratio h_{ef}/t_{ef} or l_{ef}/t_{ef}	β
0	1·00
6	1·00
8	1·00
10	0·97
12	0·93
14	0·89
16	0·83
18	0·77
20	0·70
22	0·62
24	0·53
26	0·45
27	0·40

2.12.12. Capacity reduction factor

The taller a wall, column or other compressive structural element becomes the greater is its tendency to buckle and the lower is the potential design strength of the element because it will fail by buckling before failing by crushing.

The design strength of a compressive element must therefore be reduced by a factor (called the 'capacity reduction factor β') depending on the slenderness ratio of the compressive element. Values of β given in BS 5628,[1] and based on effective thickness, are quoted in Table 2.16 for concentric loading.

Tests have shown that the values of β are the worst case values and are operative over about the central fifth of the height of the element (the area of maximum bending). Over the remaining height of the member, β varies between this value and unity at the restraints.

2.12.13. Short and long walls, and columns: limiting slenderness ratio

The graph of design stress against slenderness ratio for a steel column (an elastic material) is shown in Fig. 2.24.

The graph for steel shows some resemblance to the graph for masonry (Fig. 2.25), which is a brittle material. Fig. 2.25 plots β against slenderness ratio (based on effective thickness).

For a slenderness ratio of less than 8 the column or wall is short, i.e. the full characteristic strength of the masonry and reinforcement can be used to determine its design strength and the strut is unlikely to fail by buckling. When the slenderness ratio exceeds 8 the column is long and the longer it gets the greater will be the reduction of the characteristic strength, and hence the lower its design strength.

The limit to the magnitude, for design purposes, of the slenderness ratio is 27. It has been found in research that columns with a slenderness ratio of greater than 27 still have some loadbearing capacity, but it would be prudent in design not to exceed this value.

2.12.14. Design compressive strength of a wall

The design compressive strength of a wall per unit area is the product of the capacity reduction factor β and the characteristic strength $f_{\mathbf{k}}$. This

Fig. 2.24 (below). Design stress plotted against slenderness ratio for steel column

Fig. 2.25 (right). Design stress plotted against slenderness ratio for masonry

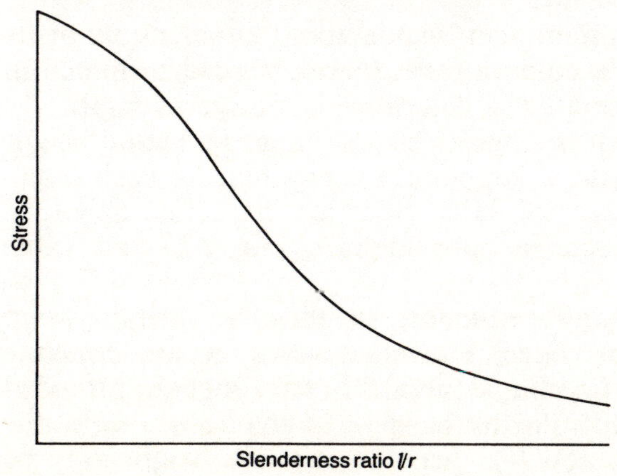

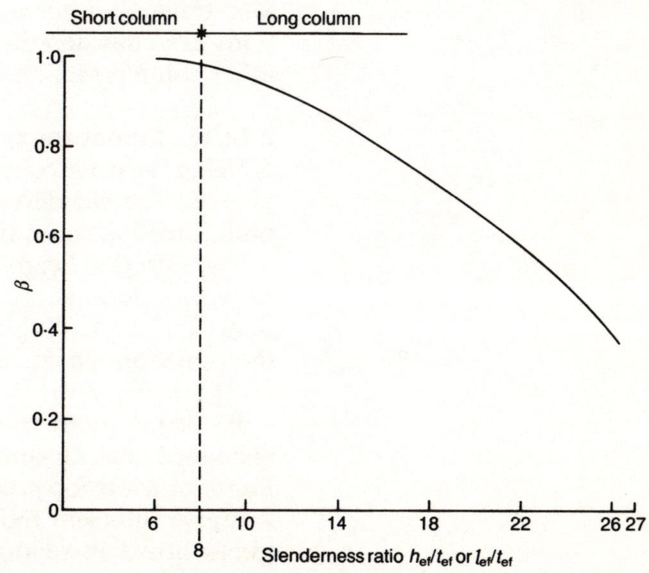

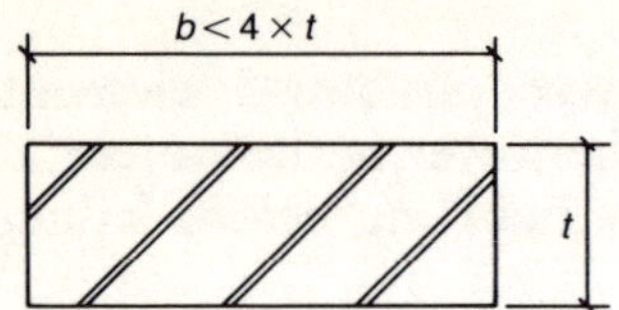

Fig. 2.26. Column definition

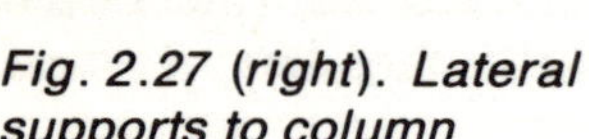

Fig. 2.27 (right). Lateral supports to column

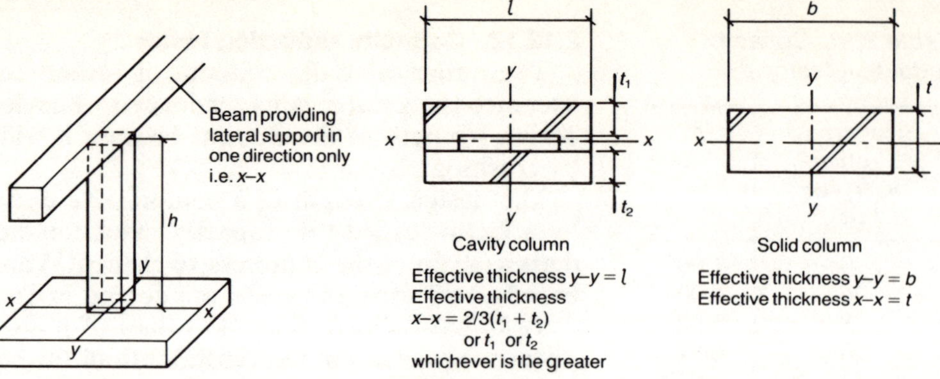

Fig. 2.28. Solid and cavity columns: effective thicknesses

strength must be divided by the partial safety factor for materials γ_{mm}. Hence the design compressive strength is $\beta f_k / \gamma_{mm}$.

The design compressive strength of the wall, per unit length, is determined by multiplying this term by the actual thickness of the wall t. Therefore the design compressive strength of masonry per unit length is $\beta f_k t / \gamma_{mm}$.

The addition of reinforcement increases axial loadbearing capacity. The added design compressive strength of the reinforcement is again the product of β and the characteristic strength of steel f_s, divided by the partial safety factor for reinforcement γ_{ms}, multiplied by the area of reinforcement A_s per unit length of the wall. Therefore the design strength of reinforcement is $\beta f_s A_s / \gamma_{ms}$.

The design compressive strength of a reinforced masonry wall is thus

$$\beta \left[\left(\frac{f_k t}{\gamma_{mm}} \right) + \left(\frac{f_s A_s}{\gamma_{ms}} \right) \right]$$

This assumes, reasonably, that the ratio of area of steel to area of masonry is small and that no reduction in masonry area is necessary.

2.12.15. Columns

Columns are designed in a similar manner to walls. A column is defined in BS 5628[1] as a 'vertical loadbearing member whose width is not greater than four times its thickness' (see Fig. 2.26) and only rectangular or square sections are considered. The Authors consider this to be restrictive. Advice on other column sections is given in section 4.3 and chapter 7.

2.12.16. Slenderness ratio of columns

Being an isolated member, a column can buckle about either its xx or its yy axis. The slenderness ratio of a column must therefore be determined in both directions, and the worst case used to determine its design strength.

The effective height of a column is equal to or less than the actual height between lateral supports (when lateral supports are provided in both directions and depending on their degree of fixity) or twice the actual height in the direction where lateral support is not provided (see Fig. 2.27 and Table 2.17).

BS 5628[1] does not suggest any reduction in effective height when enhanced lateral supports are provided. However, when in situ concrete floors or roofs (or precast slabs providing equivalent restraints) are provided and give sufficient rigidity perpendicular to the span to give resistance to the forces noted in section 2.12.4 (*a*) and (*b*), then the effective height may be assumed as 0·75 times the clear distance between such lateral supports only at the designer's discretion, based on engineering judgment and experience.

The effective thickness of a solid column is equal to the actual thickness t relative to the direction being considered. The effective thickness of a cavity column (perpendicular to the cavity) is, as with walls, taken as two-thirds of the sum of the thickness of the leaves or the actual thickness of the thicker leaf, whichever is the greater. In the other direction the effective thickness is equal to the plan length of the leaves (see Fig. 2.28).

2.12.16.1. *Types of cavity columns*

Where the cavity is properly tied, reinforced and fully grouted up, as in a reinforced masonry column, the effective thickness may be taken as the thickness of the two leaves plus the thickness of the grouted cavity, i.e. the overall thickness.

When the cavity is used to position prestressing rods and the leaves are fully bonded to the return ends, so as to act structurally integrally, then the section may be treated as a hollow box section. The effective thickness may be considered as either the actual thickness or the thickness of a solid column of the same radius of gyration.

For hollow box sections, cruciforms, tubes and similar, the effective thickness may be taken as the overall thickness or the thickness of a solid column of the same radius of gyration as the geometric section column (and axes xx, yy and vv may need consideration). (Columns formed by openings are dealt with in reference 13.)

2.12.17. Design compressive strength of columns

The design strength of a column is given by the product of the capacity reduction factor β, the column area (bt for solid columns) and the characteristic strength of the masonry f_k, divided by the partial safety factor for materials γ_{mm}.

Table 2.17. Effective heights of columns

End condition		Type of restraint	Effective height h_{ef}
Column restrained at least against lateral movement top and bottom		Floor or roof of any construction spanning on to column from both sides at the same level	h in respect of both axes
	Bearing	Concrete floor or roof, irrespective of direction of span, which has a bearing of at least $2r/3$ but but not less than 90 mm	h in respect of both axes
Column restrained against lateral movement at top and bottom by at least two ties 30 × 5 mm min. at not more than 1·25 m centres		No bearing or bearing less than case above	h in respect of minor axis
	Ties	Floor or roof of any construction irrespective of direction of span	$2h$ in respect of major axis

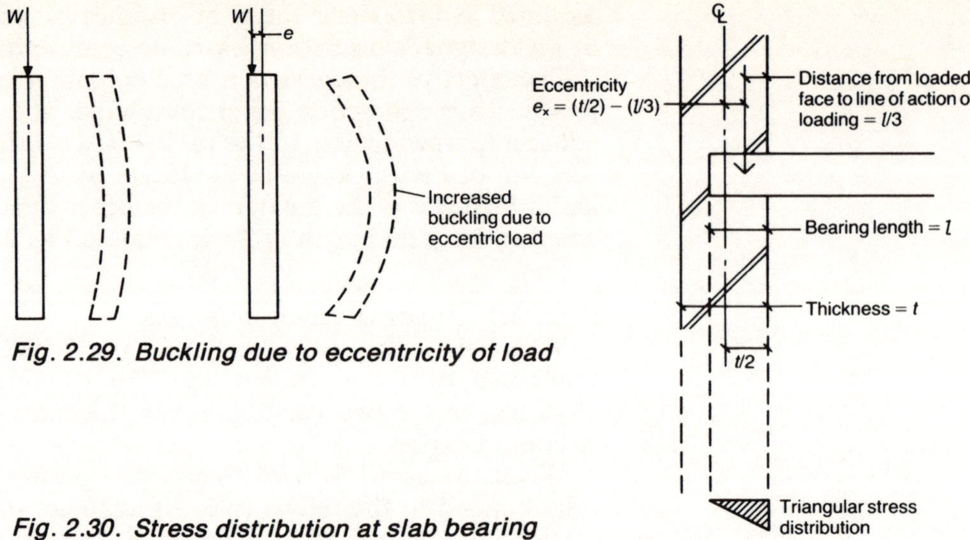

Fig. 2.29. Buckling due to eccentricity of load

Fig. 2.30. Stress distribution at slab bearing

Hence the design compressive strength of a column is given by $\beta b t f_k / \gamma_{mm}$.

The additional loadbearing capacity provided by the reinforcement is similarly $\beta A_s f_s / \gamma_{ms}$.

Thus the design compressive strength of a reinforced masonry column is

$$\beta\left[\left(\frac{b t f_k}{\gamma_{mm}}\right) + \left(\frac{A_s f_s}{\gamma_{ms}}\right)\right]$$

2.12.18. Eccentric loading and bending moments

It is difficult to obtain true concentric loading in a laboratory, and even more so in construction. The values for β given in Table 2.18 are only for eccentricities of up to $0.05t$. Such a minor eccentricity is, for all practical purposes, considered as a concentric load.

Larger eccentricities are common in practice. Typical examples are internal walls supporting floors of unequal span and loading, and gable walls supporting floors on one side of the wall only. In prestressed masonry it can be structurally efficient to apply the load at an eccentricity of Z/A. (For a solid rectangle

$$\frac{Z}{A} = \frac{b d^2/6}{b d} = \frac{d}{6}$$

i.e. at the edge of the middle third or kern.)

The application of a load eccentrically induces a moment into the column of We (where W is load and e is eccentricity) which can increase its tendency to buckle (see Fig. 2.29).

In floors and beams it is generally assumed (unless there is evidence to the contrary) that the load transmitted to a wall by a single floor, or to a column by a single beam, acts at one third of the length of the bearing from the loaded face. A triangular stress distribution is generally assumed under the bearing (see Fig. 2.30).

Where floors (or beams) are continuous over walls (or columns), each span of the floor or beam is generally assumed as being supported individually on half the total bearing area (see Fig. 2.31). If the reactions from the slabs or beams are equal then a concentric load results, but if the common variations in loading conditions and spans occur then an effective eccentric load on the wall will result.

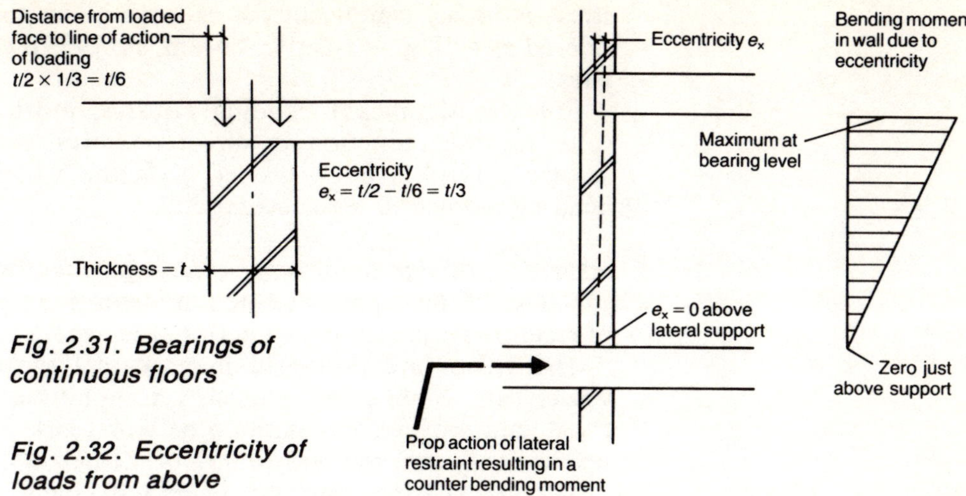

Fig. 2.31. Bearings of continuous floors

Fig. 2.32. Eccentricity of loads from above

Tests have shown that the effect of the eccentricity decreases down the height until finally it is zero at the base of the member. The Authors suggest that this phenomenon, which has not been fully considered in practical masonry design for many years, is dependent on the stiffness of the lateral support propping the top and bottom of the walls.

Thus the eccentric vertical load at the top of the member may be considered as a concentric load at the base (at the lateral supports of the base, see Fig. 2.32).

2.12.19. Combined effect of slenderness and eccentric loading

As discussed in section 2.12.18, the eccentric loading increases the tendency of the strut to buckle, and thus can reduce its loadbearing capacity.

The eccentric load may be considered as a concentric load plus a couple We_x (see Fig. 2.33(a)). The downwards concentric load W is cancelled by the upward force of the couple so that the net force is W at an eccentricity of e_x. The stress in the strut comprises a direct compressive stress W/A plus a bending stress $\pm We_x/Z$. The combined stress is $W/A \pm (We_x)/Z$ (see Fig. 2.33(b)).

The combined stress is a combination of direct and bending stress. The tendency of the strut to buckle is increased by the bending stress which can further reduce its compressive loadbearing capacity. When the bending

Fig. 2.33. Combined stress due to eccentric load

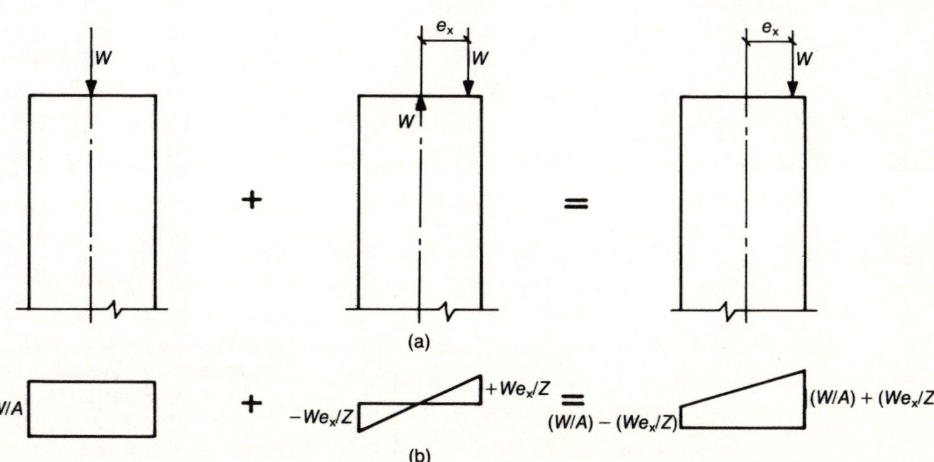

stress is highly significant (i.e. is large in magnitude) the strut can be considered as acting as a vertical beam, subject to combined bending and direct stress.

For less significant bending stresses, most codes (including BS 5628[1]) cater for the reduction in direct compressive loadbearing resistance (see section 2.12.19.1) by a revised β factor which makes allowance for the bending moment due to eccentricity.

2.12.19.1. Design compressive strength: walls

Values of the capacity reduction factor β are given in Table 2.18 for values of eccentricity e_x ranging from 0 to $0\cdot3t$, where t is the thickness of the wall.

The values are produced in graphical form in Fig. 2.34. Intermediate values can be obtained in linear interpolation. As discussed in section 2.11 these values of β are maximum or worst case values and are considered to operate only over the central fifth of the height of the element.

The loadbearing resistance of reinforcement must also be reduced by the same capacity reduction factor as for the column or wall.

The design compressive strength of walls is given as

$$\frac{\beta f_k t}{\gamma_{mm}} + \frac{\beta f_s A_s}{\gamma_{ms}}$$

per unit length.

2.12.19.2. Design compressive strength of eccentrically loaded columns

Since the application of loading to a column can be eccentric relative to two axes (i.e. biaxial, as opposed to that for a wall where the eccentricity is usually, except for some shear walls, related to the centre line axis only), the treatment of eccentricity is more complicated.

A rectangular column has a major axis, about which the column has the

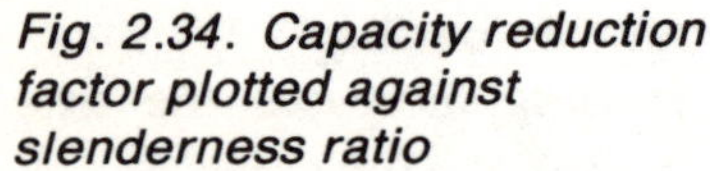

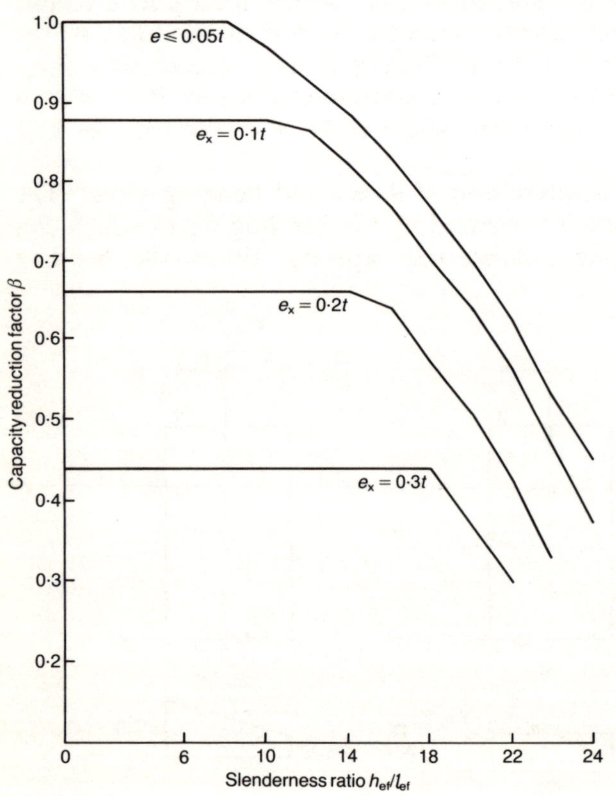

Fig. 2.34. Capacity reduction factor plotted against slenderness ratio

Table 2.18. Capacity reduction factor β (from BS 5628[1])

Slenderness ratio h_{ef}/t_{ef}	Eccentricity at top of wall e_x			
	Up to $0\cdot05t$*	$0\cdot1t$	$0\cdot2t$	$0\cdot3t$
0	1·00	0·88	0·66	0·44
6	1·00	0·88	0·66	0·44
8	1·00	0·88	0·66	0·44
10	0·97	0·88	0·66	0·44
12	0·93	0·87	0·66	0·44
14	0·89	0·83	0·66	0·44
16	0·83	0·77	0·64	0·44
18	0·77	0·70	0·57	0·44
20	0·70	0·64	0·51	0·37
22	0·62	0·56	0·43	0·30
24	0·53	0·47	0·34	
26	0·45	0·38		
27	0·40	0·33		

* It is not necessary to consider the effects of eccentricities up to and including $0\cdot05t$.
† Linear interpolation between eccentricities and slenderness ratios is permitted.
‡ The derivation of β is given in appendix B of BS 5628.[1]

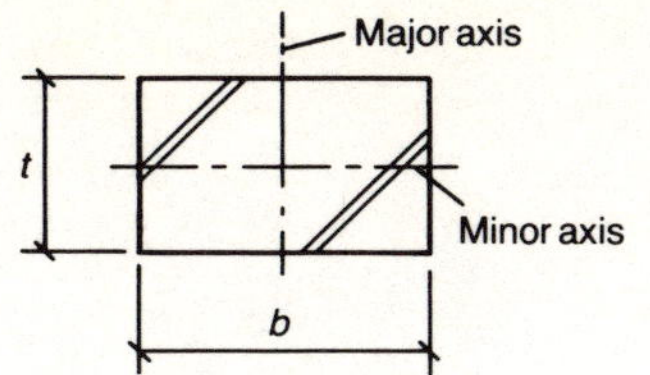

Fig. 2.35. Major and
minor axes of a
rectangular column

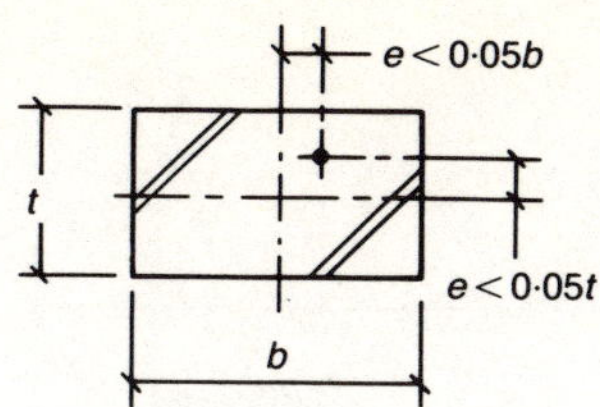

Fig. 2.36. Case 1

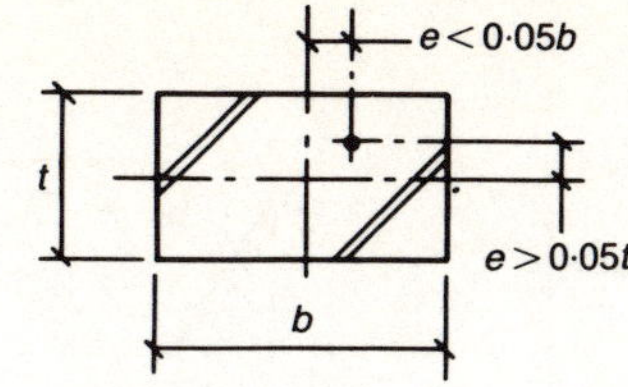

Fig. 2.37. Case 2

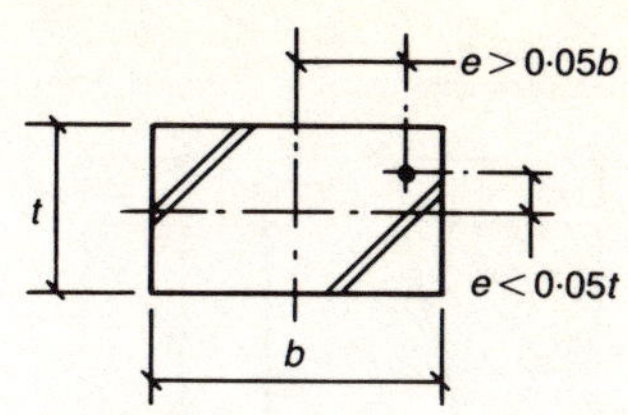

Fig. 2.38. Case 3

larger second moment of area, and a minor axis perpendicular to the major axis (see Fig. 2.35).

There are four different loading cases

- case 1: nominal eccentricity about both axes (i.e. e_x less than $0.05t$ or $0.05b$)—a common case is an internal column supporting main beams of equal loading on both axes
- cases 2 and 3: nominal eccentricity about one axis and significant about the other axis—typical is an external column supporting edge beams about one axis and a main beam about the other axis
- case 4: significant eccentricity about both axes—such cases of biaxial bending occur in external corner columns.

2.12.19.2.1. Case 1: nominal eccentricity about both axes. When the eccentricities about both axes are less than $0.05t$ and $0.05b$, β is taken from Table 2.16 as the value for e_x up to $0.05t$, with the slenderness ratio based on the value of t_{ef} of the minor axis (see Fig. 2.36). In this case the slenderness ratio SR is based on effective height (determined relative to the minor axis)/ effective thickness (based on t), or h_{ef} (relative to the major axis)/b_{ef}, whichever is the greater.

2.12.19.2.2. Case 2: nominal eccentricity about major axis, significant eccentricity about minor axis. When the eccentricities about both axes are less than $0.05b$ but greater than $0.05t$, the value of β is taken from Table 2.18 using the eccentricity values and slenderness ratio appropriate to the minor axis (see Fig. 2.37). In this case the slenderness ratio SR is based on effective height (minor axis)/effective thickness (minor axis).

2.12.19.2.3. Case 3: nominal eccentricity about the minor axis, significant eccentricity about the major axis. When the eccentricities about both axes are greater than $0.05b$ but less than $0.05t$, β is taken from Table 2.18, using the value of eccentricity appropriate to the major axis, but the value of the slenderness ratio is again appropriate to the minor axis (see Fig. 2.38). In this case the slenderness ratio SR is based on effective height (minor axis)/ effective thickness (minor axis).

2.12.19.2.4. Case 4: significant eccentricity about both axes. When the eccentricities about both axes are greater than $0.05t$ and $0.05b$, β is calculated by deriving additional eccentricities and then substituting them in the appropriate formula (based on appendix A of BS 5628[1]) as follows.

(a) The effect of the eccentricity of the applied load is assumed to vary from e_x at the point of application to zero above the lateral support at the base (see Fig. 2.39).

(b) An additional eccentricity e_a is used to allow for slenderness effect (the tendency to buckle). The additional eccentricity is, again, considered to vary linearly from zero at the lateral support to a value over the central fifth of the height of the element given by

$$e_a = t[(1/2400)(h_{ef}/t_{ef})^2 - 0.015]$$

(see Fig. 2.39).

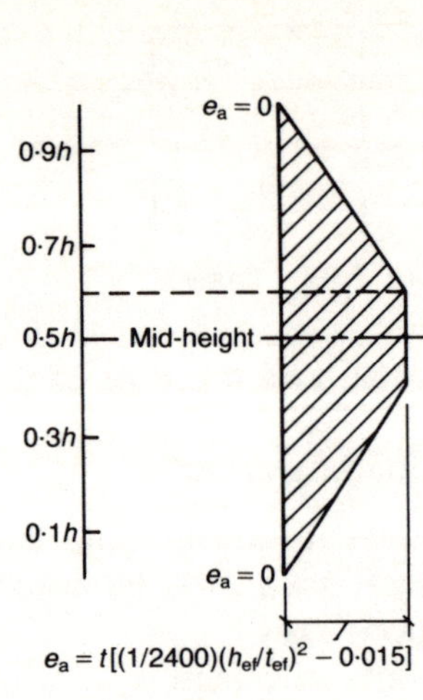

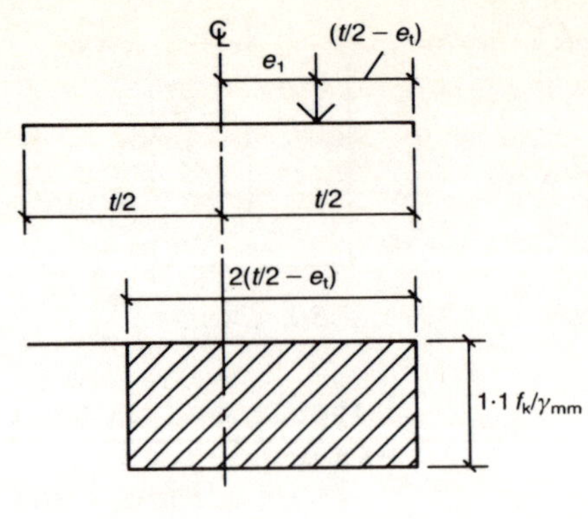

Fig. 2.40. Ultimate stress block under eccentrically loaded wall

$$e_a = t[(1/2400)(h_{ef}/t_{ef})^2 - 0.015]$$

Fig. 2.39. Variation of e_a and e_x over height of member

Table 2.19. Columns: maximum slenderness ratios

Condition	Maximum slenderness ratio
Multi-storey	27
Single storey	18
Free end	9

Table 2.20. Basic span/effective depth ratios for rectangular beams with spans of up to 10 m

Support conditions	Ratio
Cantilever	7
Simply supported	20
Continuous	26

(c) The total design eccentricity e_t (to determine β) is the sum of e_x and e_a at the point being considered. When considering the mid-height section (where e_a is a maximum) the maximum value of e_t will be

$$e_t = 0.6e_x + e_a$$

at mid-height. When considering the top of the element, e_a will be zero and e_x at its maximum. Thus e_t will be equal to e_x.

The ultimate stress block for an eccentrically loaded section is then assumed to be as shown in Fig. 2.40.

In Fig. 2.40 the characteristic strength f_k has been multiplied by a factor of 1·1, i.e. there is a 10% increase. By implication BS 5628[1] recognises that the stress is not direct alone but also partly due to bending. The Authors suggest that the approach in BS 5628 is not entirely satisfactory and propose a sliding scale for the multiplying factor of from 1 to 1·20, depending on the magnitude of the bending stress.

From Fig. 2.40 it can be shown that the design vertical loading is equal to the area of the stress block multiplied by the design stress, i.e.

$$2\left(\frac{t}{2} - e_t\right)\frac{1\cdot1f_k}{\gamma_{mm}} = \frac{1\cdot1(1 - 2e_t)tf_k}{\gamma_{mm}}$$

per unit length, which may be written as $\beta tf_k/\gamma_{mm}$ because β is, in effect, $1\cdot1(1 - 2e_t/t)$.

For the columns in case 4 the value of β may be calculated for each axis and the minimum design capacity calculated. This method has a more general application and may be used to determine β for any member at any position.

2.12.19.3. Eccentricity for geometric sections

The eccentricity and its effect as discussed in section 2.12.19.2 are solely for solid rectangular sections and as such they can be inappropriate for application to geometric sections.

Consider the diaphragm wall section shown in Fig. 2.22. For maximum efficiency of prestress, the eccentricity of the prestressing force from the central axis should be Z/A.

For the section shown in Fig. 2.22 $Z = 56·67/2 = 28·53$ units[3] and $Z/A = 28·33/26 = 1·09$ units.

Table 2.21. Special span/effective depth ratios for rectangular beams with spans greater than 10 m

Span, m	Cantilever	Simply supported	Continuous
10	Value to be	20	26
12	justified by	18	23
14	calculation	16	21
16		14	18
18		12	16
20		10	13

Table 2.22. Limiting ratios of span to effective depth for walls and columns

End condition	Ratio
Simply supported	35
Continuous	45
Cantilever with up to 0·5% reinforcement	18*

* May be increased by 30% for free-standing walls, with no finishes which may be damaged by deflection, subjected mainly to wind pressure.

If the section is considered to have an effective thickness, equal to its overall depth, of 4 units then $e_x = 1·09/4·0 = 0·2725$ and, from interpolation of Table 2.18, for a short column $\beta = 0·49$.

If the section is considered to have an effective thickness of a solid section of equal radius of gyration, then the effective thickness is 5·1 and $e_x = 1·09/5·1 = 0·214$.

In this case, by interpolation, for a short column $\beta = 0·63$. This results in an increase in design strength of about 25%.

The Authors are of the opinion that e_x should be related to the effective thickness of a solid section of equal radius of gyration. Cautious designers may prefer to use the overall depth of the geometric section. In design for bending, designers are likely to employ the β factor for zero eccentricity, taking the additional stresses due to bending into account by designing for the bending moment produced by resulting eccentricity and applied loads.

2.12.20. Vertical free and propped cantilevers subject to lateral load

Vertical free and propped cantilevers are a special case in that, when subject to lateral load resulting in bending stresses greater than direct compression, they are treated not as struts but as cantilevers and beams. However, before lateral loading is applied their strut behaviour must be considered.

For reinforced free and propped cantilevers this is usually no problem for there is relatively light compressive loading, but heavily prestressed sections are often subject to significant direct compressive loading, from the prestress, before the lateral load is applied. A typical example is a prestressed earth retaining wall. Before the backfilling (producing the lateral pressure) is applied the wall will act as a free-ended strut and the strut behaviour discussed in chapter 7 must be examined. When the lateral pressure is applied, then the section should be analysed as a beam section under bending. Generally it is the bending action which is the more critical design case.

2.13. Limiting dimensions for columns and beams

2.13.1. Introduction

To reduce detailed calculations in checking that the limit states of buckling, deflection and cracking are not reached, the limiting ratios from Tables 2.19–2.22 may be used.

2.13.2. Columns

The maximum slenderness ratios for multi-storey, single storey and free end conditions are given in Table 2.19.

2.13.3. Beams

The basis of the design of reinforced masonry beams is dealt with in detail in chapter 3 where it is shown to have similarities with the design of reinforced concrete beams. There is considerable scope for widening the application of structural masonry by reinforcing it. Reinforced masonry has been used extensively in earthquake zones in America and Japan, and in countries such as India where concrete is expensive. Its application in western Europe and elsewhere has been limited owing to lack of appreciation of the potential of the technique and lack of guidance on its design.

Detailed guidance is given in chapter 3. The following sections deal only with spans and limiting dimensions.

2.13.3.1. Effective span

The effective span of a simply supported member may be taken as whichever is the lesser of the distance between the centres of supports and the clear distance between supports plus the effective depth (the effective depth is the distance between the top of the compression flange of the member to the centre line of the reinforcement). These distances are shown in Fig. 2.41.

The effective span of a continuous member may be taken as the distance between the centres of supports (see Fig. 2.42).

The effective span of a cantilever may be taken as the length from the free end to the centre of support, where it forms the end of a continuous member or the length from the free end to the face of the support plus half the effective depth (see Fig. 2.43).

2.13.3.2. Limiting dimensions: beams

To ensure the lateral stability of a masonry beam (i.e. to prevent the compression flange from buckling) it is advisable to restrict the ratio beam effective span/beam breadth b_c.

For simply supported and continuous beams, the clear distance between lateral restraints should not exceed the lesser of $60b_c$ and $250b_c/d$, where b_c is the width of the compression face midway between restraints and d is effective depth.

For cantilevers with lateral restraint at the support only, the clear distance from the end of the cantilever to the face of the support should not exceed the lesser of $25b_c$ and $100b_c/d$.

Fig. 2.41 (right). Effective span of simply supported beams

Fig. 2.42 (below). Effective span of continuous beams

Fig. 2.43 (below right). Effective span of cantilever beam

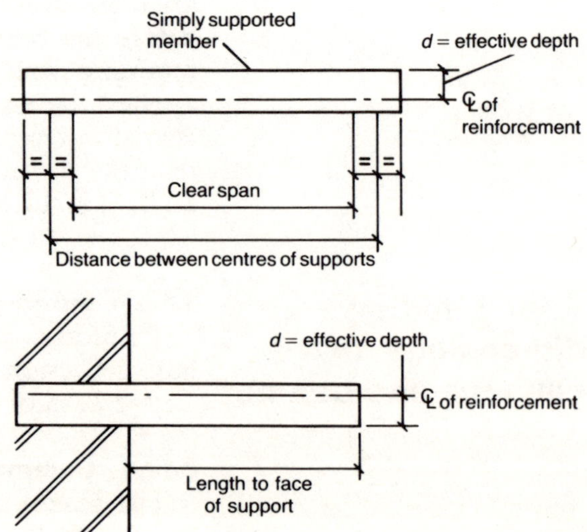

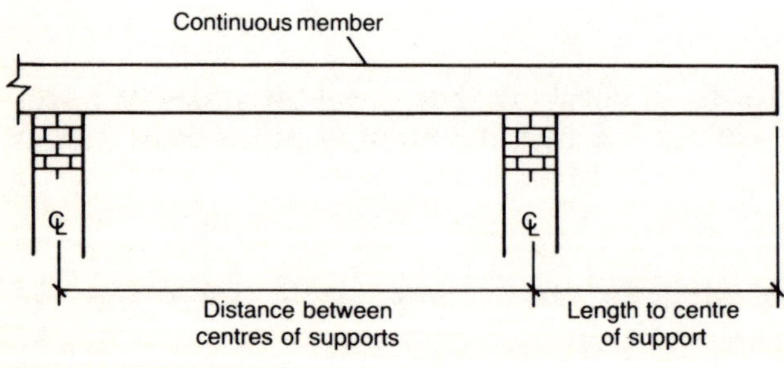

The lateral restraints (i.e. intersecting beams, walls and so on) should be horizontal lateral restraints perpendicular to the direction of span of the beam.

To prevent undue deflection and to ensure that the serviceability limit state of deflection is not exceeded, it is advisable to restrict the ratio beam effective span/effective depth d. Generally, for beams up to 10 m span, the limiting deflection should not exceed span/250 to avoid damage to finishes and partitions. The values quoted in Table 2.20 are for beams of up to 10 m span and those in Table 2.21 for beams exceeding 10 m.

2.13.3.3. Limiting dimensions: walls (and columns in bending)

To ensure compliance with the serviceability limit state for deflection the slenderness ratio of walls and columns should not exceed 27.

When walls are reinforced to resist lateral loading the ratio span/effective depth should comply with the values given in Table 2.22.

2.13.4. Sections with high proportion of concrete infill

There are occasions (particularly when using concrete hollow blockwork) when the proportion of concrete infill is so high, compared with the masonry, that the section could be designed as reinforced concrete. Since the factors of safety are more realistic in reinforced concrete design than the excessive factors in the masonry code, this usually results in a more economical design for such sections.

3 | Basis of reinforced masonry design

3.1. Introduction

For all structural materials, most present-day design theories are based on the stress–strain relationship of the material. Masonry (and particularly reinforced and prestressed masonry) is no different and account must be taken, in considering the behaviour and performance of the section under load, of the stress–strain curves and the limits being set at working and ultimate load. Fig. 3.1 shows a typical stress–strain curve for steel reinforcement.

Within the elastic range (i.e. the lower part of the graph), where stress is proportional to strain, the elastic theory is applied. Approaching failure, however, the upper portion of the graph is applied and the material layers yield. Fig. 3.2(a) indicates a typical theoretical stress block for a medium stressed uncracked masonry wall, subjected to bending and compression. Stress block (a) shows that the stress–strain relationship is directly proportional, i.e. the section is assumed to be behaving elastically. However, in Fig. 3.2 (b), where the wall has cracked and is approaching ultimate load, the tensile stress resistance is taken as zero, because the section is cracked on the tensile side and compressive stress is assumed not to be proportional to strain, i.e. the material has yielded.

The outer fibres, as they reach the yield stress, progressively transfer the excess stress into the body of the material as the strain in the outer fibres increases for little or no increase in stress.

3.2. Resistance to bending

In the design of sections to resist bending moments, full advantage of the high compressive strength of masonry is not always taken. For example, the design of plain masonry elements to resist bending often restricts the compressive stress to a relatively low value so as to limit the tensile stress to that permissible for plain uncracked sections (see Fig. 3.3).

In order to improve the performance of masonry under such loading conditions, it is necessary to exploit the compressive strength of the masonry and overcome its tensile weakness. This can be achieved by reinforcing the section to improve its tensile resistance or by precompressing the section, using prestressing methods such as the post-tensioned technique (see chapter 6).

In order to calculate the required sections for reinforced or prestressed masonry, certain assumptions and theories need to be adopted. The principles of limit state design philosophy as applied to plain masonry and used as a basis for BS 5628: Part 1[1] can be applied to reinforced and prestressed masonry with some adjustments and additions. This has been attempted in BS 5628: Part 2 and where appropriate it has been used in this book. Other additional variations are also highlighted in chapters 4 and 7.

The basic principle of the structural design philosophy is that under the most onerous loading conditions the various design strengths of the materials comprising a particular element should not be exceeded (see chapter 2).

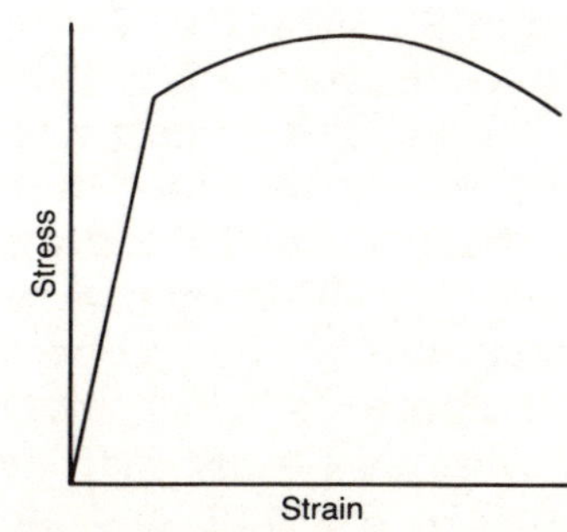

Fig. 3.1. Stress–strain diagram

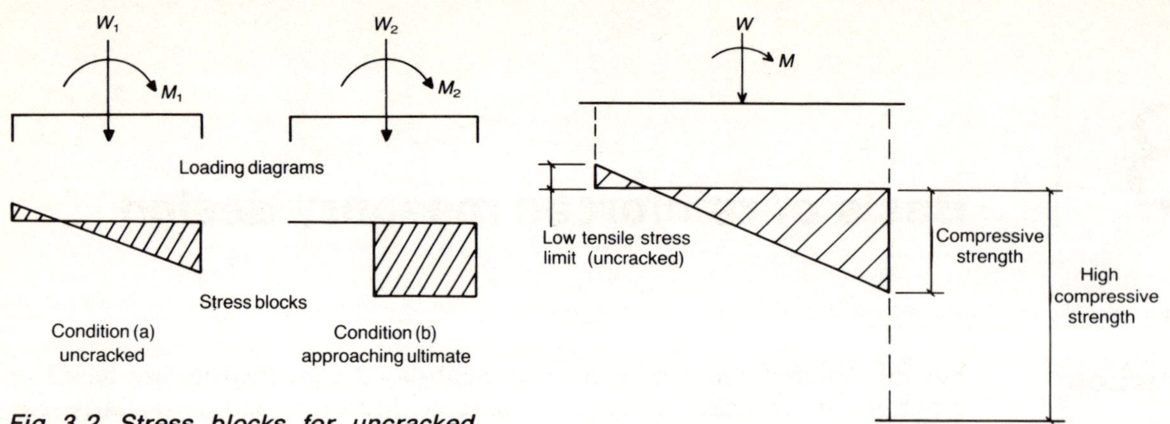

Fig. 3.2. Stress blocks for uncracked and cracked section

Fig. 3.3. Stresses due to combined axial and flexural loading

In addition, the element must remain serviceable, i.e. it must not have excessive deflection, cracking and so on. This generally means that sections are analysed for the ultimate state of collapse and are checked using either a working load analysis or a rule of thumb guide for the serviceability limit state.

In the limit state design of plain masonry, in BS 5628: Part 1 the masonry strengths are expressed in terms of characteristic compressive strength (i.e. direct compression) and characteristic flexural tensile strength (i.e. tensile strength in bending).

However, for masonry in bending, flexural compression occurs as well as direct compression, and both need to be considered in the assessment of material strength.

The relationship between direct and flexural compression values in masonry is currently under discussion and in BS 5628: Part 2 the same value is adopted for both conditions. In the Authors' opinion, when purely flexural compression is being considered a value 1·5 times that of the direct compressive strength, based on a simplified rectangular stress block, could be used and where direct and flexural compression is combined a sliding scale could be adopted, varying from 1·0 to $1·5f_k$, depending on the ratio of direct to flexural compression. Such an approach, however, would need to be supported by appropriate research. In the absence of such research, the Authors proposed that a global value for flexural compression f_f of $1·2f_k$ should be used for the general situation, with adjustment up or down, at the design engineer's discretion, where appropriate. This proposal varies slightly from that of BS 5628. The proposed stress blocks are summarised in Fig. 3.4.

The alternative approach which has been adopted in BS 5628: Part 2 is to allow no increase in flexural compression but to reduce the partial factors of safety on material strengths from 2·5 and 2·8 to 2·0 and 2·3, respectively. The Authors are of the opinion that there is good reason to adopt both lower values for the partial factors of safety on material strengths and the increased value of flexural compressive strength of $1·2f_k$. This opinion is based on many years of design and practical experience but has not yet been universally accepted. Therefore for a conservative approach which gives results compatible with BS 5628: Part 2, the Authors would suggest that the $1·2f_k$ value should be used for flexural compressive strength and no reduction made to the partial safety factors for material strength, i.e. the original γ_m factors from table 4 of BS 5628: Part 1 should be used. In this book in many examples both the increased stress and lower γ_m values have been adopted but at present these must be used with caution.

3.3. Resistance to axial load

For reinforced masonry subjected to axial loading, the load is shared between the masonry, the grout and the reinforcement. In considering this load-sharing, a number of factors need to be taken into account in deciding the proportion of load on each of the materials or in selecting the most suitable combination.

In addition to the more general considerations used in reinforced concrete (e.g. bond, anchorage and compatibility of elasticity) the differential moisture expansion and shrinkage combinations can be critical. For example, Fig. 3.5 shows a column constructed with a clay brick box section, reinforced and grouted, and indicates the likely movement and resulting effects.

The expansion of the clay brickwork would have the effect of taking up a larger proportion of the load from the combined section and the grout would thus carry less load.

The Authors recommend that the following points be observed in design.

- Large grout pockets in clay brickwork should be avoided where pos-

Fig. 3.4. Proposed flexural compressive stress blocks

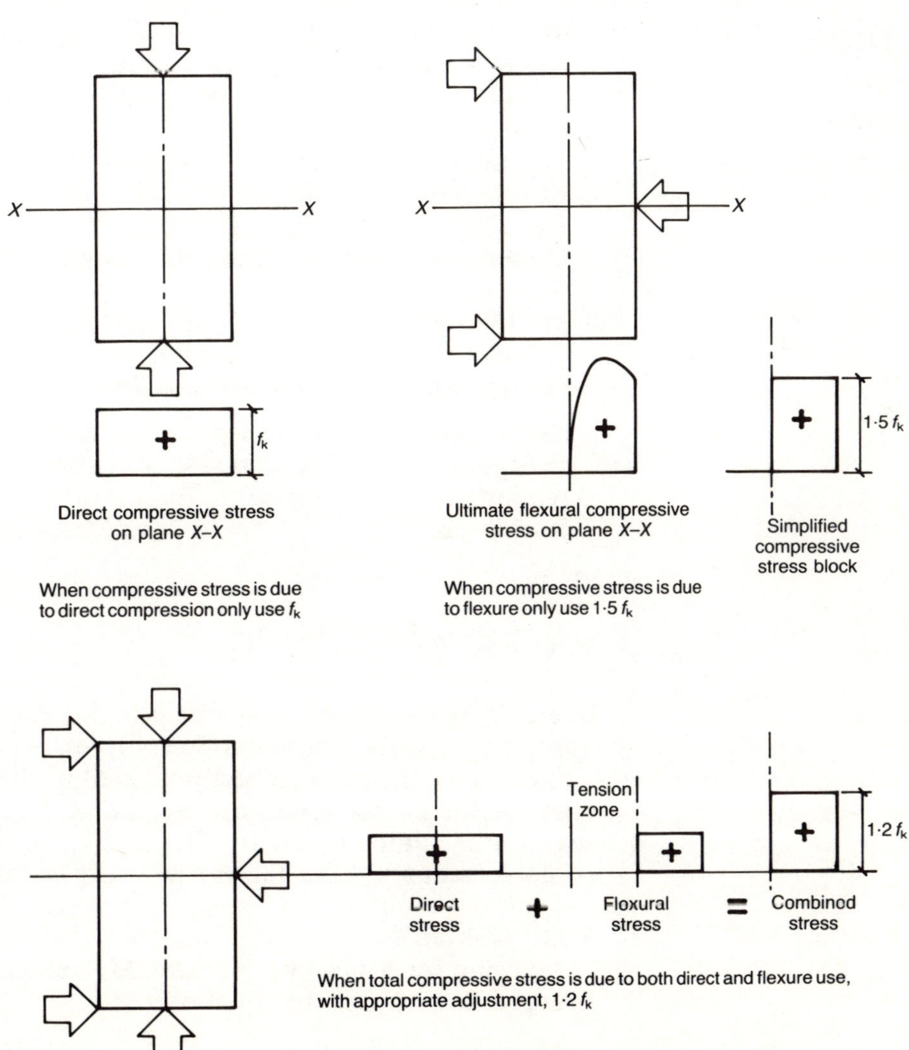

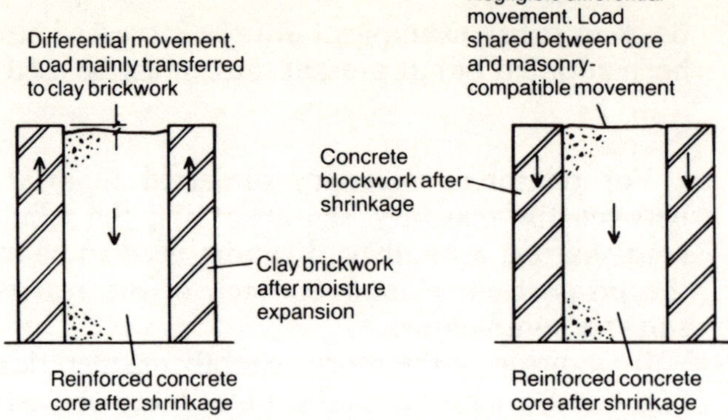

Fig. 3.5. Column sections

sible, but when used should be considered to transfer bond and shear and should not be assumed to share the direct compressive stresses.

- Where large grout pockets are used in concrete brickwork/blockwork or calcium silicate brickwork (these have similar movement characteristics to those of the grout) load-sharing can be taken into account, at the engineer's discretion, and assumed to act as an extra area of masonry.

3.4. Design formulae: compressive load resistance

After deciding on the section to resist compressive loads, the design strength of the axially loaded member is taken as the sum of the axial strength of the masonry and the axial strength of the reinforcement in compression.

The reduction of area of the masonry displaced by reinforcement is generally neglected.

The effect of slenderness is taken into account by method A or method B described in sections 3.4.1 and 3.4.2, respectively.

3.4.1. Method A

Calculate the capacity reduction factor (see chapter 2) and apply this to the design strength calculation as follows

| design axial strength of reinforced brickwork | = | capacity reduction factor | × | design strength of brickwork | + | design strength of reinforcement |

or

$$N_d = \beta\left(\frac{f_k A_m}{\gamma_{mm}} + \frac{0{\cdot}83 f_y A_s}{\gamma_{ms}}\right) \tag{3.1}$$

where N_d is the design axial strength, β is the capacity reduction factor (see Table 3.1), f_k is the characteristic compressive strength of the brickwork (see Table 2.3), A_m is the cross-sectional area of the brickwork, γ_{mm} is the partial safety factor for the brickwork (see Table 3.2), f_y is the characteristic tensile strength of reinforcement (see Table 2.4), A_s is the cross-sectional area for reinforcement and γ_{ms} is the partial safety factor for the steel (see Table 3.3).

3.4.2. Method B

Allowing for a bending moment M_a induced by the vertical load due to lateral deflection, use the equation

$$M_a = N(h_{ef})^2/2000t$$

where t is the width of the column in the plane of bending, h_{ef} is the effective height of the column and N is the design axial load. This puts the design into the category of design theory for bending and compression (see designs in chapter 4).

For geometric sections such as the diaphragm or fin wall, the concept of 'effective thickness' used in BS 5628 is inappropriate for arriving at a slenderness ratio, and in the Authors' opinion the radius of gyration should be used in a similar way to that for other materials. In the meantime the effective thickness approach, with adjustments, can be made to work.

3.5. Design formulae: bending resistance

The assumptions generally accepted for the design of reinforced masonry are similar to those commonly adopted for reinforced concrete and are as follows.

(a)   For the purposes of determining stress it is assumed that plane sections remain plane.
(b)   The strain in the outermost compressed fibre at failure is 0·0035.
(c)   The reinforcement is just about to yield at a strain of 0·002.
(d)   The tensile strength of brickwork and grout may be ignored.
(e)   The ultimate stress distribution in brickwork is taken to be rectangular with a uniform value of f_f/γ_{mm} taken over the whole of the compression zone, where f_f is the characteristic compressive strength of brickwork in bending (see section 3.11).
(f)   The depth of the compressive stress block does not exceed one half of the effective depth.

The Authors proposed in section 3.2 that a global figure for the characteristic flexural compressive strength f_f of masonry of $1·2f_k$ be adopted for the general case of combined bending and axial load. However, they also recommend that adjustments to this be made at the designer's discretion. In this

*Table 3.1. Capacity reduction factor β**

Slenderness ratio h_{ef}/t_{ef}	Eccentricity at top of wall e_x			
	Up to 0·05t†	0·1t	0·2t	0·3t
0‡	1·00	0·88	0·66	0·44
6‡	1·00	0·88	0·66	0·44
8‡	1·00	0·88	0·66	0·44
10‡	0·97	0·88	0·66	0·44
12‡	0·93	0·87	0·66	0·44
14	0·89	0·83	0·66	0·44
16	0·83	0·77	0·64	0·44
18	0·77	0·70	0·57	0·44
20	0·70	0·64	0·51	0·37
22	0·62	0·56	0·43	0·30
24	0·53	0·47	0·34	
26	0·45	0·38		
27	0·40	0·33		

* Linear interpolation between eccentricities and slenderness ratios is permitted.
† It is not necessary to consider the effects of eccentricities up to and including 0·05t.
‡ Short columns.

Table 3.2. Partial safety factors γ_{mm} for strength of reinforced masonry in direct compression and bending: ultimate limit state

Category of manufacturing control of structural units	Value of γ_{mm}
Special	2·0
Normal	2·3

Table 3.3. Partial safety factors γ_{mv}, γ_{mb} and γ_{ms}; ultimate limit state

Partial safety factor	Value
Shear strength of masonry γ_{mv}	2·0
Bond strength between concrete infill or mortar and steel γ_{mb}	1·5
Strength of steel γ_{ms}	1·15

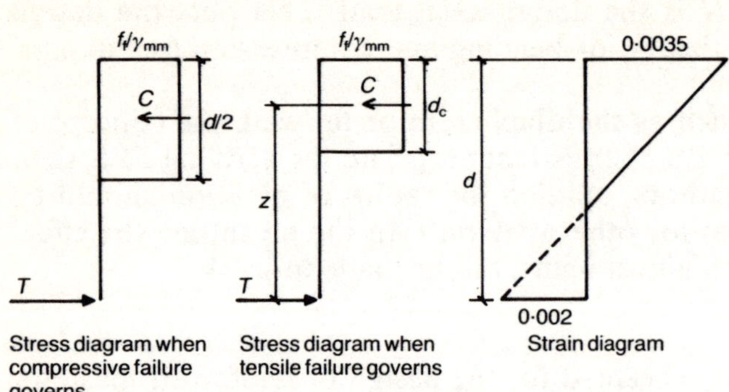

Fig. 3.6. Stress and strain diagrams due to bending

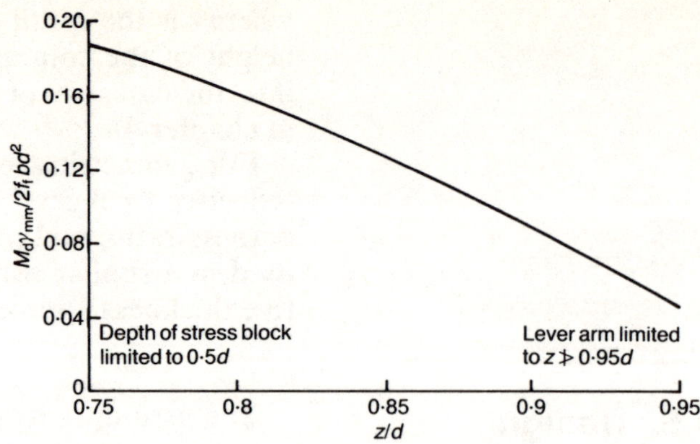

Fig. 3.7. Lever arm curve: reinforced brickwork based on rectangular stress box

recommendation the Authors suggest that for columns subject to vertical load and bending, where the vertical load is dominant, a value of $1 \cdot 1 f_k$ should be adopted, as noted in appendix B of BS 5628: Part 1.[1]

By adopting f_f as the characteristic strength of brickwork in bending, and applying this to a rectangular section subjected to bending as shown in Fig. 3.6 (where d is the effective depth, d_c is the depth of the compression zone, C is the compressive force, T is the tensile force and z is the lever arm) the maximum compressive force is given by

$$C = (f_f/\gamma_{mm})0 \cdot 5db$$

(i.e. using the maximum allowable depth of $d/2$ when the section is fully stressed) where b is the breadth of the section.

Thus, based on the masonry in bending compression, the maximum design moment of resistance M_d is given by

$$M_d = C \times \text{lever arm}$$
$$= (f_f/\gamma_{mm}) \times 0 \cdot 5d \times b \times 0 \cdot 75d$$
$$= 0 \cdot 375(f_f/\gamma_{mm})bd \tag{3.2}$$

and the maximum tensile resistance force T is given by

$$T = A_s \times 0 \cdot 83 f_y/\gamma_{ms}$$

where A_s is the cross-sectional area of the tensile steel reinforcement. The moment of resistance M_d based on the reinforcement resistance is given by

$$M_d = T \times \text{lever arm}$$
$$= (A_s \times 0 \cdot 83 f_y/\gamma_{mm})z \tag{3.3}$$

The lesser value of the design moment of resistance as obtained from equations (3.2) and (3.3) will thus be the design strength of the section in bending. By equating the tensile and compressive forces on the section, which must be equal and opposite in accordance with the laws of statics, the lever arm value can be obtained as follows.

Since $T = C$ or

$$\frac{A_s \times 0 \cdot 83 f_y}{\gamma_{ms}} = \frac{f_f}{\gamma_{mm}} d_c b$$

as $Z = d - d_c/2$

or $d_c = 2(d - z)$.

Substituting for this value of d_c and rearranging gives

$$z = d[1 - (0\cdot5A_s \times 0\cdot83f_y\gamma_{ms}/bdf_f\gamma_{mm})] \tag{3.4}$$

The lever arm is usually limited to a maximum value of $0\cdot95d$. A graphical method for determining the lever arm factor z/d is given in Fig. 3.7.

3.6. Vertical and horizontal shear resistance

The design of reinforced masonry in shear is treated generally in the same way as plain brickwork, with the exception of those shear zones in beams and other elements subjected mainly to bending, which may require supplementary shear reinforcement (which is dealt with in section 3.15).

The general expression for shear stress, v_h is

$$v_h = VA\bar{y}/Ib \tag{3.5}$$

where v_h is the shear stress, V is the horizontal or vertical shear force, A is the area of cross-section to one side of the position where shear stress is being checked, $\bar{y}$ is the distance from the neutral axis to the centroid of the area, I is the second moment of the area and b is the width at the point being checked.

The design shear strength is equal to the characteristic shear strength f_v divided by the partial factor of safety for the material strength in shear γ_{mv}. The characteristic horizontal shear strength given in BS 5628: Part 2[1] is as for racking shear and shear in prestressed sections, i.e.

$$f_v = (0\cdot35 + 0\cdot6g_B)\ \text{N/mm}^2 \tag{3.6}$$

for mortar designations (i) and (ii), with a maximum value of $1\cdot75$ N/mm², where g_B is the design vertical load in the wall. The values in BS 5628: Part 2 vary for different situations. Clauses 19.1.3.1, 19.1.3.2, and 22.5 are reproduced in this section. The vertical shear stress is always equal to the horizontal shear stress because only alternate units are bonded across the vertical joints and CP 110[14] allows a maximum of only $0\cdot35$ N/mm² for 20–40 N/mm² lightly reinforced concrete. The Authors therefore recommend caution in the use of these values, particularly if low strength units are to be used.

For reinforced masonry, the Authors advise that in general only mortar designations (i) and (ii) should generally be used and that a sensible approach to shear strength be adopted.

In masonry sections designed for flexure on the basis of an uncracked section, the horizontal shear resistance of the section may be taken as occurring over the whole plan cross-sectional area. When the cracked section analysis is applicable, only the uncracked section of masonry will provide resistance to horizontal shear. However, the vertical stress on this reduced area will be increased considerably and thus the horizontal shear resistance will be likely to be maintained at a comparably high level.

19.1.3 *Charactistic shear strength of masonry, f_v*
19.1.3.1 *Shear in bending (reinforced masonry)*
19.1.3.1.1 For sections in which the reinforcement is placed in bed or vertical joints, including Quetta bond and other sections where the reinforcement is wholly surrounded with mortar designation (i) or (ii) (see table 1), the characteristic shear strength, f_v, may be taken as $0\cdot35$ N/mm².

For simply supported beams or cantilevers where the ratio of the shear span (see **2.6**) to the effective depth is less than 2, f_v may be increased by a factor:

$2d/a_v$

where d is the effective depth; a_v is the distance from the face of the support to the nearest edge of a principal load; provided that f_v is not taken to be greater than 0.7 N/mm^2.

At sections in certain laterally loaded walls there may be substantial compressive stresses from vertical loads. In such cases the shear may be adequately resisted by the plain masonry (see clause **25** of BS 5628: Part 1: 1978).

19.1.3.1.2   For reinforced sections in which the main reinforcement is placed within pockets, cores or cavities filled with concrete infill as defined in **13.1**, the characteristic shear strength of the masonry, f_v, may be obtained from the following equation:

$$f_v = 0.35 + 17.5\rho$$

where $\rho = A_s/bd$; A_s is the cross-sectional area of primary reinforcing steel; b is the width of section; d is the effective depth (see **2.4**); provided that f_v is not taken to be greater than 0.7 N/mm^2.

For simply supported reinforced beams or cantilever retaining walls where the ratio of the shear span, a, (see **2.6**) to the effective depth, d, is six or less, f_v may be increased by a factor $\{2.5 - 0.25(a/d)\}$ provided that f_v is not taken to be greater than 1.75 N/mm^2.

19.1.3.2 *Racking shear in reinforced masonry shear walls*. When designing reinforced masonry shear walls the characteristic shear strength of masonry, f_v, may be taken to be:

$$0.35 + 0.6g_B, \text{ with a maximum of } 1.75 \text{ N/mm}^2$$

where g_B is the design load per unit area normal to the bed joint due to the loads calculated for the appropriate loading condition detailed in clause **20**.

Alternatively, for reinforced sections in which the main reinforcement is placed within pockets, cores or cavities filled with concrete infill as defined in **13.1**, the characteristic shear strength of masonry, f_v, may be taken to be 0.7 N/mm^2 provided that the ratio of height to length of the wall does not exceed 1.5.

Designers should consider the effect of damp-proof courses on shear strength of masonry (see **17.1**).

22.5   Shear resistance of elements

22.5.1 *Shear stresses and reinforcement in members in bending*. The shear stress, v, due to design loads at any cross section in a member in bending should be calculated from the equation:

$$v = V/bd$$

where b is the width of the section; d is the effective depth (or for a flanged member the actual thickness of the masonry between the ribs if this is less than the effective depth as defined in **2.4**); V is the shear force due to design loads.

Where the shear stress calculated from this equation is less than the characteristic shear strength of masonry, f_v, divided by the partial safety factor, γ_{mv}, shear reinforcement is not generally needed. In beams, however, the designer should consider the use of nominal links, bearing in mind the sudden nature of shear failure. If required, they should be provided in accordance with **26.5.2**.

Where the shear stress, v, exceeds f_v/γ_{mv}, shear reinforcement should be provided. The following requirement should be satisfied:

$$\frac{A_{sv}}{s_v} \geqslant \frac{b(v - f_v/\gamma_{mv})\gamma_{ms}}{f_y}$$

where A_{sv} is the cross-sectional area of reinforcing steel resisting shear forces; b is the width of the section; f_v is the characteristic shear strength of masonry obtained from **19.1.3**; f_y is the characteristic tensile strength of the reinforcing steel resisting shear forces obtained from table 4; s_v is the spacing of shear reinforcement along the member, provided that it is not taken to be greater than $0 \cdot 75d$ (see **26.4**); v is the shear stress due to design loads, provided that it is not taken to be greater than $2 \cdot 0/\gamma_{mv}$ N/mm^2; γ_{ms} is the partial safety factor for strength of steel given in **20.2.2**; γ_{mv} is the partial safety factor for shear strength of masonry given in **20.2.2**.

Horizontal and vertical shear stress and/or principal tensile stress can govern the design in certain instances, e.g. heavily loaded retaining walls and reinforced beams. BS 5628:[1] Part 2 recommends the use of $v = V/bd$ for the calculation of such shear stresses.

3.7. Columns in combined bending and compression

To save design office time column design charts, as given in Appendix C, are normally used in design. The theory is given in this section.

A column is defined in BS 5628: Part 1[1] as an isolated vertical loadbearing member whose width is not more than four times its thickness.

The length of a column is defined in the same way as for walls (see section 3.15). A short column is a column whose slenderness ratio does not exceed 12 and a slender column is one which has a slenderness ratio greater than 12, where the slenderness ratio is the effective length divided by the effective thickness.

The critical axis of a column will vary depending on its plan dimensions, shape and restraints. Generally for long columns consideration of the slenderness effect due to deflection about each axis is required (see Fig. 3.8).

However, for short columns only single axis loadings need generally be considered. Fig. 3.9 shows a basic reinforced masonry column section subjected to bending and compression, and indicates the stress blocks.

For equilibrium of the vertical forces in Fig. 3.9

$$N_d = F_{BC} + F_{sc} + F_s \tag{3.7}$$

Fig. 3.8 (below). Bending and buckling axes

Fig. 3.9 (right). Stress block for column subject to bending and compression

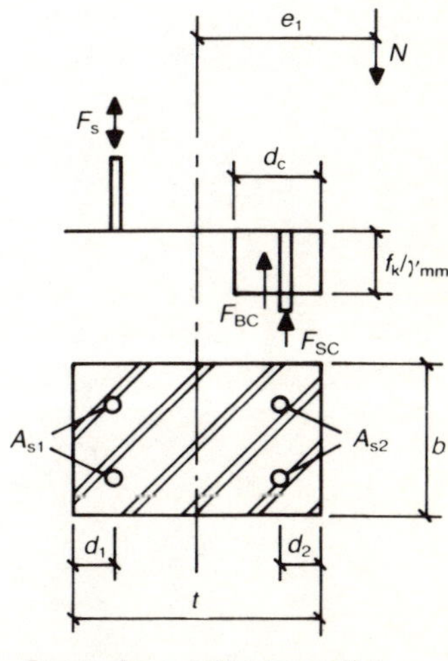

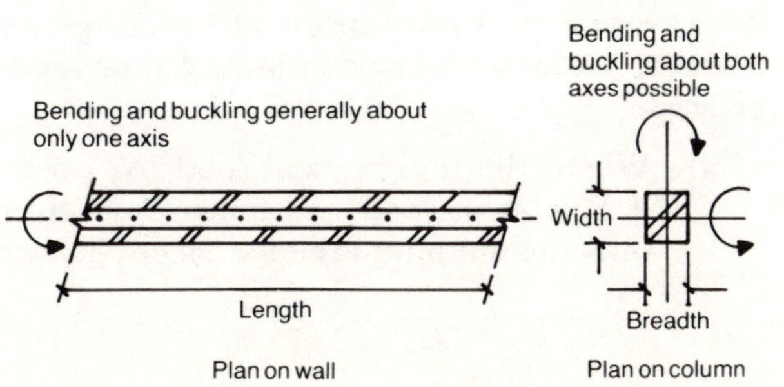

where N_d is the design vertical load resistance, F_{BC} is the compressive force in the masonry, F_{sc} is the compressive force in the reinforcement and F_s is the force in the reinforcement on the opposite face to F_{sc}, which may be tensile compressive or zero.

Equation (3.7) can be expanded as follows

$$N_d = \left(\frac{f_k}{\gamma_{mm}}\, bd_c \right) + \frac{0{\cdot}83 f_y\, A_{s1}}{\gamma_{ms}} + \frac{f_{s2}\, A_{s2}}{\gamma_{ms}} \tag{3.8}$$

where A_{s1} is the area of reinforcement in the main compressive zone, A_{s2} is the area of reinforcement in the opposite face to A_{s1} and f_{s2} is the design stress in the reinforcement A_{s2}.

f_{s2} will be equal to $+f_y/\gamma_{ms}$ in tension and $0{\cdot}83 f_y/\gamma_{ms}$ in compression.

Taking moments about the centreline of the section

$$N_d e_x = M_d = F_{BC}\left(\frac{t}{2} \times \frac{d_c}{2} \right) + F_{sc}\left(\frac{t}{2} \times d_1 \right) + F_s\left(\frac{t}{2} \times d_2 \right) \tag{3.9}$$

where M_d is the design moment of resistance which may be expressed as

$$M_d = \frac{f_k}{\gamma_{mm}}\, bd_c\left(\frac{t - d_c}{2} \right) + \frac{0{\cdot}83 f_y\, A_{s1}}{\gamma_{ms}}\left(\frac{t}{2} - d_1 \right) + \frac{f_{s2}\, A_{s2}}{\gamma_{ms}}\left(\frac{t}{2} - d_2 \right) \tag{3.10}$$

The design procedure is thus one of trial and error. For example, let $d_c = t$ and substitute the areas of reinforcement A_{s1} and A_{s2} into equation (3.10). If this results in N_d or M_d being less than the applied load or moment, respectively, a revised value of d_c should be tried, e.g. let $d_c = t - d_2$ and/or the areas of reinforcement be adjusted. This trial and error process is thus repeated until the values of N_d and M_d are greater than or equal to the values derived from the loaded conditions.

Where d_c is chosen as equal to t it is recommended that f_{s2} varies linearly between 0 and $-0{\cdot}83 f_y$. Where d_c is chosen between t and $t - d_2$ then $f_{s2} = 0$. Where d_c is chosen between $t - d_2$ and $t/2$ then f_{s2} is varied linearly between 0 and f_y. Where d_c is chosen between $t/2$ and $2d$ then f_{s2} may be taken as $+f_y$. The parameter d_c should not be chosen as less than $2d$ (see clause 23.3.1.1 of BS 5628: Part 2[1] reproduced below).

23.3 Design

23.3.1 *Columns subjected to a combination of vertical loading and bending*

23.3.1.1 *Short columns.* Where the slenderness ratio of a column does not exceed 12, only single axis bending generally requires consideration. Even where it is possible for significant moments to occur simultaneously about both axes, it is usually sufficient to design for the maximum moment about the critical axis only. However, where biaxial bending has to be considered reference should be made to **23.3.1.2**.

Either the cross section of the column may be analysed to determine the design moment of resistance and the design axial resistance, using assumptions (a), (c), (d) and (e) given in **22.4.1**, or the following design method may be used.

(a) Where the design axial load, N, does not exceed the value of the design axial load resistance, N_d, given in the following equation only the minimum reinforcement given in **26.1** or **26.3** is required:

$$N_d = \frac{f_k}{\gamma_{mm}}\, b(t - 2e_x)$$

where b is the width of the section; e_x is the resultant eccentricity; f_k is the characteristic compressive strength of the masonry; t is the overall thickness of the section in the plane of bending; γ_{mm} is the partial safety factor for strength of masonry

Note. This formula does not cover cases where the resultant eccentricity $e_x = M/N$ exceeds $0\cdot5t$, where M is the bending moment due to design load.

(*b*)   Where the design axial load, N, is greater than that given by the equation in (*a*) the strength of the section may be assessed by using the following equations and the relation $f_{s1} = 0\cdot83f_y$.

$$N_d = \frac{f_k}{\gamma_{mm}}\, bd_c + \frac{f_{s1}A_{s1}}{\gamma_{ms}} - \frac{f_{s2}A_{s2}}{\gamma_{ms}}$$

$$M_d = \frac{0\cdot5f_k}{\gamma_{mm}}\, bd_c(t - d_c) + \frac{0\cdot83f_y}{\gamma_{ms}}\, A_{s1}(0\cdot5t - d_1) + \frac{f_{s2}}{\gamma_{ms}}A_{s2}(0\cdot5t - d_2)$$

where A_{s1} is the area of compression reinforcement in the more highly compressed face; A_{s2} is the area of the reinforcement nearer the least compressed face; this may be considered as being in compression, inactive or in tension, depending on the resultant eccentricity of the load; b is the width of the section; d_1 is the depth from the surface to the reinforcement in the more highly compressed face; d_c is the depth of masonry in compression; d_2 is the depth to the reinforcement from the least compressed face; f_k is the characteristic compressive strength of the masonry; f_{s1} is the stress in the reinforcement in the most compressed face; f_{s2} is the stress in the reinforcement in the least compressed face, equal to $-0\cdot83f_y$ in compression or $+f_y$ in tension; f_y is the characteristic tensile strength of the reinforcement nearer the least compressed face; M_d is the design moment of resistance; N_d is the design axial load resistance; t is the overall thickness of the section in the plane of bending; γ_{mm} is the partial safety factor for strength of masonry given in clause **20**; γ_{ms} is the partial safety factor for strength of steel given in clause **20**.

The designer should choose a value of d_c which ensures that both the design axial load resistance, N_d and the moment of resistance, M_d, obtained from these equations exceed the design axial load, N, and the design bending moment, M. The choice of d_c establishes the assumed strain distribution in the section and appropriate values for the stresses in the reinforcement may be determined from the stress-strain relationship given in figure 2 or as follows:

(1)   where d_c is chosen as t, then f_{s2} varies linearly between 0 and $-0\cdot83f_y$;

(2)   where d_c is chosen between $(t - d_2)$ and t, then $f_{s2} = 0$;

(3)   where d_c is chosen between $(t - d_2)$ and $t/2$, then f_{s2} varies linearly between 0 and $+f_y$;

(4)   where d_c is chosen between $t/2$ and $2d_1$, f_{s2} may be taken as $+f_y$;

(5) d_c should not be chosen as less than $2d_1$.

(*c*)   As an alternative to (*b*) when the resultant eccentricity is greater than $(t/2 - d_1)$, the axial load may be ignored and the section designed to resist an increased moment, M_a, given by:

$$M_a = M + N(t/2 - d_1)$$

The area of tension reinforcement necessary to provide resistance to this increased moment may be reduced by $N\gamma_{ms}/f_y$.

As can be seen from clause 23.3.1.1 of BS 5628: Part 2, for columns subject to combined bending and axial loading the calculation method is an iterative process and tends to be cumbersome and tedious, and thus totally unsuited to the busy design office. In practice, therefore, design charts are used in which the various design formulae are plotted to form graphical curves, suitably non-dimensional in form and incorporating the numerous variable values for materials, geometric proportions, reinforcement proportions and partial safety factors.

A column subject to both vertical loading and bending could fail either in compression (where the masonry crushes), or in tension (where the reinforcements yields). The condition between these two extremes is termed a balanced failure (where the masonry crushes as the reinforcement yields).

The column design charts are obtained by plotting a series of points which define the three zones of failure. First, the point relating to the maximum axial vertical load which the column can carry is calculated and plotted. Second, the point relating to the balanced failure condition is calculated and plotted. It can be assumed that there is a linear relationship between these two failure conditions and hence the corresponding points plotted are connected together with straight lines, as shown in Fig. 3.10.

The remaining lengths of the curves are obtained from equations relating to the tensile failure condition. The equation for balanced and tensile failure is derived in sections 3.9 and 3.10.

3.8. Condition 1: axial loading failure

The maximum axial loading which a column can support is discussed in section 3.4 and the design formula is given as

$$N_d = \beta\left(\frac{f_k A_m}{\gamma_{mm}} + \frac{0{\cdot}83 f_y A_s}{\gamma_{ms}}\right) \tag{3.1}$$

If the term r is equal to A_s/bt, then $A_s = rbt$ and this can be substituted into equation (3.1). Also if A_m is replaced by bt equation (3.1) may be rewritten as

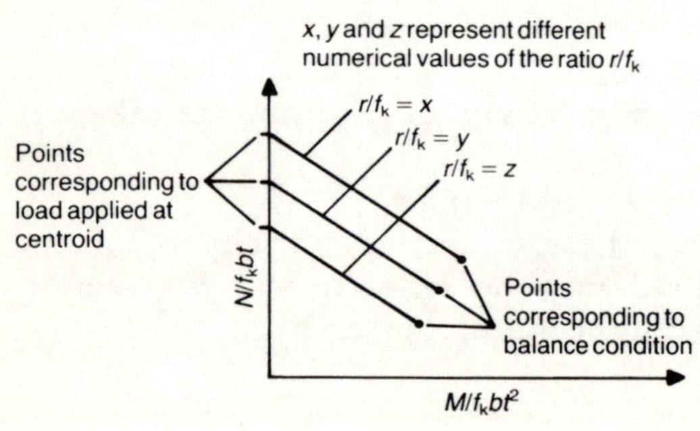

Fig. 3.10 (below). Balanced failure curves for columns

Fig. 3.11 (right). Stress conditions at balanced failure

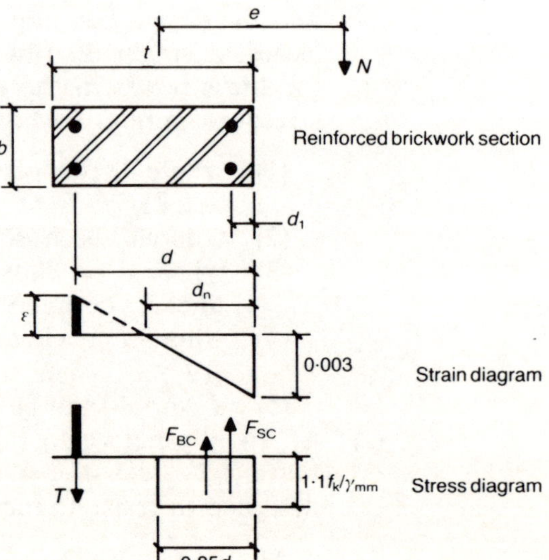

$$\frac{N_d}{f_k \, bt} = \frac{\beta}{\gamma_{mm}} + \frac{0 \cdot 83 f_y}{\gamma_{ms}} \frac{\beta r}{f_k} \tag{3.11}$$

This is in non-dimensional form and is the equation used to plot the first set of points on the column charts.

3.9. Condition 2: balanced failure

The stress conditions assumed to be applicable at the balanced failure are shown in Fig. 3.11 for a symmetrical section with an eccentric load, where N_b is the vertical load in the balanced condition, e is the eccentricity, d_1 is the depth from the compression face to compression steel, d is the effective depth, ε is the strain in the reinforcement, d_n is the depth to the neutral axis, T is the tension force in the tensile reinforcement, F_{BC} is the compression force in the masonry and F_{SC} is the compression force in the compression reinforcement.

The depth of the stress block is assumed to be $0 \cdot 85 d_n$ and the maximum strain is assumed to be $0 \cdot 003$. The term $1 \cdot 1 f_k/\gamma_{mm}$ is discussed in chapter 2 and is taken from figure 9 in appendix B of BS 5628: Part 1.[1]

From the strain diagram

$$\frac{d_n}{0 \cdot 003} = \frac{d - d_n}{\varepsilon} \quad \text{or} \quad d_n = \frac{1}{1 + \varepsilon/0 \cdot 003} \, d$$

As the strain varies for particular grades of steel, this may conveniently be expressed as

$$d_n = sd$$

where

$$s = \frac{1}{1 + \varepsilon/0 \cdot 003}$$

Equating forces on the vertical section gives

$$N_b = F_{BC} + F_{SC} - T$$

i.e.

$$N_b = \frac{1 \cdot 1 f_k}{\gamma_{mm}} \times b \times 0 \cdot 85 d_n + \frac{A_s}{2} \frac{0 \cdot 83 f_y}{\gamma_{ms}} - \frac{A_s \, f_y}{2 \gamma_{ms}} \tag{3.12}$$

Substituting for d_n and $A_s/bt = r$, equation (3.11) may be rewritten as

$$\frac{N_b}{f_k \, bt} = \left(\frac{0 \cdot 935 s}{\gamma_{mm}}\right) \frac{d}{t} - \left(\frac{0 \cdot 085 f_y}{\gamma_{ms}}\right) \frac{r}{f_k} \tag{3.13}$$

In a non-dimensional form this is essentially

$$\frac{N_b}{f_k \, bt} = A \times \frac{d}{t} - B \times \frac{r}{f_k}$$

where A and B are constants for any given situation.

Taking moments about the tensile steel gives

$$N_b \left(e + \frac{d - d_1}{2}\right) = F_{BC}\left(d - \frac{0 \cdot 85 d_n}{2}\right) + F_{SC}(d - d_1) \tag{3.14}$$

Substituting M_b for $N_b e$, equation (3.14) becomes

$$M_b + N_b\left(\frac{d - d_1}{2}\right) = F_{BC}\left(d - \frac{0 \cdot 85 d_n}{2}\right) + F_{SC}(d - d_1)$$

Substituting the value of N_b obtained previously, and expanding the F_{BC} and F_{SC} terms, equation (3.14) becomes

$$M_b + \left[\left(\frac{1 \cdot 1 f_k}{\gamma_{mm}} \times b \times 0.85 d_n \right) + \left(\frac{A_s}{2} \frac{0.83 f_y}{\gamma_{ms}} \right) - \left(\frac{A_s f_y}{2 \gamma_{ms}} \right) \right] \frac{d - d_1}{2}$$

$$= \left(1 \cdot 1 \frac{f_k}{\gamma_{mm}} \times b \times 0.85 d_n \right) \left(d - \frac{0.85 d_n}{2} \right) + \left(\frac{A_s}{2} \frac{0.83 f_y}{\gamma_{ms}} \right) (d - d_1)$$

$$(3.15)$$

If equation (3.15) is rearranged with $d_n = sd$ and $A_s/bt = r$ it becomes

$$M_b = \left(\frac{1 \cdot 1 f_k}{\gamma_{mm}} b \frac{0.85}{2} sd \right) (d - 0.85 sd + d_1) + \frac{rbt}{2} \frac{f_y}{\gamma_{ms}} 0.915(d - d_1)$$

But $d_1 = t - d$ and therefore

$$M_b = \left(\frac{1 \cdot 1 f_k}{\gamma_{mm}} b \frac{0.85 sd}{2} \right) (t - 0.85 sd) + \frac{rbt}{2} \frac{f_y}{\gamma_{ms}} 0.915(2d - t)$$

Dividing by $f_k bt^2$ gives

$$\frac{M_b}{f_k bt^2} = \frac{0.47 sd}{\gamma_{mm} t^2} (t - 0.85 sd) + 0.46 \frac{r f_y}{\gamma_{ms} f_k t} (2d - t)$$

$$= \frac{0.47 s}{\gamma_{mm}} \left(\frac{d}{t} \right) - \frac{0.4 s^2}{\gamma_{mm}} \left(\frac{d}{t} \right)^2 + \frac{0.9 r f_y}{\gamma_{ms} f_k} \left(\frac{d}{t} - 0.5 \right) \qquad (3.16)$$

Thus, for each value of d/t there is for each value of r/f_k a value of $N_b/f_k bt$ given by equation (3.13) and a corresponding value of $M_b/f_k bt^2$ given by equation (3.16).

If the ratio of effective depth to overall depth (i.e. for columns the ratio d/t) is kept constant, values of $N_b/f_k bt$ and $M_b/f_k bt$ can be plotted for different areas of reinforcement as given by the term r/f_k.

3.10. Condition 3: tensile failure

When considering the tensile failure condition it is not necessary to consider strains in the reinforcement since it is at yield point. The depth of the masonry in compression d_c is less than the assumed maximum value of $0.85 d_n$ as the compressive stress has not yet reached its maximum value. Fig. 3.12, on which the analysis is based, is a simplified version of Fig. 3.11.

Equating vertical forces gives

$$N = F_{BC} + F_{SC} - T$$

but

$$N = \left(\frac{1 \cdot 1 f_k}{\gamma_{mm}} d_c b \right) + \left(\frac{A_s}{2} \frac{0.83 f_y}{\gamma_{ms}} \right) - \left(\frac{A_s}{2} \frac{f_y}{\gamma_{ms}} \right)$$

which may be rewritten as

$$\frac{N}{f_k bt} = \frac{1 \cdot 1}{\gamma_{mm}} \left(\frac{d_c}{t} \right) - \frac{0.085 f_y r}{f_k \gamma_{ms}} \qquad (3.17)$$

where $A_s/bt = r$.

Taking moments about the tensile steel gives

$$N \left(\frac{e + d - d_1}{2} \right) = F_{BC} \left(d - \frac{d_c}{2} \right) + F_{SC} (d - d_1) \qquad (3.18)$$

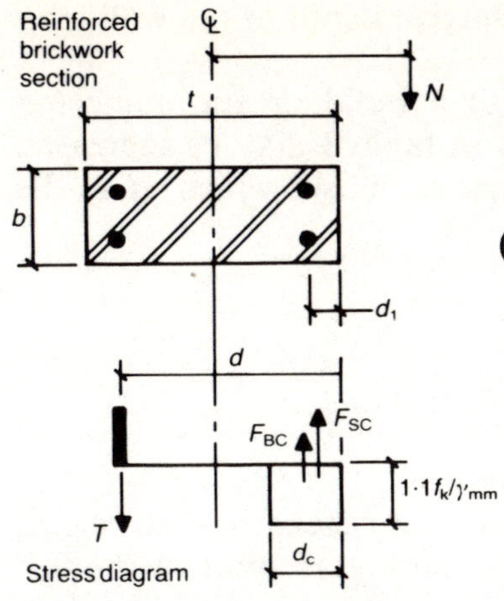

Fig. 3.12. Stress conditions before maximum values

Substituting $M = Ne$ into equation (3.18) gives

$$M + N\left(\frac{d - d_1}{2}\right) = F_{BC}\left(d - \frac{d_c}{2}\right) + F_{SC}(d - d_1) \tag{3.19}$$

Expanding the F_{BC} and F_{SC} terms and substituting $r = A_s/bt$ equation (3.19) becomes

$$M + N\left(\frac{d - d_1}{2}\right) = \left[\frac{1 \cdot 1 f_k}{\gamma_{mm}}\, d_c\, b\left(d - \frac{d_c}{2}\right)\right] + \left[\frac{rbt}{2}\frac{0 \cdot 83 f_y}{\gamma_{ms}}(d - d_1)\right]$$

Dividing by $f_k bt^2$ and substituting $d_1 = t - d$ gives

$$\left(\frac{M}{f_k bt^2}\right) + \frac{N}{f_k bt}\left(\frac{d}{t} - 0 \cdot 5\right) = \left[\frac{1 \cdot 1}{\gamma_{mm}}\frac{d_c}{t}\left(\frac{d}{t} - \frac{d_c}{2t}\right)\right] + \left[\frac{0 \cdot 83 f_y}{\gamma_{ms}}\frac{r}{f_k}\left(\frac{d}{t} - 0 \cdot 5\right)\right]$$
$$\tag{3.20}$$

A curve may then be plotted on the design chart by taking a value of r/f_k and assuming a value of $N/f_k bt$, and then a value of d_c/t may be found from equation (3.17) and substituted into equation (3.20) to determine the corresponding value of $M/f_k bt^2$.

The respective points may then be plotted to complete the curves in Fig. 3.10. Suitable design curves are given in Appendix C. Engineers should exercise their judgment in accepting values from these curves and, if necessary, carry out a design check of their own.

3.11. Beams in bending

The design of members subject to pure bending can be carried out by direct calculation analysis. For combined bending and axial loading a different approach is required. This has been dealt with in section 3.7.

The design formula for bending resistance and the assumptions generally accepted for design are derived in section 3.5. However, there are other factors affecting the design which need to be taken into account. For example, beams subjected to pure bending may fail by buckling of the compression flange if the slenderness of the flange is critical.

To ensure lateral stability, therefore, limits need to be put on the proportions of the beam which influence this failure. BS 5628:[1] Part 2 sets out limits for these proportions as follows.

To ensure lateral stability of simply supported or continuous beams they should be proportioned so that the clear distance between lateral restraints does not exceed the lesser of $60b_c$ and $250b_c^2/d$, where d is the effective depth and b_c is the width of the compression face midway between the restraints.

For a cantilever with lateral restraint provided at the support, the clear distance from the end of the cantilever to the face of the support should not exceed the lesser of $25b_c$ and $100b_c^2/d$.

In addition, limits need to be set to limit deflection and cracking. Generally these items are set by restriction on span to depth ratios unless more stringent requirements are to be met than those normally applicable. Clauses 22.3 and 16.2.2 of BS 5628: Part 2[1] follow.

22.3 Limiting dimensions

22.3.1 *General*. To avoid detailed calculations to check that the limit states of deflection and cracking are not reached, the limiting ratios given in tables 8 and 9 may be used, except when the serviceability requirements are more stringent than those given in **16.2.2**.

22.3.2 *Walls subjected to lateral loading*. When walls are reinforced to resist lateral loading, the ratio of span to effective depth of the wall may be taken from table 8.

For free-standing walls not forming part of a building and subjected predominantly to wind loads, the ratios given in table 8 may be increased by 30%, provided such walls have no applied finish which may be damaged by deflection or cracking.

16.2.2 *Serviceability limit states*

16.2.2.1 *Deflection.* The deflection of the structure or any part of it should not adversely affect the performance of the structure or any applied finishes, particularly in respect of weather resistance.

The design should be such that deflections are not excessive, with regard to the requirements of the particular structure, taking account of the following recommendations.

(*a*) The final deflection (including the effects of temperature, creep and shrinkage) of all elements should not, in general, exceed length/125 for cantilevers or span/250 for all other elements.

(*b*) Consideration should be given to the effect on partitions and finishes of that part of the deflection of the structure taking place after their construction. A limiting deflection of span/500 or 20 mm, whichever is the lesser, is suggested.

(*c*) If finishes are to be applied to prestressed masonry members, the total upward deflection, before the application of finishes, should not exceed span/300 unless uniformity of camber between adjacent units can be ensured.

In any calculation of deflections (see appendix C) the design loads and the design properties of materials should be those recommended for the serviceability limit state in clauses **18** to **20**. For reinforcement, stresses lower than the characteristic strengths given in table 4 may need to be used to reduce deflection or control cracking.

Also, where the ratio of the span to the depth of a beam is less than 1·5, the beam should be treated as a wall beam. Tension reinforcement should be provided to take the whole of the tensile force calculated on the basis of the moment arm equal to two thirds of the depth with a minimum value equal to 0·7 times the span.

The effective span is shown in Fig. 3.13.

The design loading on a particular structural element is determined from the combination of the characteristic loadings from the relevant codes of practice and the partial factors of safety appropriate to the case being considered.

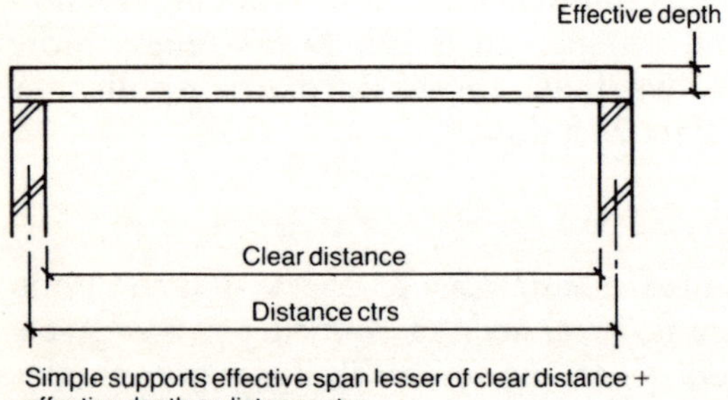

Simple supports effective span lesser of clear distance + effective depth or distance ctrs

Fig. 3.13. Effective span of beam

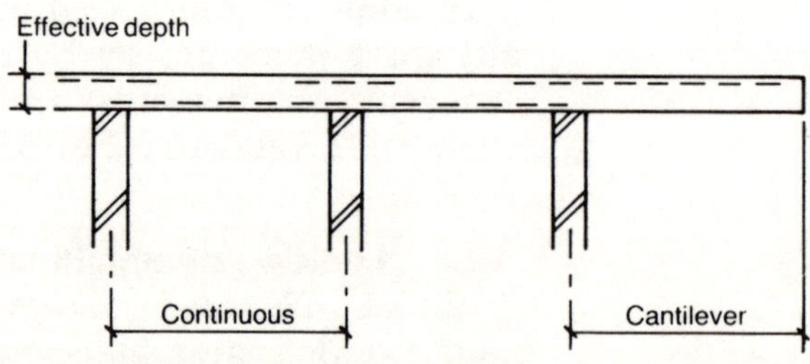

Fig. 3.14. Effective span of continuous beams and cantilevers

In the case of elements subject to bending (e.g. beams and retaining walls) the following information should be used in assessing the design loads as illustrated in Figs 3.13 and 3.14.

The effective span of a simply supported member should be taken as the lesser of the distance between the centres of bearing and the clear distance between supports plus the effective depth.

The effective span of a continuous member should be taken as the distance between the centres of supports.

The effective span of a cantilever should be taken as its length to the face of the support plus half of its effective depth, except where it forms the end of a continuous beam, in which case the length to the centre of the support should be used.

3.12. Lever arm: calculation and graphical curve

The design equations derived in chapter 2 involve the use of the lever arm z. It is convenient to present lever arm values in graphical form rather than to calculate the value for each individual case. The rectangular stress block shown in Fig. 3.15 may be used to derive the curve given in Fig. 3.7. However, it is more convenient to adopt an equivalent rectangular stress block in place of the semi-parabolic curve noted in Fig. 3.4, the resulting resistance moment is similar.

The moment of resistance M_d is given by

$$M_d - \frac{f_f}{\gamma_{mm}}\, d_c\, bz \tag{3.20}$$

but

$$z = d - \frac{d_c}{2}$$

and therefore

$$d_c = 2(d - z).$$

Substituting this value for d_c into equation (3.20) gives

$$M_d = \frac{f_f}{\gamma_{mm}}\, bz \times 2(d - z)$$

or

$$M_d = \frac{2f_f}{\gamma_{mm}}\, bd^2\left(1 - \frac{z}{d}\right)\frac{z}{d}$$

or

$$\frac{M_d\,\gamma_{mm}}{2f_f\,bd^2} = \frac{z}{d}\left(1 - \frac{z}{d}\right)\frac{z}{d} < 0\cdot95$$

A curve of z/d may be plotted against $M_d\,\gamma_{mm}/2f_f\,bd^2$ (see Fig. 3.17).

Column design charts compiled in accordance with these procedures are given in Appendix C. These charts include those required for the analysis of symmetrical reinforced columns subjected to combined bending and axial loading. However, when the resulting eccentricity exceeds $t/2 - d_1$ the section may be designed to resist bending only, but using an increased M_a given by

$$M_a = M + N\left(\frac{t}{2} - d_1\right)$$

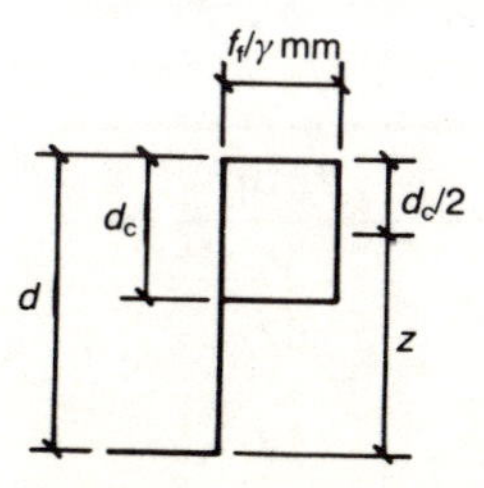

Fig. 3.15. Stress diagram

where M is the design moment, N is the vertical load and d_1 is the depth from the surface to the reinforcement in the more highly compressed face.

To allow for the cancelling effect of the axial loading on the tensile stresses, the area of tensile reinforcement required to resist the increased moment may be reduced by $\gamma_{ms} N/f_y$.

When significant moments occur simultaneously about both axes, columns should be designed to resist biaxial bending, and for slender columns it may be necessary to consider the increased moments resulting from the effects of lateral deflections.

3.13. Walls in combined bending and compression

A distinction is usually made for design purposes between short and slender walls in a manner similar to that discussed in section 3.7 for columns. 'Short' is the term used to describe walls with a slenderness ratio not exceeding 12 (where the slenderness ratio is effective height divided by effective thickness).

For slender walls the additional bending moment which is induced by lateral deflection of the member under vertical load should be taken into account in the design. For masonry the buckling tendency is generally allowed for by applying a capacity reduction factor to the design strength rather than by increasing the applied moment. For short walls the effect is small and is generally ignored.

Masonry under combined vertical loading and bending is analysed using the basic assumptions discussed in this chapter and chapter 2.

The effective resultant eccentricity is determined by dividing the total applied bending moment by the axial load (see Fig. 3.16).

For the wall shown in Fig. 3.16 the eccentricity will be

$$e_x = (M + Ne_t)/N \tag{3.21}$$

The failure of such a section may result from either compressive failure of the masonry or tensile yielding of the reinforcement. Design formulae for these conditions would be similar to those for columns.

For the purpose of design, a resulting eccentricity e_x of less than $0{\cdot}05t$ may be analysed for axial loading only. However, when a wall is subjected to a large moment compared with the vertical load (i.e. the resulting eccentricity e_x is greater than $0{\cdot}5t$) the vertical load produces only a marginal modification to the bending stresses and the wall may be analysed for bending only. The definition of this in BS 5628[1] is that in the analysis of a cross-section which has to resist a small axial thrust, the effect of the axial force of

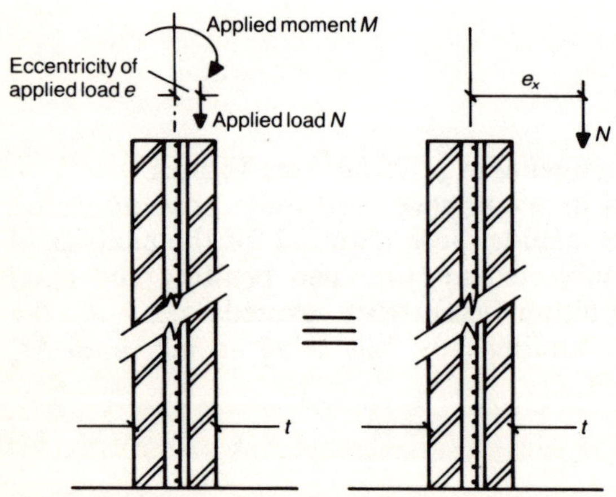

Fig. 3.16. Resultant eccentricity: section through reinforced brickwork wall

Table 3.4. Limiting ratios of span to effective depth for laterally-loaded walls

End condition	Ratio
Simply supported	35
Continuous or spanning in two directions	45
Cantilever with values of ρ up to and including $0{\cdot}005$	18

the design may be ignored if it does not exceed $0{\cdot}1f_k\,A_m$ where A_m is the cross-sectional area of the masonry, i.e. the member may be designed for bending only.

In cases where it is felt necessary to calculate the additional moment due to the deflection under vertical load it may be taken as

$$M_a = N(h_{ef})^2/2000t \tag{3.22}$$

where N is the design axial load, h_{ef} is the effective height, t is the thickness of the wall and M_a is the moment due to lateral deflection.

The alternative to this approach is to use a reduction factor β which makes allowance for the deflections based on the slenderness ratio of the section (see design examples in chapter 4). The factor effectively reduces the allowable stress remaining after account has been taken of stress due to deflections. Either approach is reasonable for the analysis of sections.

3.14. Laterally loaded walls and shear walls

Laterally loaded reinforced masonry walls can be designed as cantilevers, simply supported, spanning one direction, continuous or two-way spanning, or a combination of these support conditions.

In order to limit the calculation checks necessary to achieve the limit states for deflection and cracking criteria, limiting ratios of span to effective depths are given in BS 5628: Part 2[1] as shown in Table 3.4, where $\rho = A_s/bd$.

Where such walls do not form part of a building the deflection and cracking limitations are less stringent and an increase of 30% in the values given in Table 3.4 is allowed, providing that such walls have no applied finishes, which may be damaged by deflection or cracking.

Shear walls are subjected to racking shear. BS 5628: Part 2[1] gives the following information for design limits for shear for reinforced members.

25 Reinforced masonry subjected to horizontal forces in the plane of the element

25.1 Racking shear

25.1.1 Where a vertically reinforced wall resists horizontal forces acting in its plane, adequate provision against the ultimate limit state in shear being reached may be assumed if the following relationship is satisfied:

$$v \leqslant f_v/\gamma_{mv}$$

where f_v is the characteristic shear strength of masonry; γ_{mv} is the partial safety factor for shear strength of masonry; v is the shear stress due to design loads given by: $v = V/tL$ where t is the thickness of the wall; L is the length of the wall; V is the horizontal shear force due to design loads.

25.1.2 Where the relationship given in **25.1.1** is not satisfied, horizontal shear reinforcement should be provided but in no case should v exceed $2{\cdot}0/\gamma_{mv}$ N/mm^2.

Where horizontal reinforcement is provided, the following requirement should be satisfied:

$$\frac{A_{sv}}{s_v} \geqslant \frac{t(v - f_v/\gamma_{mv})}{f_y/\gamma_{ms}}$$

where A_{sv} is the cross-sectional area of reinforcing steel resisting shear forces; t is the thickness of the wall; f_v is the characteristic shear strength of masonry obtained from **19.1.3.2**; f_y is the characteristic tensile strength of the reinforcing steel resisting shear forces obtained from table 4; s_v is the spacing of shear reinforcement along member; γ_{mv} is the partial safety factor for shear strength of masonry given in **20.2.2**; γ_{ms} is the partial safety factor for strength of steel given in clause **20**.

25.2 Bending

When the bending is in the plane of the wall, the analysis and design of the wall should follow the recommendations for beams given in clause **22**. Where the slenderness ratio exceeds 12 in any direction, it is essential also to take account of the slenderness at right angles to the plane of the wall by calculating the maximum compressive stress in the wall and checking that the recommendations for slender columns described in **23.3.1.3** are satisfied.

Shear walls acting about their major axis and cantilevering vertically provide stability to the structure to resist wind forces or other laterally applied loads (see Fig. 3.17).

Figure 3.18 shows a typical loading and bending moment diagram applicable to a shear wall.

Many shear walls are adequate in plain masonry. However, where the length of the wall available is restricted or where vertically applied loads are small relative to the applied lateral load, it may be necessary to reinforce the section. In such cases the reinforced section should be designed to resist both bending and shear. The section shown in Fig. 3.17 is checked for bending as a cantilever, taking into account the restrictions on restraints and slenderness, and a check is made on compressive stresses due to direct loads.

The design moment of resistance is first checked against the formula

$$M = \left(\frac{f_{kx}}{\gamma_{mm}} + g_B\right)Z \tag{3.23}$$

in which the limiting condition is the flexural tensile strength f_{kx}. The check should be made using the minimum design load for the calculation of g_B, and the Authors recommend that for unreinforced sections f_{kx} should be taken as zero.

Fig. 3.17 (below). Shear walls in multi-storey structure

Fig. 3.18 (right). Shear wall loading diagram

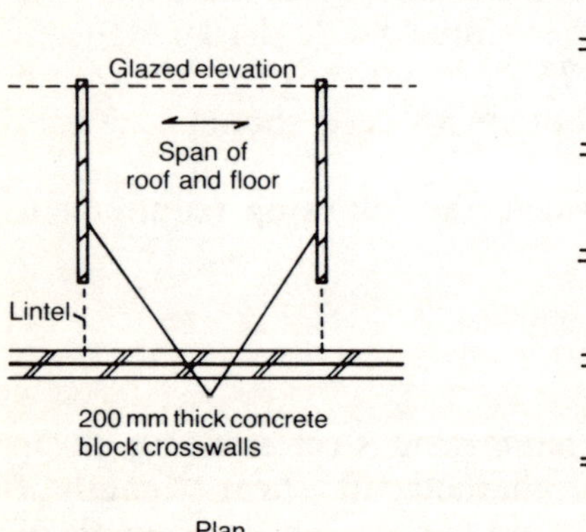

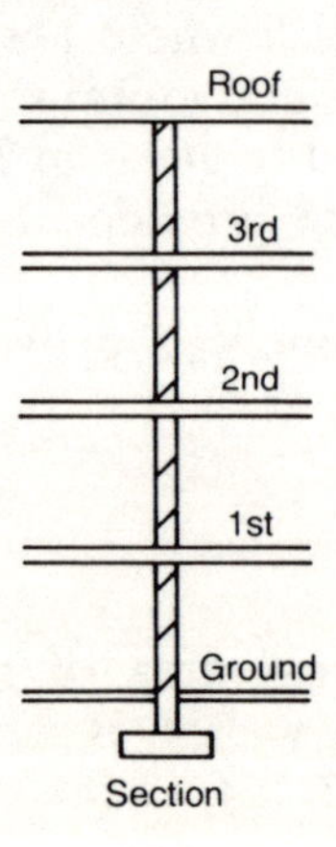

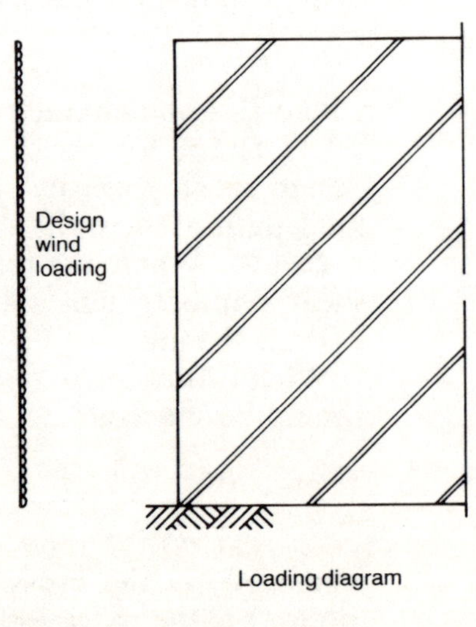

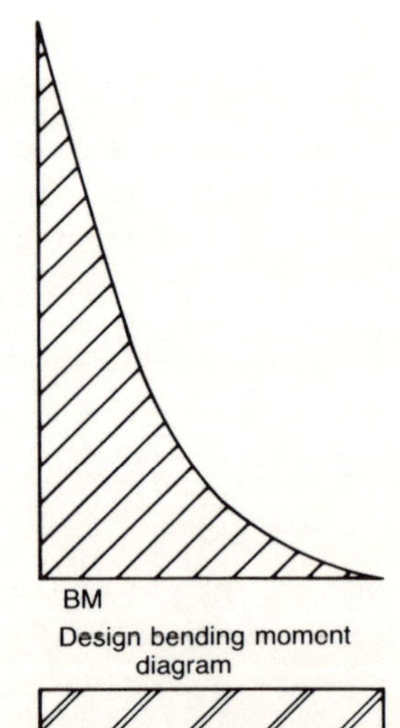

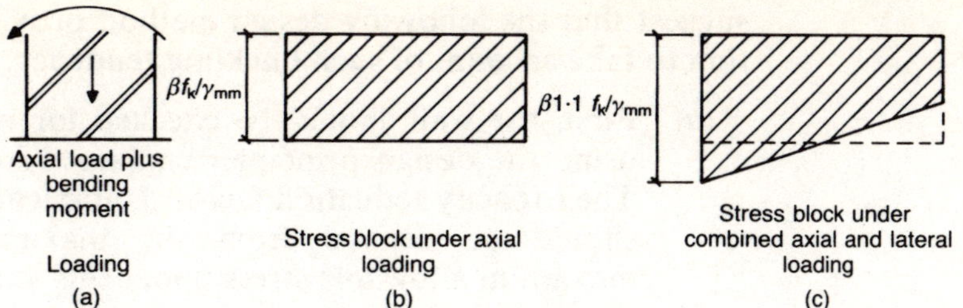

Fig. 3.19. Loading distribution and stress block

In addition, flexural compression stress should be checked. Where a section is checked on this basis, and is found to exceed the limit with regard to flexural tension but has spare capacity with regard to flexural compression, the use of reinforced masonry could be advantageous, particularly if the shear forces are also well within the design strength.

In the case of compression and bending compression in crosswalls, gables and other walls acting as shear walls, the intensity of loading at any particular position should be assessed on the basis of the load distribution shown in Fig. 3.19. The strength of the wall should then be determined, taking into account the shear strength or shear loading.

Two limiting conditions can be said to apply to the maximum allowable stresses which are given as f_k/γ_{mm} for flexural stresses and $1 \cdot 1 f_k/\gamma_{mm}$ for flexural compressive stresses, the latter having been adopted in appendix B of BS 5628:[1] Part 1 for the derivation of β. The flexural tensile stresses will sometimes be the limiting factor under such loading conditions and reinforcement can be added. However, when the axial compressive stresses already in the wall are added to the flexural compressive stresses, this may produce a more critical design condition. Consideration must be given to the need for limiting the flexural compressive stresses because of the possibility of buckling of the section under the application of such stress. Buckling of a section, due to flexural compressive stresses, will occur perpendicular to the direction of application of the bending. Two common plan sections are shown in Fig. 3.20.

Clearly, the shear wall section shown in Fig. 3.20(b) is the more critical with respect to buckling due to flexural compressive stresses. The Authors

Fig. 3.20. Buckling of shear walls (axial compressive stresses omitted for clarity; t_{ef} is the effective thickness of section resisting tendency to buckle under flexural compressive stresses): (a) bending about minor axis of wall; (b) bending about major axis of wall (as in shear walls)

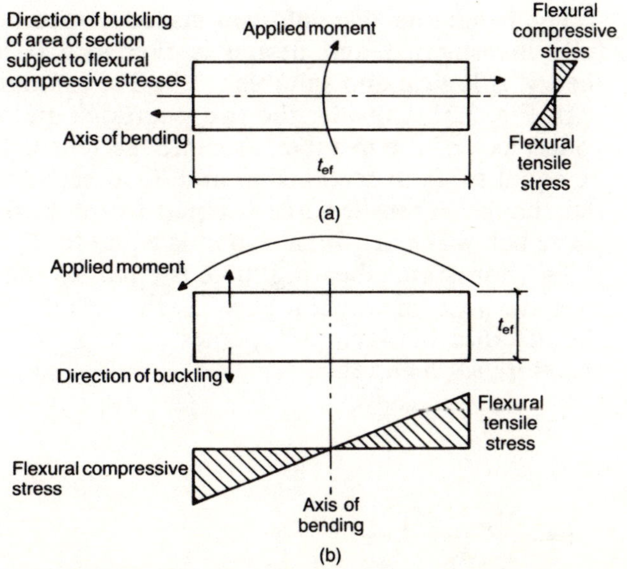

suggest that the following design method provides a safe and practical solution to take account of such buckling tendency.

(a) First, the wall should be checked for maximum axial loading only, using the design principles explained in chapter 2 and this chapter. The capacity reduction factor β, applicable to this stage of the design, should be derived from the maximum slenderness ratio. The maximum allowable stress under this loading condition is $\beta f_k / \gamma_{mm}$.

(b) The additional compressive stress resulting from the bending due to lateral loading is then considered, and the maximum allowable combined compressive stress is $\beta \times 1 \cdot 1 f_k / \gamma_{mm}$, in which a 10% increase has been applied to the flexural aspect of the stress in a similar manner to appendix B of BS 5628: Part 1.[1] The capacity reduction factor β should be derived from the slenderness ratio which incorporates the effective thickness appropriate to the direction of buckling tendency (i.e. perpendicular to the direction of application of the bending) as shown in Fig. 3.20.

In checking the compression zone, the wall section should be checked at two critical levels

(a) at foundation level where the effect of axial loading and design bending moments are greatest

(b) at a point $0 \cdot 4\,h$ above the ground floor slab where the design bending moment will be less, but where the capacity reduction factor β reduces the compressive resistance of the wall under combined axial and lateral loading.

The check for condition (a) should be carried out in two stages, the first being a check on the axial loading condition only and the second on the combined axial and lateral loading.

For the design of reinforcement, the worst condition must be adopted, i.e. the maximum tensile stress condition for the design of tensile steel and the maximum compressive stress condition for compressive steel if this should be required. In addition a check must be made on shear stress and, where appropriate, principal tensile stress should also be considered.

The design procedure for the section is that considered for deep beams and dealt with in section 3.11.

3.15. Analysis for shear

The behaviour of reinforced masonry members in shear has not yet been fully investigated and design is therefore based on experience and basic theory. A logical and suitable method of estimating shear stress follows.

In Fig. 3.21, consider the two planes ab and cd perpendicular to the horizontal axis of the member. Assume the tensile forces in the reinforcement to be equal to T_0 at sections ab and T_1 at section cd. Over the short length δ_2 the change in tensile force is equal to the horizontal shearing force on any plane below the neutral axis and is equal to $T_0 - T_1$.

The horizontal shearing force is equal to the product of the shear stress v over the area on which it acts, i.e. the breadth of the section b multiplied by the distance between two planes ab and cd. The horizontal shear force is therefore $v\delta_2 b$ and thus

$$T_0 - T_1 = v\delta_2 b$$

or

$$v = \frac{M_0 - M_1}{\delta_2} \frac{1}{zb} \tag{3.24}$$

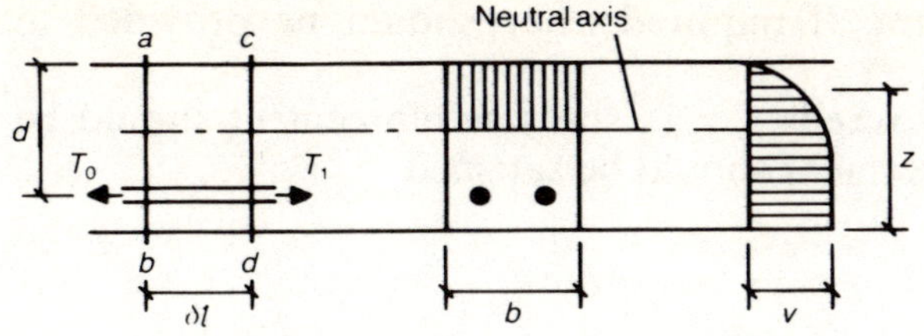

Fig. 3.21 (above). Shear stress distribution

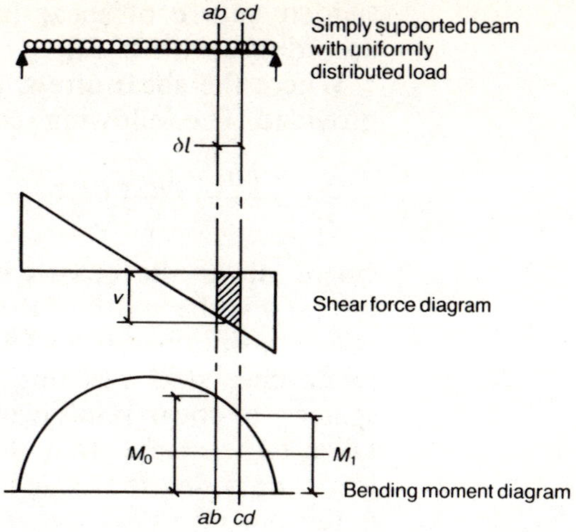

Fig. 3.22 (right). Shear force distribution along beam

As is well known the bending moment at any position on a member is equal to the area of the shear force diagram to one side of the position being considered.

Figure 3.22 shows a simply supported beam loaded with a uniformly distributed load. The area of the shear force diagram between two positions on the beam will be equal to the difference in bending moment between the same positions. Thus the difference in bending between planes ab and cd will be equal to the area of the shear force diagram between them, i.e.

$$M_0 - M_1 = V\delta_2$$

where V is the average value of the shear forces between ab and cd. Assuming a unit length

$$\frac{M_0 - M_1}{\delta_2} = V$$

and substituting this in equation (3.24) gives

$$v = V/zb$$

In most recent codes, including BS 5628: Part 2,[1] this expression has been modified by replacing the lever arm z by the effective depth d. Thus the applied shear stress v is obtained using the equation $v = V/bd$. However, the Authors see no justification for this.

Clause 22.5 of BS 5628: Part 2 is as follows.

22.5   Shear resistance of elements

22.5.1   *Shear stresses and reinforcement in members in bending.* The shear stress, v, due to design loads at any cross section in a member in bending should be calculated from the equation:

$$v = V/bd$$

where b is the width of the section; d is the effective depth (or for a flanged member the actual thickness of the masonry between the ribs if this is less than the effective depth as defined in **2.4**); V is the shear force due to design loads.

Where the shear stress calculated from this equation is less than the characteristic shear strength of masonry, f_v, divided by the partial safety factor, γ_{mv}, shear reinforcement is not generally needed. In beams, however, the designer should consider the use of nominal links, bearing in mind the sudden nature of shear failure. If required, they should be provided in accordance with **26.5.2**.

Where the shear stress, v, exceeds f_v/γ_{mv}, shear reinforcement should be provided. The following requirement should be satisfied:

$$\frac{A_{sv}}{s_v} \geqslant \frac{b(v - f_v/\gamma_{mv})\gamma_{ms}}{f_y}$$

where A_{sv} is the cross-sectional area of reinforcing steel resisting shear forces; b is the width of the section; f_v is the characteristic shear strength of masonry obtained from **19.1.3**; f_y is the characteristic tensile strength of the reinforcing steel resisting shear forces obtained from table 4; s_v is the spacing of shear reinforcement along the member, provided that it is not taken to be greater than $0{\cdot}75d$ (see **26.4**); v is the shear stress due to design loads, provided that it is not taken to be greater than $2{\cdot}0/\gamma_{mv}$ N/mm^2; γ_{ms} is the partial safety factor for strength of steel given in **20.2.2**; γ_{mv} is the partial safety factor for shear strength of masonry given in **20.2.2**.

The Authors disagree with the substitution of d for z, which gives a lower shear stress and for which there appears to be little justification. The design shear strength of masonry v is equal to f_v/γ_{mv}, where f_v is the characteristic shear strength and γ_{mv} is the partial safety factor for shear (see table 3.6).

Characteristic shear strength is discussed in clause 19.1.3 of BS 5628: Part 2.[1]

19.1.3   *Characteristic shear strength of masonry, f_v*
19.1.3.1   *Shear in bending (reinforced masonry)*

19.1.3.1.1   For sections in which the reinforcement is placed in bed or vertical joints, including Quetta bond and other sections where the reinforcement is wholly surrounded with mortar designation (i) or (ii) (see table 1), the characteristic shear strength, f_v, may be taken as $0{\cdot}35$ N/mm^2.

For simply supported beams or cantilevers where the ratio of the shear span (see **2.6**) to the effective depth is less than 2, f_v may be increased by a factor:

$$2d/a_v$$

where d is the effective depth; a_v is the distance from the face of the support to the nearest edge of a principal load; provided that f_v is not taken to be greater than $0{\cdot}7$ N/mm^2.

At sections in certain laterally loaded walls there may be substantial compressive stresses from vertical loads. In such cases the shear may be adequately resisted by the plain masonry (see clause **25** of BS 5628: Part 1: 1978).

19.1.3.1.2   For reinforced sections in which the main reinforcement is placed within pockets, cores or cavities filled with concrete infill as defined in **13.1**, the characteristic shear strength of the masonry, f_v, may be obtained from the following equation:

$$f_v = 0{\cdot}35 + 17{\cdot}5\rho$$

where $\rho = A_s/bd$; A_s is the cross-sectional area of primary reinforcing steel; b is the width of section; d is the effective depth (see **2.4**); provided that f_v is not taken to be greater than $0{\cdot}7$ N/mm^2.

For simply supported reinforced beams or cantilever retaining walls where the ratio of the shear span, a, (see **2.6**) to the effective depth, d, is six or less, f_v may be increased by a factor $\{2{\cdot}5 - 0{\cdot}25(a/d)\}$ provided that f_v is not taken to be greater than $1{\cdot}75$ N/mm^2.

19.1.3.2 *Racking shear in reinforced masonry shear walls*. When designing reinforced masonry shear walls the characteristic shear strength of masonry, f_v, may be taken to be:

$$(0\cdot35 + 0\cdot6g_B)\ \text{N/mm}^2, \text{ with a maximum of } 1\cdot75\ \text{N/mm}^2$$

where g_B is the design load per unit area normal to the bed joint due to the loads calculated for the appropriate loading condition detailed in clause **20**.

Alternatively, for reinforced sections in which the main reinforcement is placed within pockets, cores or cavities filled with concrete infill as defined in **13.1**, the characteristic shear strength of masonry, f_v, may be taken to be $0\cdot7\ \text{N/mm}^2$ provided that the ratio of height to length of the wall does not exceed $1\cdot5$.

Designers should consider the effect of damp-proof courses on shear strength of masonry (see **17.1**).

19.1.3.3 *Shear in prestressed sections*. For prestressed sections with bonded or unbonded tendons the characteristic shear strength of masonry, f_v, may be obtained from the following formula:

$$f_v = (0\cdot35 + 0\cdot6g_B)\ \text{N/mm}^2$$

where g_B is the design load per unit due to the loads acting at right angles to the bed joints, including prestressing loads (in N/mm^2).

Note. In elements prestressed parallel to the bed joints $g_B = 0$, giving $f_v = 0\cdot35\ \text{N/mm}^2$.

For simply supported prestressed beams or cantilever retaining walls where the ratio of the shear span, a, to the effective depth, d, is six or less, f_v may be increased by a factor $\{2\cdot5 - 0\cdot25(a/d)\}$.

In all cases f_v should not be taken to be greater than $1\cdot75\ \text{N/mm}^2$.

The characteristic shear strength values given in BS 5628 seem high, particularly for the lower strength units. The Authors recommend caution, particularly when low strength units are used. Comparison of the shear strength of concrete with the values given in BS 5628 will clarify the Authors' opinions, particularly when the vertical shear occurring on bonded joints is considered.

The Authors also recommend caution in the use of the increased shear strength allowance stated in BS 5628 for the effect of vertical stress on the bed joints, particularly since the shear stress on the bedding plane is repeated in the vertical direction where compressive stress normal to the shear force is likely to be much less than in the horizontal direction. It can be argued that, in general, the vertical shear resistance of masonry is greater than its horizontal shear resistance. This is not necessarily true of the weaker units, however, and this aspect is not defined in BS 5628: Parts 1 or 2. In addition it must be remembered that in the vertical plane, at perpendicular joint locations, only alternate units pass through the plane of possible vertical shear failure.

The contribution of main reinforcement in resisting shear still requires additional research but clause 19.1.3 of BS 5628: Part 2 gives the increases in characteristic shear stresses for certain conditions.

The apparent encouragement offered in the code to use large percentages of reinforcement should be resisted because quantities in excess of 1% may present difficulties in grout/mortar compaction, and hence bond and durability may be impaired.

Even where reinforcement is provided the design shear stress should not exceed $2\cdot0/\gamma_{mv}\ \text{N/mm}^2$.

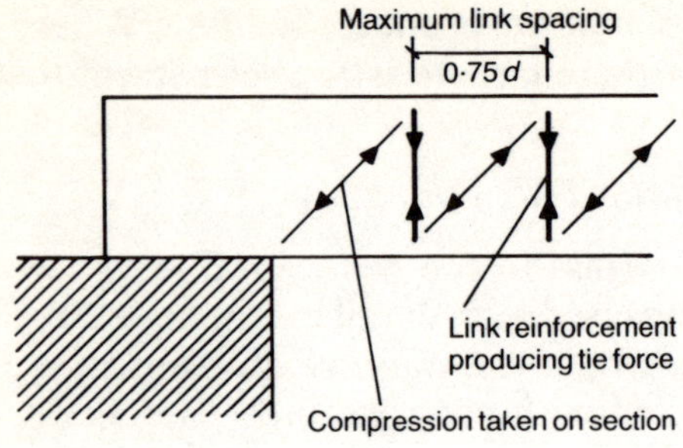

Fig. 3.23. Truss analogy

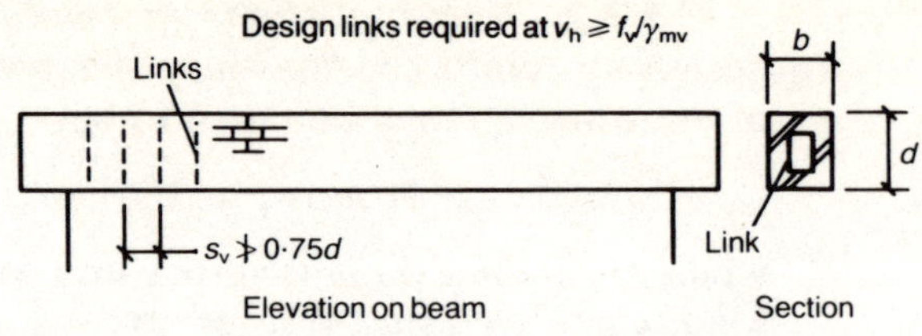

Fig. 3.24. Spacing of links

3.15.1. Shear reinforcement

Most shear reinforcement used for reinforced masonry takes the form of links and is generally designed on the commonly used truss analogy (see Fig. 3.23).

The shear reinforcement is designed to resist the shear force in excess of that which could be resisted on the masonry alone.

Consider a cross-section through the beam shown in Fig. 3.23 required to resist a design shear force V. The tie force provided by the reinforcement is equal to the area of the shear reinforcement times the design shear strength, i.e.

$$A_{sv} \times f_{yv}/\gamma_{ms}$$

where A_{sv} is the area of shear reinforcement and f_{yv} is the characteristic strength of the shear reinforcement.

The shear strength of the masonry is the area of the masonry times the design shear strength, i.e.

$$bd \times f_{v}/\gamma_{mv}$$

The total shear resistance is therefore

$$\frac{A_{sv} f_{yv}}{\gamma_{ms}} + \frac{bd \times f_{v}}{\gamma_{mv}}$$

In order to ensure that the 45° line of potential diagonal tension always passes through a link, the links must be spaced at less than the lever arm. The limit for such spacing is generally kept to a maximum of $0.75d$ (see Fig. 3.24).

The contribution of the link reinforcement is increased by the ratio of the actual spacing S_v to the effective depth, i.e. the number of ties effective at any one section is given by the ratio d/S_v.

Using a link spacing of S_v the shear resistance becomes

$$\frac{d}{S_v}\left[\left(A_{sv} \times \frac{f_{yv}}{\gamma_{ms}}\right) + \frac{bdf_{v}}{\gamma_{mv}}\right]$$

Expressed in terms of the design shear stress v it becomes

$$vbd \leqslant \frac{d}{S_v}\left[\left(A_{sv} \times \frac{f_{yv}}{\gamma_{ms}}\right) + \frac{bdf_{v}}{\gamma_{mv}}\right]$$

Rearranged for area and spacing of links this becomes

$$\frac{A_{sv}}{S_v} \geqslant b\left(v - \frac{f_{v}}{\gamma_{mv}}\right)\frac{\gamma_{ms}}{f_{yv}}$$

3.16. Analysis for bond

In order that the masonry and reinforcement act compositely under load it is essential that the bond between the materials remains intact.

Under conditions of applied bending, the bond stresses change in relation to the rate of change in bending moment and are termed local bond stresses. Under direct tension or compression the bond stresses offer anchorage to such direct forces and are checked on an average bond stress.

3.16.1. Local bond

It is shown in section 3.15.1 that the change in tensile force is equal to the horizontal shearing force on any plane below the neutral axis.

For the planes *ab* and *cd* it was shown that the change in the tensile force between the two planes for a member subjected to bending was equal to the differences in bending moments between the two planes divided by the lever arm, i.e.

$$(M_0 - M_1)\frac{1}{z}$$

The change in tensile force is transferred to the reinforcement via the bond stress at the bar interface, and is equal to the horizontal shear force within the masonry over the unit of length considered (see Fig. 3.21).

The bond must therefore be adequate to resist the change in force. The strength of this bond may be defined, for a very short length, as the product of the bond resistance stress and the contact area and may be expressed for design purposes as

$$\frac{f_{bs}}{\gamma_{mb}} \delta_1 \Sigma_u$$

where f_{bs} is the characteristic local bond stress, γ_{mb} is the partial safety factor for the bond, δ_1 is the length of the bar under consideration and Σ_u is the sum of the perimeters of bars providing the reinforcement.

The characteristic local bond strength f_{bs} is $1\cdot4f_b$. BS 5628: Part 2[1] does not now include local bond; however, the Authors do not agree with the approach to bond stresses in BS 5628. Clause 19.1.6 of BS 5628 is as follows.

19.1.6 *Characteristic anchorage bond strength*, f_b. The characteristic anchorage bond strength, f_b, between mortar and steel in tension or compression should be taken as $1\cdot5$ N/mm² for plain bars and $2\cdot0$ N/mm² for deformed bars of types 1 and 2 as defined in BS 4449 and BS 4461.

The characteristic anchorage bond strength between concrete infill and steel in tension or compression should be taken as $1\cdot8$ N/mm² for plain bars and $2\cdot5$ N/mm² for deformed bars of types 1 and 2 as defined in BS 4449 and BS 4461.

Note. The recommendations in this clause may not apply to walls incorporating bed joint reinforcement to enhance lateral load resistance (see appendix A).

Where austenitic stainless steel reinforcement other than types 1 and 2 is used tests as described in appendix A of BS 4449: 1978 should be carried out.

The resistance must be greater than or equal to the force being transferred, i.e.

$$\frac{f_{bs}}{\gamma_{mb}} \delta_1 \Sigma_u \geqslant (M_0 - M_1)\frac{1}{z}$$

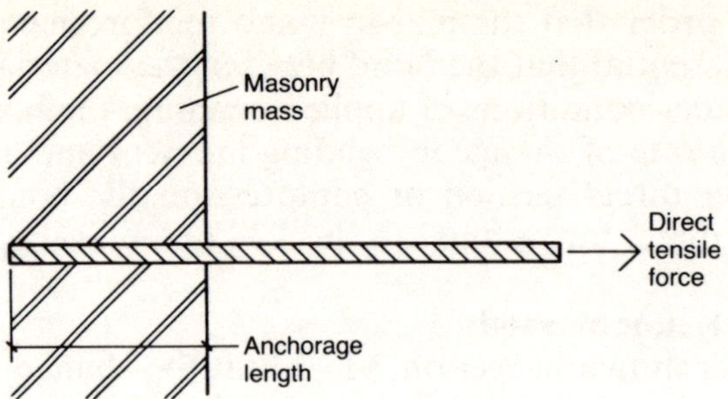

Fig. 3.25. Anchorage length

Rearranged this becomes

$$\frac{f_{bs}}{\gamma_{mb}} \geqslant M_0 - M_1 \times \frac{1}{2\delta_1 \times \Sigma_u}$$

and since over a short unit length $(M_0 - M_1)/\delta_1$ is equal to the shear force V

$$\frac{f_{bs}}{\gamma_{mb}} \geqslant \frac{V}{\Sigma_u d}$$

As noted in section 3.15, this has been modified in most recent codes by replacing the lever arm by the effective depth d and is expressed as

$$\frac{f_{bs}}{\gamma_{mb}} \geqslant \frac{V}{\Sigma_u d}$$

However, the Authors are of the opinion that this change may have inadequate justification.

13.16.2. Anchorage bond

The anchorage or average bond is activated by a direct force, as shown in Fig. 3.25. As the term suggests it is calculated from the average bond stress on the bar, i.e.

$$\frac{\text{force in the bar}}{\text{anchorage length} \times \text{perimeter}}$$

The force in the bar will be equal to the actual stress f in the bar times the cross-sectional area of the bar, i.e. $f\pi D^2/4$.

The anchorage bond stress is the force in the bar divided by the surface area of the bar, i.e. $\pi D^2 f \times \pi D/4$ and

$$\frac{\pi D^2 f}{\text{anchorage length} \times 4D\pi}$$

i.e.

$$\text{anchorage bond stress} = \frac{fD}{\text{anchorage length} \times 4}$$

Rearranging this gives

$$\text{anchorage length} = \frac{fD}{4 \times \text{anchorage bond stress}}$$

As limit state principles are relevant, partial safety factors apply to both loads and strengths so this equation should be multiplied by the factors γ_f

and γ_{mb} so that the appropriate total safety factors can be applied as follows

$$\frac{fD\gamma_f}{4f_b/\gamma_{mb}} = \frac{fD}{4f_b}(\gamma_f \times \gamma_{mb})$$

where f_b is the characteristic anchorage bond stress (see clause 19.1.6 of BS 5628: Part 2[1] in section 3.16.1) and γ_{mb} is the partial safety factor for bond (see Table 3.3).

To prevent bond failure, the tension or compression in any bar due to design loads should be developed on each side of the section by the appropriate anchorage bond strength divided by the partial safety factor for bond. The cover of concrete infill or mortar should not be less than the bar diameter.

4

Design examples: reinforced masonry

In the following design examples of reinforced masonry, the Authors have used the lower values of γ_{mm} and the increased values of f_f for flexural compression checks (see section 3.2). In other respects the design follows the recommendations of BS 5628: Part 2[1] where appropriate. However, the designer may wish to consider substituting f_k for f_f if stricter compliance with the code is thought appropriate. In all cases the Authors would recommend a cautious approach, particularly by inexperienced designers.

4.1. Reinforced masonry walls

4.1.1. Quetta bond wall: axial loading

The clay brick wall shown in Fig. 4.1 is required to support a design axial load of 1100 kN/m. The specification is for bricks with a compressive strength of 27·5 N/mm² laid in a designation (ii) mortar. Special manufacturing and construction control will apply to materials and workmanship and the grout proposed for cavities of the Quetta bonded wall will be grade 25.

The following design follows the basic principles recommended in clause 22.4.1 of BS 5628: Part 2,[1] adapted as felt appropriate by the Authors.

The characteristic compressive strength of brickwork is given by (from table 3(A) of BS 5628: Part 2)

$$f_k = 7\cdot9 \text{ N/mm}^2$$

For special control (from table 6 of BS 5628: Part 2)

$$\gamma_{mm} = 2\cdot0$$

The effective height of the wall (as the floor slabs provide enhanced lateral restraint) is

$$0\cdot75 \times \text{clear height} = 0\cdot75 \times 8\cdot0 \text{ m}$$
$$= 6\cdot0 \text{ m}$$

The effective thickness of the wall is the actual thickness (i.e. 0·328 m) and therefore the slenderness ratio

$$SR = 6\cdot0/0\cdot328 = 18\cdot3$$

and (from table 7 of BS 5628: Part 1)

$$\beta = 0\cdot76$$

The designer should first check if the wall section, without reinforcement, can support the design loading, using the formula from clause 32.2 of BS 5628

$$\text{design strength of plain wall} = \beta t f_k/\gamma_m = 0\cdot76 \times 328 \times 7\cdot9/2\cdot5$$
$$= 787\cdot7 \text{ kN/m}$$

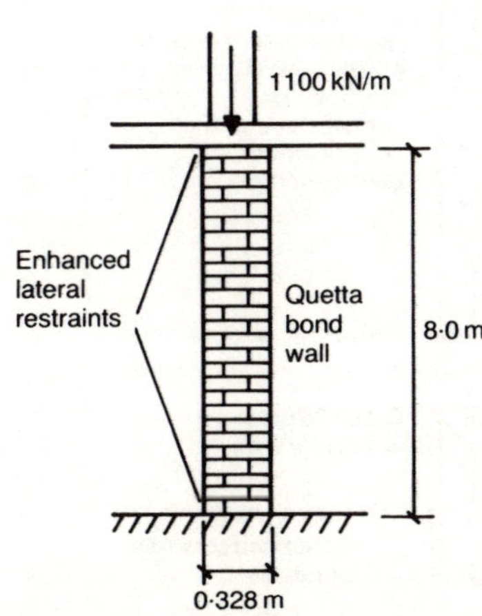

Fig. 4.1. Quetta bond wall: elevation

which is inadequate. Hence it is necessary either to increase the brickwork strength or, as demonstrated in this example, to reinforce the wall (see Fig. 4.2).

Using mild steel reinforcement of grade 250

$f_y = 250$ N/mm^2 from table 4 of BS 5628: Part 2

$\gamma_{ms} = 1\cdot15$ from table 7 of BS 5628: Part 2

The design vertical load resistance of an axially loaded member is given by

$$N_d = \beta\left(\frac{f_k A_m}{\gamma_{mm}} + \frac{0\cdot83 f_y A_s}{\gamma_{ms}}\right)$$

Considering a metre run of wall and a design axial load of 1100 kN/m

$$1100 = 0\cdot76\left(\frac{7\cdot9 \times 328 \times 10^3}{2\cdot0 \times 10^3} + \frac{0\cdot83 \times 250}{1\cdot15 \times 10^3} A_s\right)$$

$$1100 = 0\cdot76(1295\cdot6 + 0\cdot18 A_s)$$

$$1100 = 984\cdot67 + 0\cdot137 A_s$$

and so

$$A_s = \frac{1100 - 984\cdot67}{0\cdot137}$$

$$= 842 \text{ mm}^2 \text{ per m length of wall}$$

Therefore 16 mm dia. mild steel bars at 169 mm centres grouted into voids in Quetta bond wall using grade 25 grout should be used.

Clause 24 of BS 5628: Part 2 on axial compressive loading also allows the use of clause 23.3.1.3 on combined vertical load and bending for the design

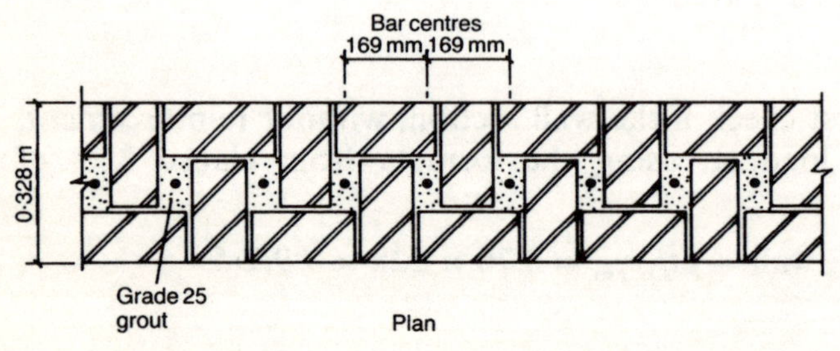

Fig. 4.2 (below). Quetta bond wall: sectional plan

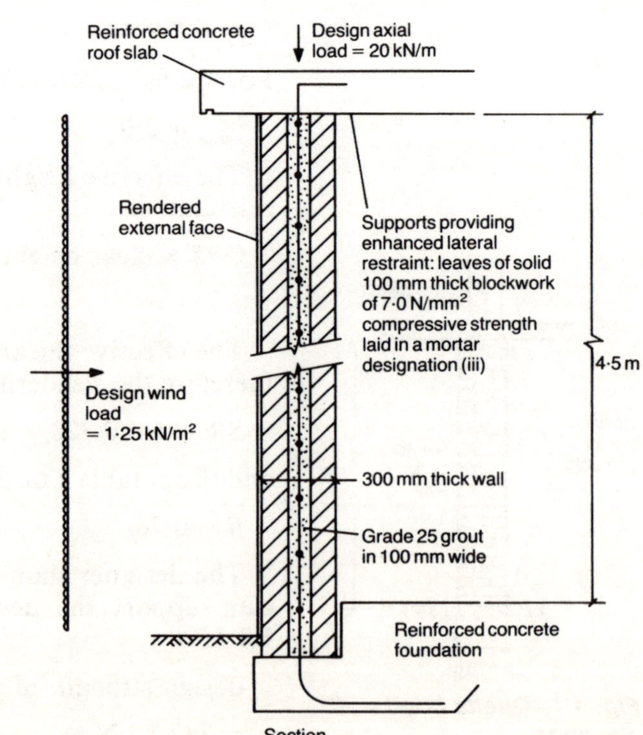

Fig. 4.3 (right). Grouted cavity wall: section

of slender columns. Clause 23.3.1.3 allows the use of clause 23.3.1.1 for the same design but includes the additional bending moment M_a given in clause 23.3.1.3 as follows.

$$M_a = \frac{N(h_{ef})^2}{2000t}$$

$$= \frac{1100 \times 6^2}{2000 \times 0.328}$$

$$= 60.36 \text{ kNm}$$

Therefore the total M is 60.36 kNm. Checking against clause 23.3.1.1(a) shows that

$$N \not> N_d$$

$$= \frac{f_k}{\gamma_{mm}} b(t - 2e_x)$$

$$= \frac{7.9}{2.0} \times 10^3 \left(328 - 2 \times \frac{60.36}{1100} \right)$$

$$= 1295 \text{ kN}$$

This clause is intended for short columns and takes no account of buckling.

Checking against clause 23.3.1.1(b) is inappropriate to Quetta bond (and similar) walls as it involves sections with two layers of reinforcement, i.e. A_{s1}: A_{s2}: f_{s1}: f_{s2}: d_1: d_2 (see Figs 3.9 and 3.11). Also buckling is again ignored. Checking against clause 23.3.1.1(c) is also inappropriate because the axial load dominates the design in this example.

Hence the design method adopted uses the basic principles of BS 5628: Part 1 in which the capacity reduction factor β makes a correction for the tendency of the section to buckle.

4.1.2. Grouted cavity wall: wind and axial loading

The grouted cavity wall shown in Fig. 4.3 is to be constructed of concrete blockwork and rendered externally. The wall is to support a design axial load of 20 kN/m and a design wind load of 1.25 kN/m². Manufacturing control is normal and construction control is special for these particular circumstances. In this example, the reinforcement required is to be designed.

BS 5628: Part 2 recommends the use of mortar designations (i) and (ii) generally, although mortar designation (iii) is allowed for panel walls. In this example, a mortar designation (iii) has been specified and is considered more appropriate for use with concrete blockwork particularly as, in this case, the reinforcement is required only to resist wind forces.

The wall will be considered to span vertically as a propped cantilever although, with the wall reinforcement tied into the roof slab, fixity could be considered at both the top and the bottom of the wall. However, rotation of the roof slab from its own bending may affect the top condition and so the propped cantilever analysis is considered to be more appropriate.

The span/effective depth ratio is 35 for simply supported walls and 45 for continuous walls. If for the propped cantilever condition a span/effective depth ratio of 40 is assumed, the effective depth required will be 4500/40 = 112.5 mm.

The effective depth provided is in excess of 150 mm and therefore the wall satisfies the serviceability limit state for deflection. At the base of the wall the design moment is

$$\frac{1\cdot25 \times 4\cdot5^2}{8} = 3\cdot16 \text{ kN m per m run}$$

and the design shear force is

$$\frac{5 \times 1\cdot25 \times 4\cdot5}{8} = 3\cdot51 \text{ kN per m run}$$

Using clause 22.4.1 of BS 5628: Part 2 the proportion of axial thrust in the wall is checked as follows

$$0\cdot1 f_k\, A_m = 0\cdot1 \times 6\cdot4 \times 0\cdot3 \times 1 \times 10^3 = 192 \text{ kN/m}$$

which exceeds the design axial force of 20 kN/m. Hence the member may be designed for bending only.

The masonry specification has been given as $7\cdot0$ N/mm² solid blocks laid in a designation (iii) mortar, and the face size of the blocks is assumed to be 440×215. Therefore for the blocks

$$\frac{\text{block height}}{\text{least horizontal dimension}} = \frac{215}{100}$$

$$= 2\cdot15$$

Hence (from table 3(c) of BS 5628: Part 2)

$$f_k = 6\cdot4 \text{ N/mm}^2$$

Therefore

$$f_f = 1\cdot2 \times 6\cdot4 = 7\cdot68 \text{ N/mm}^2$$

and (for normal/special control)

$$\gamma_{mm} = 2\cdot3$$

The design moment of resistance

$$M_d = \frac{0\cdot375 \times f_f \times b \times d^2}{\gamma_{mm}}$$

Therefore

$$d = \sqrt{\frac{M_d\, \gamma_{mm}}{0\cdot375 f_f\, b}}$$

$$= \sqrt{\frac{3\cdot16 \times 10^6 \times 2\cdot3}{0\cdot375 \times 7\cdot68 \times 10^3}}$$

$$= 50\cdot23 \text{ mm}$$

which is less than the effective depth provided.

The lever arm is obtained from

$$\frac{M \gamma_{mm}}{2 f_f\, b d^2} = \frac{3\cdot16 \times 10^6 \times 2\cdot3}{2 \times 7\cdot68 \times 10^3 \times 153^2}$$

$$= 0\cdot0205$$

and from Fig. 3.7 the lever arm factor

$$z/d = 0\cdot95 \text{ (maximum allowed)}$$

Therefore

$$z = 0\cdot95 \times 153 \text{ (see Fig. 4.4)} = 145\cdot35 \text{ mm}$$

Using mild steel reinforcement grade 250

$$f_y = 250 \text{ N/mm}^2$$

$$\gamma_{ms} = 1\cdot15$$

Hence

$$M_d = A_s f_y z / \gamma_{ms}$$

from which

$$A_s = M_d \gamma_{ms} / f_y z$$

$$= \frac{3\cdot16 \times 1\cdot15 \times 10^6}{250 \times 145\cdot35}$$

$$= 100 \text{ mm}^2 \text{ per m run of wall}$$

Use 6 mm dia. mild steel bars at 250 mm centres (113 mm^2) for both main bars and distribution bars.

Figure 4.4 shows the arrangement of the reinforcement within the wall in which 44 mm of grout cover is provided to the main reinforcement.

The shear may be checked as follows.

From clause 19.1.3.1.2 of BS 5628: Part 2 the characteristic shear strength is

$$f_v = 0\cdot35 + 17\cdot5\rho$$

where

$$\rho = A_s/bd$$

$$= 113/153 \times 10^3$$

$$= 0\cdot00074$$

Therefore

$$f_v = 0\cdot35 + (17\cdot5 \times 0\cdot00074)$$

$$= 0\cdot36 \text{ N/mm}^2$$

Now

$$\frac{f_v}{\gamma_{mv}} = \frac{V}{bd}$$

from which shear resistance

$$V = f_v bd / \gamma_{mv}$$

$$= 0\cdot36 \times 153 \times 10^3 / 2\cdot0$$

$$= 27\cdot54 \text{ kN/m}$$

Fig. 4.4. Grouted cavity wall: sectional plan

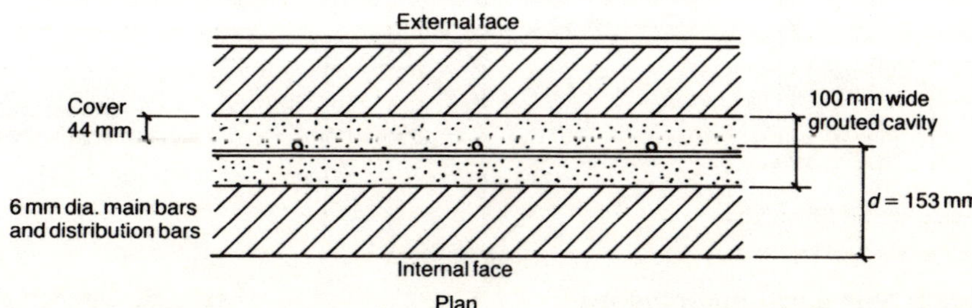

which exceeds the applied shear force due to the design loads calculated earlier as 3·51 kN/m.

Local bond may be checked as follows.

$$\frac{f_{bs}}{\gamma_{mb}} > \frac{V}{\Sigma_u d}$$

From clause 19.1.6 of BS 5628: Part 2 (quoted in section 13.16.1), for plain bars in a concrete infill grout

$$f_b = 1\cdot8 \ \text{N/mm}^2$$

$$f_{bs} = 1\cdot4 \times f_b$$

$$= 1\cdot4 \times 1\cdot8$$

$$= 2\cdot52 \ \text{N/mm}^2$$

$$\gamma_{mb} = 1\cdot5$$

Therefore

$$\frac{f_{bs}}{\gamma_{mb}} = \frac{2\cdot52}{1\cdot5} = 1\cdot68 \ \text{N/mm}^2$$

$$\frac{V}{\Sigma_u d} = \frac{3\cdot51 \times 10^3}{75\cdot41 \times 153} = 0\cdot304 \ \text{N/mm}^2$$

Hence

$$\frac{f_{bs}}{\gamma_{ms}} > \frac{V}{\Sigma_u d}$$

4.1.3. Pocket type earth retaining wall

The earth retaining wall shown in Fig. 4.5 is to be designed as a reinforced brickwork pocket wall. Manufacturing and construction controls will be taken as special. The brickwork will comprise bricks with a compressive strength of 35 N/mm² and a water absorption of less than 7% laid in a

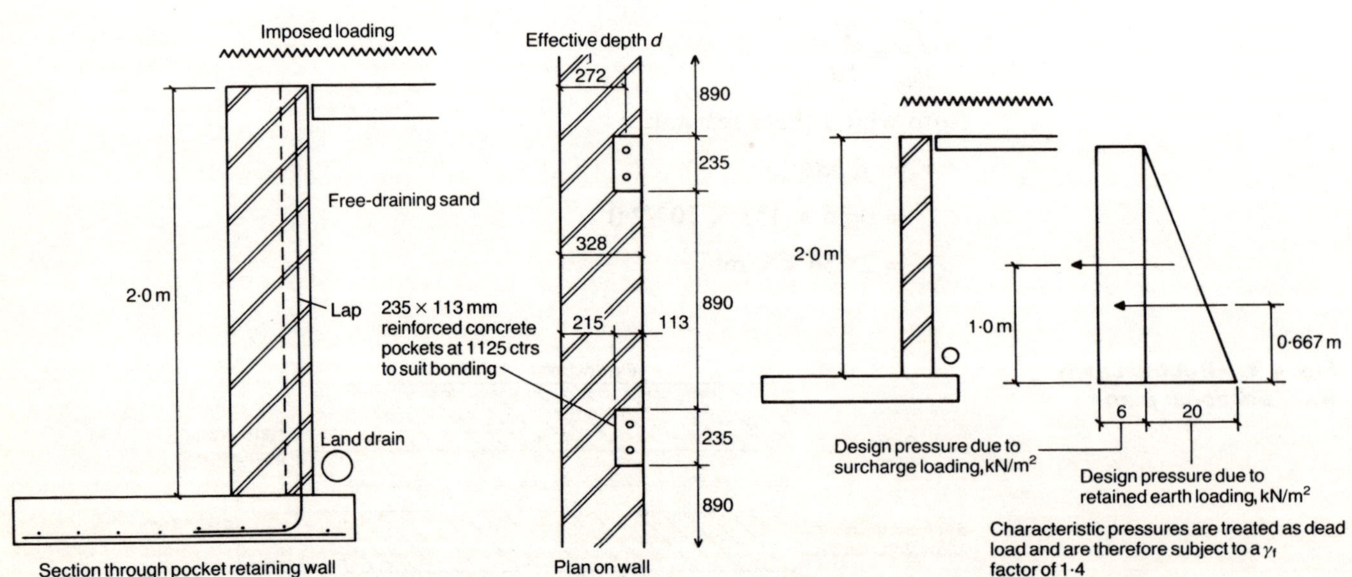

Fig. 4.5. Pocket type earth-retaining wall   *Fig. 4.6. Design pressure diagram*

designation (i) mortar. The wall retains fine dry sand with adequate provision for free drainage. The upper floor level supports an imposed loading and the combination of loads produces the design pressure diagram shown in Fig. 4.6.

The design overturning moment

$$\frac{6 \times 2^2}{2} + \frac{20 \times 2^2}{2 \times 3} = 12 + 13 \cdot 33$$

$$= 25 \cdot 33 \text{ kNm/m length of wall}$$

and for a pocket spacing of $1 \cdot 125$ m the design overturning moment is

$$25 \cdot 33 \times 1 \cdot 125 = 28 \cdot 5 \text{ kNm}$$

The check for the wall spanning horizontally between pockets would be done as for the example in section 4.1.4. From experience this example is considered adequate.

For the masonry specification given, from table 3(A) of BS 5628: Part 2

$$f_k = 11 \cdot 40 \text{ N/mm}^2$$

from table 3 of BS 5628: Part 1 for the plane of failure parallel to the bed joints

$$f_{kx} = 0 \cdot 70 \text{ N/mm}^2$$

from table 6 of BS 5628: Part 2 for special manufacture and construction control

$$\gamma_{mm} = 2 \cdot 0$$

and using high tensile reinforcement (assuming a bar diameter greater than 16 mm)

$$f_y = 460 \text{ N/mm}^2$$

$$\gamma_{ms} = 1 \cdot 15$$

The design resistance moment of the masonry may be checked as follows.

$$M_d = C \times \text{lever arm}$$

$$= 0 \cdot 375 \, \frac{f_f}{\gamma_{mm}} \, bd^2$$

$$= 0 \cdot 375 \, \frac{1 \cdot 2 \times 11 \cdot 4}{2 \cdot 0} \times 1125 \times 272^2 \times 10^{-6}$$

$$M_d = 213 \cdot 5 \text{ kNm}$$

which exceeds the applied moment.

For the design of the reinforcement

$$M_d = T \times \text{lever arm}$$

$$= A_s f_y z / \gamma_{ms}$$

which, rearranged, becomes

$$A_s = M \gamma_{ms} / f_y z$$

$$= \frac{28 \cdot 5 \times 10^6 \times 1 \cdot 15}{460z}$$

$$= 71\,250/z \text{ mm}^2 \text{ per pocket}$$

z can be obtained from the lever arm curve (Fig. 3.7) for which

$$\frac{M\gamma_{\mathrm{mm}}}{2f_{\mathrm{f}}bd^2} = \frac{28{\cdot}5 \times 10^6 \times 2{\cdot}0}{2 \times 1{\cdot}2 \times 11{\cdot}4 \times 1125 \times 272^2}$$

$$= 0{\cdot}025$$

Hence, from Fig. 3.7

$$z/d = 0{\cdot}95 \ (\text{maximum})$$

Therefore

$$z = 0{\cdot}95 \times 272 = 258{\cdot}4 \ \text{mm}$$

$$A_{\mathrm{s}} = \frac{71\,250}{258{\cdot}4}$$

$$= 276 \ \text{mm}^2 \ \text{per pocket}$$

Use two 16 mm dia. high tensile deformed bars per pocket ($A_{\mathrm{s}} = 402 \ \text{mm}^2$). The check for shear is as follows.

$$V = (6 \times 2) + (20 \times 2 \times 0{\cdot}5)$$

$$= 12 + 20$$

$$= 32 \ \text{kN/m length of wall}$$

The characteristic shear strength

$$f_{\mathrm{v}} = (0{\cdot}35 + 17{\cdot}5\rho) \ \text{N/mm}^2$$

but

$$\rho = \frac{A_{\mathrm{s}}}{bd} = \frac{402}{1125 \times 272} = 0{\cdot}0013$$

Hence

$$f_{\mathrm{v}} = 0{\cdot}35 + (17{\cdot}5 \times 0{\cdot}0013)$$

$$= 0{\cdot}373 \ \text{N/mm}^2$$

The design shear strength

$$\frac{f_{\mathrm{v}}}{\gamma_{\mathrm{mv}}} = \frac{0{\cdot}373}{2{\cdot}0}$$

$$= 0{\cdot}186 \ \text{N/mm}^2$$

The design shear stress is

$$\frac{V}{A_{\mathrm{m}}} = \frac{32 \times 10^2 \times 1{\cdot}125}{1125 \times 272}$$

$$= 0{\cdot}117 \ \text{N/mm}^2$$

which is less than the design shear strength.
The check for local bond is as follows.
The characteristic local bond strength is

$$f_{\mathrm{bs}} = (1{\cdot}4 \times f_{\mathrm{b}}) \ \text{N/mm}^2$$

For deformed bars set in concrete

$$f_{\mathrm{b}} = 2{\cdot}5 \ \text{N/mm}^2$$

Hence

$$f_{bs} = 2 \cdot 5 \times 1 \cdot 4$$

$$= 3 \cdot 5 \ \text{N/mm}^2$$

$$\gamma_{mb} = 1 \cdot 5$$

The design local bond strength is

$$\frac{f_{bs}}{\gamma_{mb}} = \frac{3 \cdot 5}{1 \cdot 5}$$

$$= 2 \cdot 33 \ \text{N/mm}^2$$

The design local bond stress is

$$\frac{V}{\Sigma_u d} = \frac{32 \times 10^3 \times 1 \cdot 125}{2 \times \pi \times 16 \times 272}$$

$$= 1 \cdot 32 \ \text{N/mm}^2$$

which is less than the design local bond strength.

The design of the concrete base into which the wall reinforcement must be anchored is outside the scope of this book.

4.1.4. Reinforced fin retaining wall

The earth retaining wall shown in Fig. 4.7 is to be designed as a reinforced brickwork fin wall. Manufacturing and construction controls will be taken as special. The brickwork will comprise bricks with a compressive strength of 35 N/mm² and a water absorption of 7–12% laid in a designation (i) mortar. The wall retains fine dry sand with adequate provision for free drainage. The upper car park level supports an imposed loading and the combination of loads produces the design pressure diagram shown in Fig. 4.8.

The design overturning moment is

$$(5 \times 3 \times 1 \cdot 5) + (26 \times 3 \times 0 \cdot 5 \times 1 \cdot 0) = 22 \cdot 5 + 39 \cdot 0$$

$$= 61 \cdot 5 \ \text{kNm}$$

Fig. 4.7. Reinforced fin retaining wall

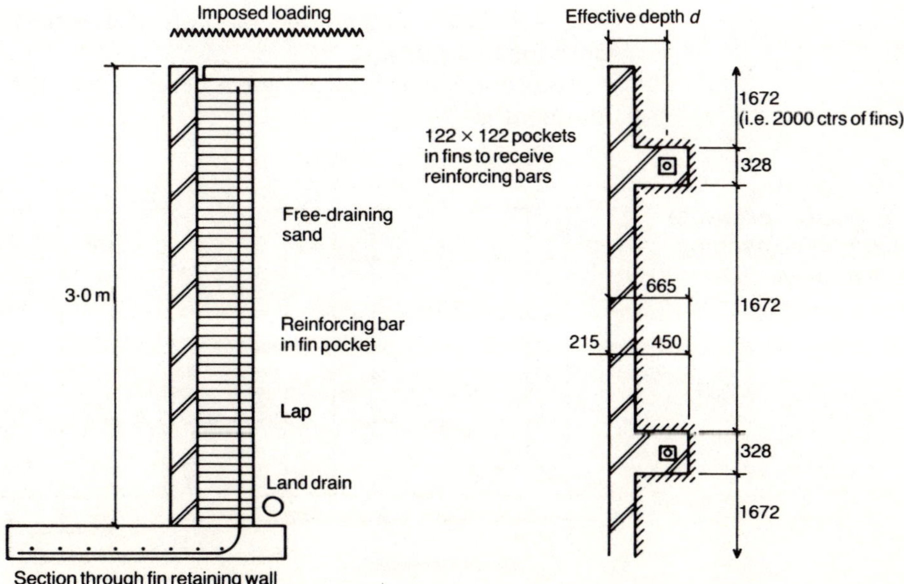

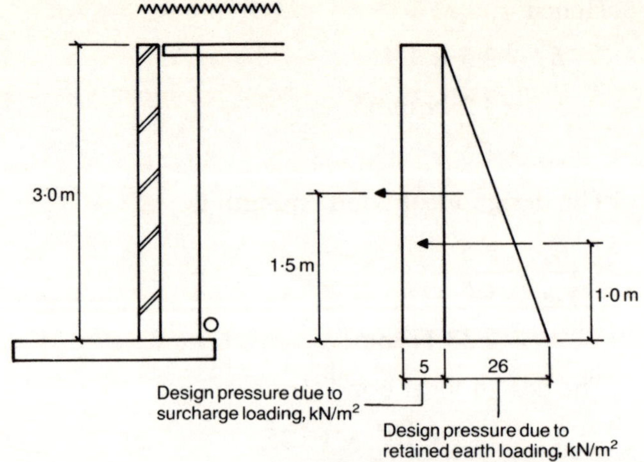

Fig. 4.8. Design pressure diagram

Try fins spaced at 2·0 m centres. Then the design moment per fin is

$61.5 \times 2 = 123.0$ kNm

For the masonry specification given

$f_k = 11.40$ N/mm²

$f_{kx} = 0.50$ N/mm² parallel and 1.5 N/mm² perpendicular

$\gamma_{mm} = 2.0$

($\gamma_m = 2.5$ for plain masonry, i.e. flange spanning horizontally)

$f_y = 460$ N/mm²

$\gamma_{ms} = 1.15$

Check a 215 mm thick flange spanning horizontally between fins as follows. It is considered unreasonable to design the horizontal span of the flange for the maximum design pressures shown in Fig. 4.8 as a considerable amount of restraint will be provided by the reinforced concrete foundation. Fig. 4.9 shows a suggested method of determining a reasonable design pressure for this purpose.

Assuming continuity across the fins, the maximum design bending moment is

Fig. 4.9. Design pressure for horizontally spanning flange of fin wall

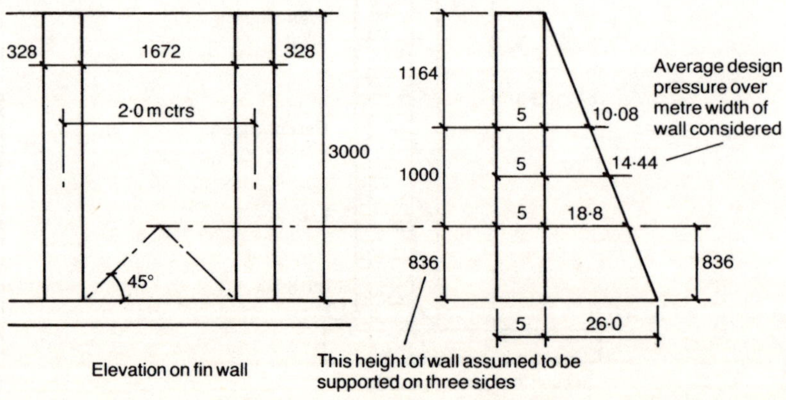

$$(5 + 14{\cdot}44) \times 2^2/12 = 6{\cdot}48 \text{ kNm}$$

which occurs on the centreline of the fin.

The critical design bending moment is 3·24 kNm and occurs, as shown in Fig. 4.10, at midspan.

The design moment of resistance is

$$f_{kx} z/\gamma_m = 1{\cdot}5 \times 1{\cdot}0 \times 0{\cdot}215^2 \times 10^3/2{\cdot}5 \times 6$$
$$= 4{\cdot}62 \text{ kNm}$$

which exceeds the midspan design bending moment of 3·24 kNm.

Now consider the vertical cantilever action of the T profile. The effective flange width is the least of

- fin centres: 2·0 m
- $1/3 \times$ effective span: $3{\cdot}0/3 = 1{\cdot}0$ m
- fin breadth + 12 times flange thickness: $0{\cdot}328 + (12 \times 0{\cdot}215) = 2{\cdot}9$ m

The effective fin section is shown in Fig. 4.11 in which a fin 665 mm × 328 mm providing an effective depth d of 500 mm is adopted for trial purposes.

The design resistance moment of the masonry is

$$M_d = C \times \text{lever arm}$$

The minimum design resistance moment will be given when the lever arm is taken from the centre of the thickness of the 215 mm flange

$$\text{minimum } M_d = (f_f/\gamma_{mm}) \times 1000 \times 215 \times 392 \times 10^{-6}$$

For $f_f = 1{\cdot}2f_k$ this becomes

$$(1{\cdot}2 \times 11{\cdot}4/2{\cdot}0) \times 1000 \times 215 \times 392 \times 10^{-6} = 576{\cdot}5 \text{ kNm}$$

which exceeds the applied design bending moment of 123·0 kNm already calculated.

Since the flange of the T section is approximately $0{\cdot}5d$ the lever arm can be calculated by analysing the T profile as a rectangular section of width 1000 mm and depth 665 mm.

The lever arm z is obtained from the lever arm curve (see Fig. 3.7) for which

$$\frac{M\gamma_{mm}}{2f_f bd^2} = \frac{123 \times 2{\cdot}0 \times 10^6}{2 \times 1{\cdot}2 \times 11{\cdot}4 \times 1000 \times 665^2}$$
$$= 0{\cdot}020$$

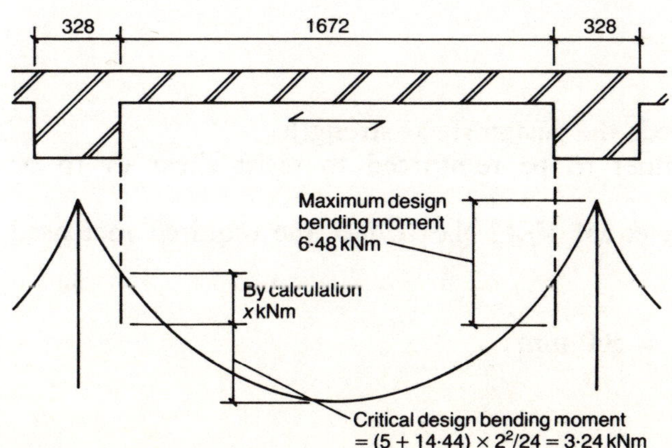

Fig. 4.10. Critical design bending moment diagram: horizontal span

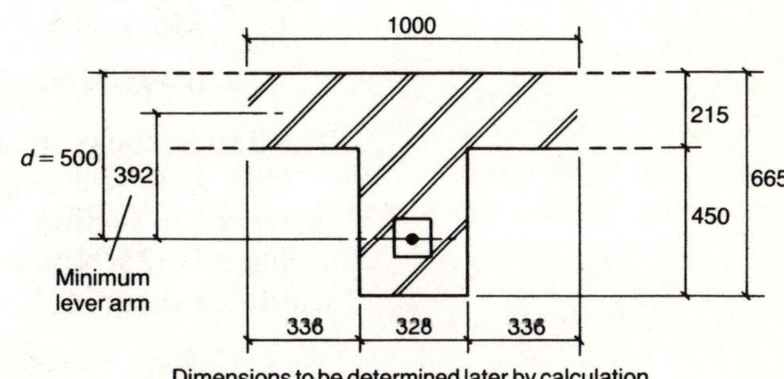

Fig. 4.11. Effective fin section

Hence, from Fig. 3.7

$$z/d = 0.95$$

Therefore

$$z = 0.95 \times 665$$

$$= 632 \text{ mm}$$

Using this lever arm the minimum A_s is designed

$$A_s = M\gamma_{ms}/f_y z$$

$$= \frac{123.0 \times 10^6 \times 1.15}{460 \times 632}$$

$$= 486 \text{ mm}^2 \text{ per fin}$$

Use two T20 high tensile bars per fin ($A_s = 628 \text{ mm}^2$).
Check for shear as follows.

$$V = (5 \times 3) + (26 \times 3 \times 0.5) \times 2.0 \text{ m centres}$$

$$= (15 + 39) \times 2$$

$$= 108.0 \text{ kN}$$

The design shear strength is f_v/γ_{mv}

$$f_v = 0.35 + 17.5\rho$$

Where

$$\rho = A_s/bd$$

$$= 628/(665 \times 328)$$

$$= 0.003$$

then

$$f_v = 0.35 + (17.5 \times 0.003)$$

$$= 0.403 \text{ N/mm}^2$$

Only the breadth of the fin is used to resist shear.
The design shear strength is

$$\frac{0.403}{2.0} = 0.202 \text{ N/mm}^2$$

$$\frac{V}{A_m} = \frac{108 \times 10^3}{328 \times 665}$$

$$= 0.495 \text{ N/mm}^2$$

The design shear stress exceeds the design shear strength.
   This wall will require either to be reinforced to resist shear or to be increased in section.
   Since 0.325 N/mm^2 must equal V/A_m (i.e. V/bd), the required increased width for shear is

$$b = d\frac{V/f_v}{\gamma_{mv}} = \frac{108 \times 10^3}{665 \times 0.325} = 500 \text{ mm}$$

say 554 for brick sizing.

Check local bond as follows.
The characteristic local bond strength is given by

$$f_{bs} = 1\cdot4 \times f_b$$

and for deformed bars set in concrete

$$f_b = 2\cdot5 \ \text{N/mm}^2$$

Hence

$$f_{bs} = 1\cdot4 \times 2\cdot5$$
$$= 3\cdot5 \ \text{N/mm}^2$$

$$\gamma_{mb} = 1\cdot5$$

The design local bond strength is given by

$$f_{bs}/\gamma_{mb} = 3\cdot5/1\cdot5$$
$$= 2\cdot33 \ \text{N/mm}^2$$

and the design local bond stress by

$$\frac{V}{\Sigma_u d} = \frac{108 \times 10^3}{2\pi \times 20 \times 665}$$
$$= 1\cdot29 \ \text{N/mm}^2$$

which is less than the design local bond strength.
The revised wall is shown in Fig. 4.12.

Fig. 4.12. *Revised wall detail*

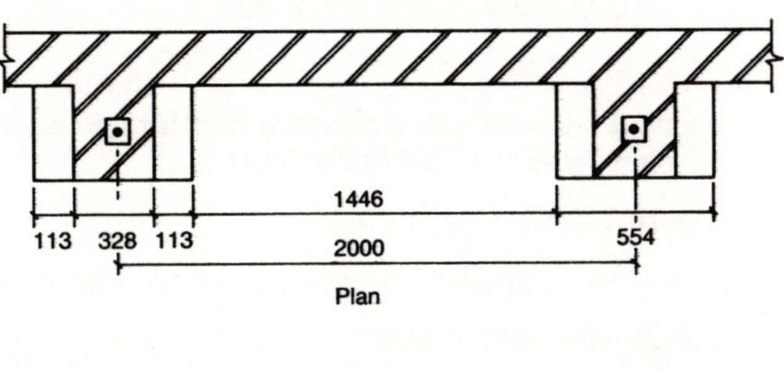

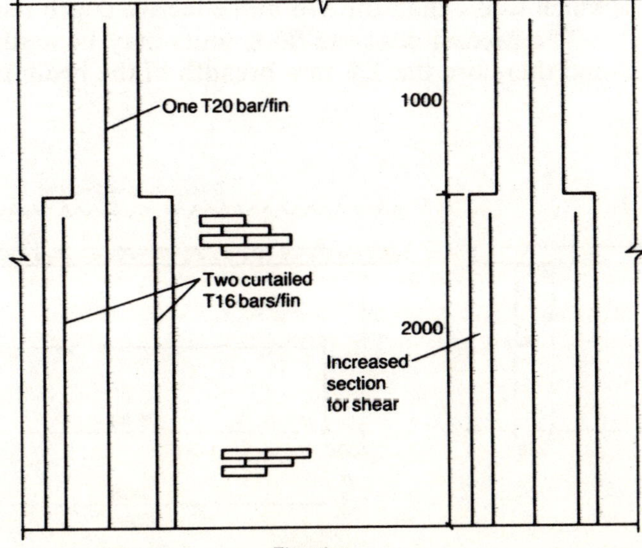

The design of the reinforced concrete base, into which the reinforcement must be anchored, is outside the scope of this book.

4.2. Reinforced masonry beams

4.2.1. Reinforced brickwork beam

A reinforced brickwork beam is required to span the 3·5 m clear opening as shown in Fig. 4.13. The design loading, including the self-weight of the beam, will be taken as 45 kN/m. The brickwork will comprise bricks with a compressive strength of 27·5 N/mm² laid in a designation (ii) mortar and the manufacturing and construction controls will be taken as special.

The design bending moment is

$$45 \times 3\cdot828^2/8 = 82\cdot43 \text{ kNm}$$

and the design shear force is

$$V = 45 \times 3\cdot5/2$$

$$= 78\cdot75 \text{ kN}$$

First, consider the brickwork in compression. The design moment of resistance is

$$M_d = \sqrt{(0\cdot375 f_f\, bd^2/\gamma_{mm})}$$

Hence, the effective depth required is

$$d = \sqrt{\frac{M_d \gamma_{mm}}{0\cdot375 f_f\, b}}$$

$$= \sqrt{\frac{82\cdot43 \times 10^6 \times 2\cdot0}{0\cdot375 \times 1\cdot2 \times 7\cdot9 \times 328}}$$

$$= 376 \text{ mm.}$$

Check against span/effective depth ratio as follows.
From table 9 of BS 5628: Part 2

$$\text{span/effective depth} = 20$$

for simply supported beams and hence minimum effective depth is

$$3828/20 = 191\cdot4 \text{ mm}$$

which is less than the 376 mm effective depth required.

The precast concrete floor units may be assumed to provide full restraint and therefore the 328 mm breadth of the beam is also satisfactory.

Fig. 4.13. Reinforced brickwork beam

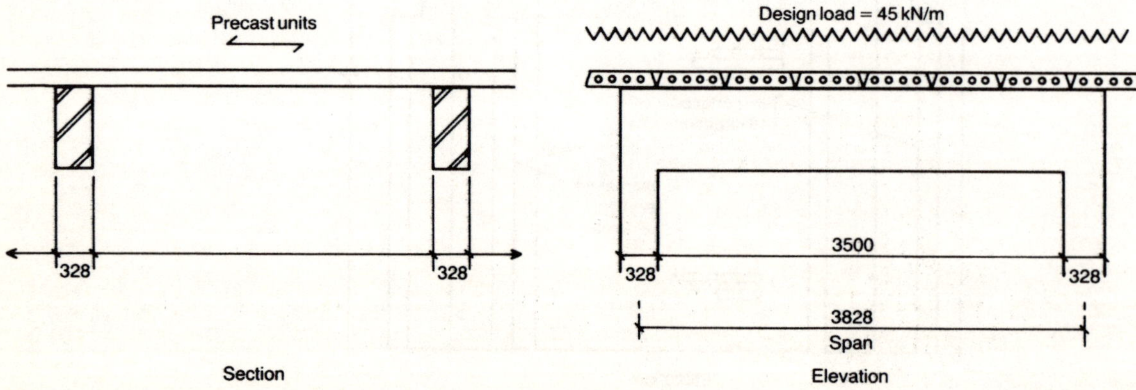

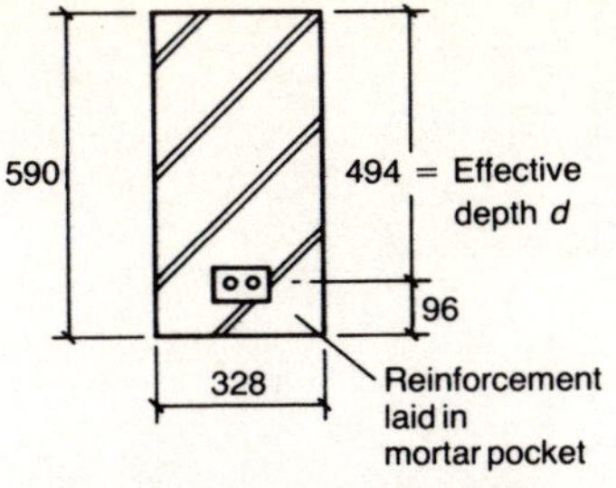

*Fig. 4.14. Section through
reinforced brickwork beam*

The minimum overall depth of the beam is the effective depth required plus half the assumed bar diameter plus the link diameter plus the grout cover and one course of brickwork, i.e.

$$376{\cdot}0 + 10{\cdot}0 + 6{\cdot}0 + 25{\cdot}0 + 65{\cdot}0 = 482{\cdot}0 \text{ mm}$$

but must be increased to a brick course dimension. Hence an overall depth of 590 mm should be used to allow for shear considerations.

The beam section is shown in Fig. 4.14.

Calculate the lever arm from Fig. 3.7 for which

$$\frac{M_d \gamma_{mm}}{2 f_f b d^2} = \frac{82{\cdot}43 \times 10^6 \times 2{\cdot}0}{2 \times 1{\cdot}2 \times 7{\cdot}9 \times 328 \times 494^2}$$

$$= 0{\cdot}109$$

Reading off from Fig. 3.7

$$z/d = 0{\cdot}87$$

Hence

$$z = 0{\cdot}87 \times 494$$

$$= 430 \text{ mm}$$

The reinforcement to be used will be high tensile deformed bars (assuming a diameter greater than 16 mm). Hence

$$f_y = 460 \text{ N/mm}^2$$

$$\gamma_{ms} = 1{\cdot}15$$

Now considering the reinforcement in tension, the design moment of resistance is

$$M_d = \frac{A_s f_y z}{\gamma_{ms}}$$

from which

$$A_s = \frac{M_d \gamma_{ms}}{f_y z}$$

$$= \frac{82{\cdot}43 \times 10^6 \times 1{\cdot}15}{460 \times 430}$$

$$= 479 \text{ mm}^2$$

Use two 20 mm dia. high tensile bars of $A_s = 628 \text{ mm}^2$.

Check for shear as follows.

The design shear strength is f_v/γ_{mv} for simply supported reinforced beams. The variable f_v may be increased by a factor of $2{\cdot}5 - 0{\cdot}25a/d$, provided that $a/d = 6$ or less.

Shear span a is given by

$$M/V = 82{\cdot}43/78{\cdot}75 = 1{\cdot}05$$

Hence f_v is increased

$$2{\cdot}5 - (0{\cdot}25 \times 1{\cdot}05) = 2{\cdot}24$$

$$f_v = 0{\cdot}35 \times 2{\cdot}24$$

$$= 0{\cdot}78 \text{ N/mm}^2$$

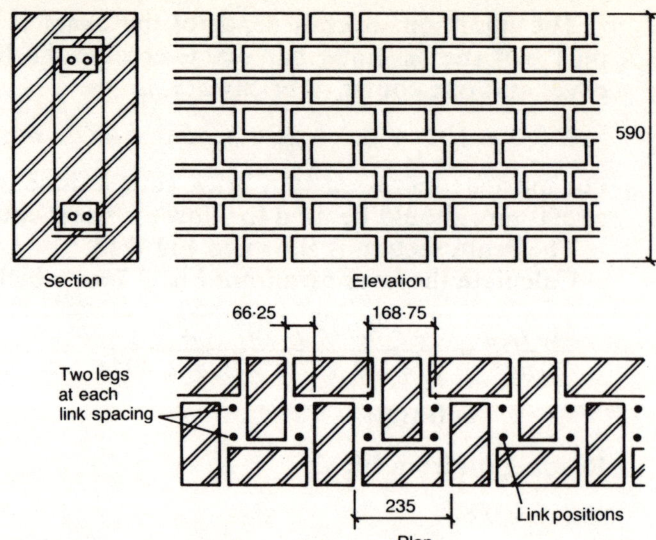

Fig. 4.15. Quetta bond for link spacing

The design shear strength is

$$0\cdot78/2\cdot0 = 0\cdot39 \text{ N/mm}^2$$

and the design shear stress is

$$V/A_m = (78\cdot75 \times 10^3)/(328 \times 494)$$

$$= 0\cdot486 \text{ N/mm}^2$$

which exceeds the design shear strength, and hence shear reinforcement is required.

Even if shear reinforcement is required the shear stress should not exceed $1\cdot75$ N/mm^2, and in this example it does not. The area of shear reinforcement is

$$A_{sv} = b\left(v - \frac{f_v}{\gamma_{mv}}\right)\frac{\gamma_{ms}}{f_y} s_v$$

where the spacing of the links s_v is determined by the bonding pattern adopted to accommodate them.

In this example it is assumed that Quetta bond will be used as shown in Fig. 4.15. This allows a link spacing of 169 mm. Care is required to ensure adequate anchorage of link ends.

For the links, mild steel will be used. Hence

$$f_y = 250 \text{ N/mm}^2$$

Therefore

$$A_{sv} = 328(0\cdot486 - 0\cdot39)(1\cdot15/250) \times 169$$

$$= 24\cdot5 \text{ mm}^2$$

$$= 12\cdot3 \text{ mm}^2 \text{ per leg of link}$$

Use 6 mm dia. mild steel links at 169 mm centres.

Check for local bond as follows.

The design local bond strength is

$$f_{bs}/\gamma_{mb} = 1\cdot4 \times 2\cdot0/1\cdot5 = 1\cdot87 \text{ N/mm}^2$$

as $f_{bs} = 1\cdot4f_b$ (see section 3.16).

For deformed bars laid in mortar pocket, the design local bond stress is

$$\frac{V}{\Sigma_u d} = \frac{78{\cdot}75 \times 10^3}{2\pi \times 20 \times 494}$$

$$= 1{\cdot}27 \ \text{N/mm}^2$$

which is less than the design local bond strength.

Check the local bearing stress at beam supports as follows.

The design local bearing stress is

$$V/\text{bearing area} = 78{\cdot}75 \times 10^3/328 \times 328$$

$$= 0{\cdot}732 \ \text{N/mm}^2$$

It is considered that the bearing detail is equivalent to a bearing type 1 as given in BS 5628: Part 1 and therefore the characteristic compressive stresses may be increased by 25%. Therefore the design bearing strength is

$$1{\cdot}25 \times f_k/\gamma_{mm} = 1{\cdot}25 \times 7{\cdot}9/2{\cdot}0$$

$$= 4{\cdot}94 \ \text{N/mm}^2$$

which exceeds the design local bearing stress.

4.3. Reinforced masonry columns

4.3.1. Axially loaded reinforced brick column

The column shown in Fig. 4.16 is required to support a design axial load of 1200 kN which may be considered to be applied axially. The floors and beams may be considered to be providing simple lateral restraint to both axes of the columns and for the purpose of this example manufacturing control is assumed to be special.

The following design follows the basic principles recommended in clause 22.4.1 of BS 5628: Part 2, adapted as felt appropriate by the Authors. The slenderness ratio $6000/440 = 13{\cdot}64$ and therefore the capacity reduction factor (from table 7 of BS 5628: Part 1) is

$$\beta = 0{\cdot}897 \ \text{based on zero eccentricity}$$

Fig. 4.16. *Reinforced brick column*

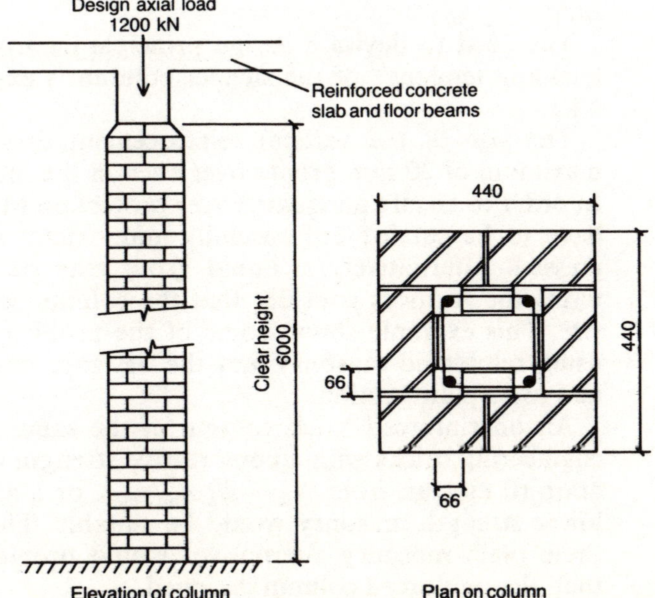

For this design example the column size of 553 mm × 440 mm has been determined from other considerations and therefore the two unknowns remaining to be determined are the masonry strength and the reinforcement quantity and type.

Initially, a brick of compressive strength 35 N/mm² set in a mortar designation (ii) is assumed and the required reinforcement is designed

$$A_s = \left(\frac{N_d}{\beta} - \frac{f_k A_m}{\gamma_{mm}} \right) \frac{\gamma_{ms}}{0 \cdot 83 f_y}$$

Given that the design axial load $N_d = 1200$ kN, $\beta = 0 \cdot 897$ (as calculated above), the assumed masonry strength $f_k = 9 \cdot 4$ N/mm², the area of a 553 mm × 440 mm column $A_m = 24 \cdot 33 \times 10^4$ mm², $\gamma_{mm} = 2 \cdot 0$ (special), $\gamma_{ms} = 1 \cdot 15$, and $f_y = 460$ N/mm² for high tensile bars of 16 mm diameter, then

$$A_s = \left(\frac{1200 \times 10^3}{0 \cdot 897} - \frac{9 \cdot 4 \times 24 \cdot 33 \times 10^4}{2 \cdot 0} \right) \frac{1 \cdot 15}{0 \cdot 83 \times 460}$$

$$= (19 \cdot 43 \times 10^4) \times 0 \cdot 003$$

$$= 583 \text{ mm}^2$$

For the column links

$$A_s = \frac{583 \times 100}{553 \times 440}$$

$$= 0 \cdot 24\% \text{ of } A_m$$

Links will be provided in spite of A_s being slightly less than $0 \cdot 25\%$ of A_m (clause 26.5.3 of BS 5628: Part 2).

With a link diameter of 6 mm and a link spacing of 440 mm (based on column dimension), or $50 \times 6 = 300$ mm (based on link diameter) or $20 \times 16 = 320$ mm (based on bar diameter), it is necessary to use four 16 mm dia. high tensile bars ($A_s = 804$ mm²) with 6 mm diameter binders spaced at 300 mm centres, all set in a 553 mm × 440 mm brick column, using bricks with a compressive strength of 35 N/mm and a designation (ii) mortar. The 66 mm square vertical reinforcement duct should be grouted solid.

The need to devise a design principle incorporating an evaluation of the buckling tendency of the slender column is explained at the end of section 4.1.1.

The size of the vertical reinforcement duct (66 mm square) permits a maximum of 20 mm grout cover even in the most favourable circumstances. In order to ensure adequate cover protection to the steel, the grouting would have to be carried out carefully and strictly supervised as the work progressed. Alternatively, a liquid grout may be considered, but in practical terms the Authors consider that the column section should be increased in size. This example shows some of the problems which can be encountered using reinforced masonry and the designer may wish to amend the design and use a plain section.

An unreinforced brick column of the same size could be constructed in engineering bricks with a compressive strength of 60 N/mm², set in a designation (i) mortar, from $N_d = \beta f_k A_m / \gamma_{mm}$, or a slightly larger cross-section of lower strength masonry would be suitable. The Authors feel that either of these plain masonry alternatives would provide a more practical solution than the reinforced column designed.

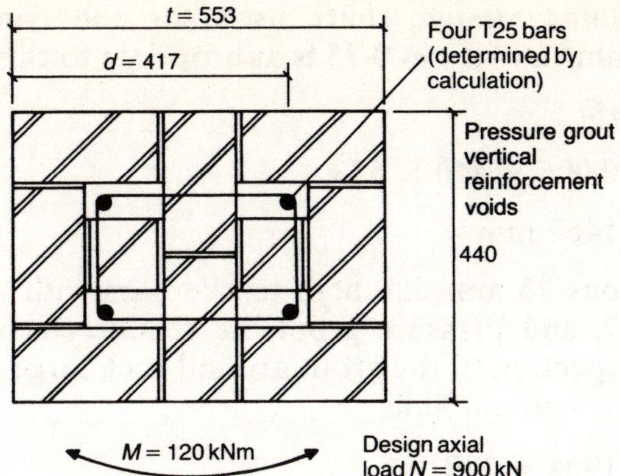

*Fig. 4.17. Plan on
reinforced brick column:
end column in a bay*

4.3.2. Reinforced brick column: combined bending/axial load using column design charts

In this example a reinforced brickwork column for a design axial load of 900 kN and a design bending moment of 120 kNm is to be designed. It is assumed that this is the end column in a bay of the same building as was considered in section 4.3.1. Hence the same conditions prevail and it is required that the column width and the brick and mortar specifications remain the same. A 553 mm × 440 mm column is designed and the required reinforcement determined, as shown in Fig. 4.17.

The slenderness ratio is 13·64 and the resultant eccentricity is

$$e_x = \text{design bending moment/design axial load}$$

$$= 120/900$$

$$= 0\cdot133 \text{ m}$$

Compare e_x with $t/2$

$$t/2 = 0\cdot553/2$$

$$= 0\cdot276 \text{ m}$$

Hence e_x is greater than $0\cdot05t$ but less than $t/2$. The column is therefore designed using the charts in Appendix C

$$d/t = 417/553$$

$$= 0\cdot75$$

$$\frac{N}{btf_k} = \frac{900 \times 10^3}{440 \times 553 \times 9\cdot4}$$

$$= 0\cdot39$$

$$\frac{M}{bt^2f_k} = \frac{120 \times 10^6}{440 \times 553^2 \times 9\cdot4}$$

$$= 0\cdot095$$

Therefore from the column design chart for HY: 2·0:0·75 (Fig. C.2)

$$r/f_k = 6\cdot5 \times 10^{-4}$$

Hence

$$r = 6\cdot5 \times 10^{-4} \times 9\cdot4$$

$$r = 0\cdot006$$

The column design chart used for the conditions of $\gamma_{mm} = 2{\cdot}0$, $f_y = 460$ N/mm^2 and $d/t = 0{\cdot}75$ is appropriate to this example.

$$A_s = rbt$$

$$= 0{\cdot}006 \times 440 \times 553$$

$$= 1460 \text{ mm}^2$$

Use four 25 mm dia. high tensile bars with $A_s = 1964$ mm^2, as shown in Fig. 4.17, and pressure grout the reinforcement voids to ensure full filling and compaction of the grout around such large diameter bars.

For the column links

$$A_s = \frac{1964 \times 100}{440 \times 553}$$

$$= 0{\cdot}81\% \text{ of } A$$

Hence links are required as A_s exceeds $0{\cdot}25\%$ of A.

With a link diameter of 6 mm and a link spacing of $50 \times 6 = 300$ mm, it is necessary to use 6 mm dia. links at 300 mm centres.

As an alternative, following the design procedure given in clause 23.3.1.3 of BS 5628: Part 2

$$M_a = \frac{N(h_{ef})^2}{2000t}$$

$$= \frac{900 \times 6^2}{2000 \times 0{\cdot}553}$$

$$= 29{\cdot}30 \text{ kNm}$$

Therefore

$$\text{total } M = 120 + 29{\cdot}30$$

$$= 149{\cdot}30 \text{ kNm}$$

Now use clause 23.3.1.1(b) (because N_a exceeds N from clause 23.3.1.1(a)). Then

$$N_d = (f_k/\gamma_{mm})b(t - 2e_x)$$

As

$$e_x = \text{total } M/N$$

$$= \frac{149{\cdot}3}{900}$$

$$= 166 \text{ mm}$$

$$N_d = (9{\cdot}4/2{\cdot}0) \times 440(553 - 2 \times 166)$$

$$N_d = 457 \text{ kN}$$

which is less than $N = 900$ kN and is therefore unacceptable and so clause 23.3.1.1(b) must be used.

$$N_d = \frac{f_k}{\gamma_{mm}} bd_c + \frac{f_{s1} A_{s1}}{\gamma_{ms}} - \frac{f_{s2} A_{s2}}{\gamma_{ms}}$$

$$M_d = \frac{0{\cdot}5f_k}{\gamma_{mm}} bd_c(t - d_c) + \frac{0{\cdot}83}{\gamma_{ms}} A_{s1}(0{\cdot}5t - d_1) + \frac{f_{s2}}{\gamma_{ms}} A_{s2}(0{\cdot}5t - d_2)$$

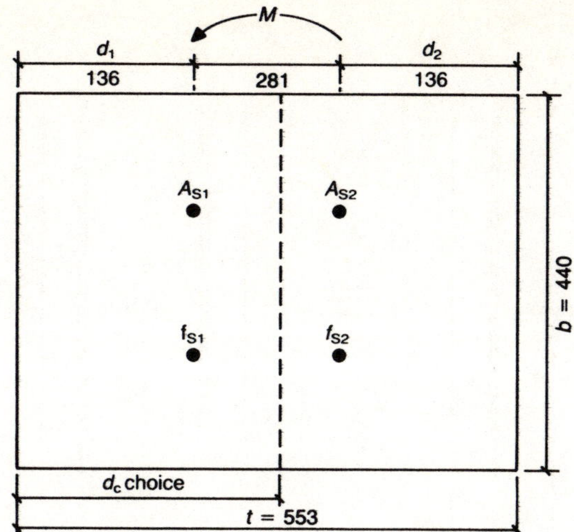

Fig. 4.18. Alternative design procedure

Choose $d_c = 350$ mm (i.e. between $t - d_2$ and $t/2$ (see Fig. 4.18)) and case 3 of clause 23.3.1.1(b) of BS 5628: Part 2. Then

$$t - d_2 = 553 - 136 = 417$$

$$t/2 = 553/2 = 276.5$$

Hence, by linear proportion, $f_{s2} = 0.523 f_y$ (i.e. linear between 0 and f_y), and therefore

$$N_d = \frac{9.4}{2.0} \times 440 \times 350 + \frac{0.83 \times 460 \times 1414}{1.15} - \frac{0.523 \times 460 \times 1414}{1.15}$$

$$= 724 + 469 - 296$$

$$= 897 \text{ kN}$$

which is virtually equal to the applied $N = 900$ kN and is therefore acceptable.

$$M_d = \frac{0.5 \times 9.4}{2.0} \times 440 \times 350(553 - 350)$$

$$+ \frac{0.83 \times 460 \times 1414}{1.15}(0.5 \times 553 - 136)$$

$$+ \frac{0.523 \times 460 \times 1414}{1.15}(0.5 \times 553 - 136)$$

$$= 73.5 + 65.95 + 41.53$$

$$= 180.98 \text{ kNm}$$

which exceeds the total applied bending moment of 149.30 kNm and is therefore acceptable.

This design method is based on a trial and error selection of column size and reinforcement. In practice, for the section shown (see Fig. 4.17) by the column charts design methods to be acceptable, it still took three choices of d_c before an acceptable one was made. This took over an hour of design time despite an acceptable answer having been started with.

The Authors therefore developed the column design charts given in Appendix C.

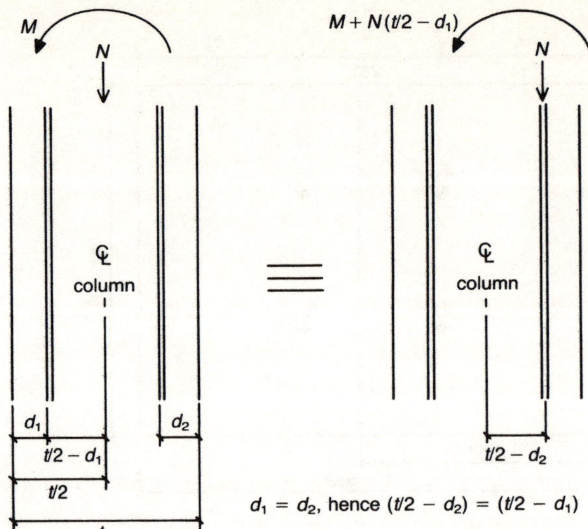

Fig. 4.19. Design procedure: bending moment dominant

4.3.3. Reinforced brick column: combined bending/axial load where bending moment is dominant design factor

In this example a reinforced brick column for a design axial load of 45 kN and a design bending moment of 125 kNm is to be designed. Manufacturing control is taken as special, and hence $\gamma_{mm} = 2{\cdot}0$. The clear height of the column is taken as 5000 mm. (The effect of increasing this clear height is considered later in this section.)

Clause 22.4.1 of BS 5628: Part 2 allows the small axial thrust to be ignored if it does not exceed $0{\cdot}1f_k A_m$. Although, at this stage of the design of this column, its A_m is not yet known, it is apparent that clause 22.4.1 will apply. Therefore, by inspection, the design bending moment dominates the design axial loading and the column is designed for bending only as tensile failure will govern the design. The axial thrust is transposed to an equivalent moment M as shown in Fig. 4.19.

For simplicity of analysis, N is moved across the section to coincide with the line of the tensile reinforcement, which creates an artificial anti-clockwise moment of value $N(t/2 - d_1)$ within the section. This is counteracted by an increase in the applied bending moment of an equal and opposite amount, which allows the section to be designed for bending only. The tensile force in the tension reinforcement may then be reduced by the value of N. Hence the tensile steel area reduction is $N\gamma_{ms}/f_y$.

The moment of resistance of the brickwork in compression is given by

$$M_{rc} = 0{\cdot}375(f_f bd^2/\gamma_{mm})$$

Hence for trial purposes only

$$d \text{ required} = \sqrt{\frac{1{\cdot}10M\gamma_{mm}}{0{\cdot}375f_f b}}$$

in which a nominal 10% increase has been applied to M

$$\sqrt{\frac{1{\cdot}1 \times 125 \times 10^6 \times 2{\cdot}0}{0{\cdot}375f_f b}}$$

Now, assuming a masonry specification of 27·5 N/mm² bricks set in designation (ii) mortar (i.e. $f_k = 7{\cdot}9$ N/mm² and therefore $f_f = 1{\cdot}2 \times 7{\cdot}9 = 9{\cdot}48$ N/mm²) and a 440 mm column width

$$d \text{ required} = \sqrt{\frac{138 \times 10^6 \times 2\cdot0}{0\cdot375 \times 9\cdot48 \times 440}}$$

$$= 419 \text{ mm}$$

Therefore try a 440 mm × 553 mm column (as was used for the design example in section 4.3.3) constructed of bricks with a compressive strength of 27·5 N/mm^2 set in a designation (ii) mortar, and design the reinforcement required ($d = 417$ mm). The slenderness ratio is

$$5000/440 = 11\cdot36$$

As this is less than 12 the column may be designed as a flexural member for a total design bending moment of

$$M + N(t/2 - d_1) = 125 + 45(0\cdot553/2) - 0\cdot136$$

$$= 125 + 6\cdot30$$

(For d see Fig. 4.17.) Therefore

total $M = 131\cdot30$ kNm

Check the compression in the brickwork as follows.

$$M_d = 0\cdot375 f_f \, bd^2/\gamma_{mm}$$

$$= 0\cdot375 \times 9\cdot48 \times 440 \times 417^2 \times 10^{-6}/2\cdot0$$

$$= 136 \text{ kNm}$$

which is greater than the total design bending moment of 137·44 kNm and is therefore acceptable.

To calculate the design reinforcement required, assuming high tensile bars, use

$$A_s = \frac{\text{total } M\gamma_{ms}}{f_y z} - \frac{N\gamma_{ms}}{f_y}$$

The lever arm z is obtained from Fig. 3.7 for which

$$\frac{M\gamma_{mm}}{2f_f bd^2} = \frac{137\cdot44 \times 10^6 \times 2\cdot0}{2\cdot0 \times 9\cdot48 \times 440 \times 417^2}$$

$$= 0\cdot19$$

From Fig. 3.17, $z/d = 0\cdot75$ and therefore $z = 0\cdot75 \times 417 = 313$ mm. Hence

$$A_s = \frac{137\cdot44 \times 10^6 \times 1\cdot15}{460 \times 313} - \frac{45 \times 10^3 \times 1\cdot15}{460}$$

$$= 986 \text{ mm}^2$$

in the tensile face, or in both faces if bending can be applied in both directions as is often the case with wind moments.

Use two 25 mm dia. high tensile bars per face and 6 mm diameter links at 300 mm centres.

Working strictly in accordance with BS 5628: Part 2 this column could be designed using clause 22.4.1 as N is less than $0\cdot1f_k A_m = 0\cdot1 \times 7\cdot9 \times 440 \times 553 = 192$ kN. Thus, ignoring the axial force of 45 kN, the column would be designed as a singly reinforced rectangular member using clause 22.4.2 in which the flexural compressive resistance moment $0\cdot4f_k b_d^2/\gamma_{mm}$ is related to f_k. This is perhaps reasonable, as clause 22.4.2 would be appropriate only where the applied axial load is limited to one tenth of the column's axial load resistance and thus the buckling tendency is of little significance. However, throughout this book the Authors have adopted the philosophy of increasing f_k for flexural compression to a value of f_f and thus buckling should always be considered. In this example buckling is taken into account by increasing the applied bending moment to $M + N(t/2 - d_1)$ as shown.

Now consider the effect of increasing the clear height to 6500 mm. The slenderness ratio is $6500/440 = 14\cdot8$, which exceeds 12.

The design approach in clause 23.3.1.3 of BS 5628: Part 2 is complicated by the fact that biaxial bending will invariably apply because the stiffer axis of the column would be placed parallel with the direction of the applied bending moment, whereas buckling from the axial load would occur perpendicular to this axis. Hence the Authors use a more practical design approach which analyses separately the actual applied axial load and bending moment but combines them in the form of a unity factor. To take account of the buckling stresses (additional M_a in clause 23.3.1.3 of BS 5628: Part 2) a reduction is made in the axial load resistance by the introduction of the capacity reduction factor β.

The design moments of resistance (both compressive and tensile) have already been shown to exceed this design bending moment.

The compressive (i.e. brickwork) design moment of resistance is less than that of the tensile reinforcement and will therefore be used in the unity calculation. The ratio

$$M/M_d = 125/136$$

$$= 0\cdot92$$

Now consider the axial load only. The design axial load

$$N = 45 \text{ kN}$$

and the slenderness ratio is $14\cdot8$. Therefore

$$\beta = 0\cdot866$$

Thus

$$N_d = \beta\left(\frac{f_k\,bd}{\gamma_{mm}} + \frac{0\cdot83f_y\,A_s}{\gamma_{ms}}\right)$$

$$= 0\cdot866\,\frac{7\cdot9 \times 440 \times 553}{2\cdot0 \times 10^3} + \frac{0\cdot83 \times 460 \times 2828}{1\cdot15 \times 10^3}$$

$$= 0\cdot866(961\cdot1 + 938\cdot9)$$

$$= 1645\cdot4 \text{ kN}$$

The ratio

$$N/N_d = 45/1645\cdot4$$

$$= 0\cdot027$$

$$\frac{M}{M_d} + \frac{N}{N_d} = 0\cdot92 + 0\cdot027$$

$$= 0\cdot947$$

which is less than unity and therefore the column is satisfactory.

5

Detailing reinforced masonry

5.1. Introduction

The translation from design to site construction is generally achieved via details assisted by specifications. Detailing is the most important link between good design and quality construction. The basic requirement for good detailing is to keep it simple, clear and practical.

5.2. Practicality

The drawings must clearly define and depict the design requirements. For reinforced masonry they should include all the detail information for the reinforcement, the units, the mortar and grout mixes and all other items necessary for the accurate and complete translation of the requirements for construction. These requirements should include durability, fire resistance and buildability. It should also be borne in mind that bricklayers may not be familiar with the common conventions used in detailing reinforced concrete. It is therefore essential to ensure that they understand the details.

The first design consideration with regard to the inclusion of reinforcement should be, 'Is it necessary?' Masonry acting alone has good resistance to many of the externally applied forces and in a plane form can resist limited amounts of flexure and shear as well as high compressive forces.

Plain masonry should be supplemented with reinforcement only where the stresses would otherwise be excessive on the masonry acting alone. The use of unnecessary reinforcement will add initial cost, increase supervision needs, add to the workmanship difficulties, lengthen construction time and extend the durability considerations. If it is necessary to use main reinforcement in a clay brick reinforced wall it may still not be necessary to follow the example of concrete and introduce distribution steel. Clay brickwork is more likely to expand than to shrink, thus reducing the need to control shrinkage cracking. Bonded masonry is often suitable, particularly in the horizontal direction, and capable of spanning and distributing the longitudinal forces between such main steel.

Much of the advice given here is similar to that given about reinforced concrete and will thus be familiar to many detailers.

5.3. Minimum areas of reinforcement and spacing

5.3.1. Minimum areas

Where the need for reinforcement has been considered and it has been decided that it should be included, minimum percentage areas are recommended. The percentage areas are based on the effective section which, for members in bending, should be taken as breadth multiplied by effective depth.

For the unusual case of reinforcement designed to act in compression, the compression reinforcement must be restrained against the tendency to buckle. This is achieved by the inclusion of secondary steel reinforcement in the form of distribution steel, links or binders.

Figure 5.1 indicates the minimum areas and/or diameters recommended.

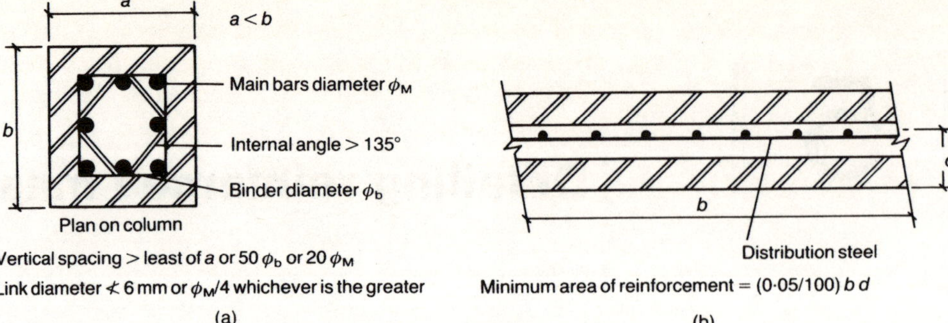

Fig. 5.1. Minimum areas of reinforcement

5.3.2. Minimum spacing

The minimum distance between individual bars and between a masonry or shutter face must be maintained. This is to ensure adequate space for the proper compaction and flow of grout or mortar in order to enclose the bar with an adequately dense cover. It is recommended that this minimum distance should be taken as the greatest of the maximum size of aggregate plus 5 mm, or the bar diameter or 10 mm (see Fig. 5.2).

The Authors recommend that this minimum should be used only when it is unavoidable; wherever possible a larger spacing or cover should be adopted.

5.3.3. Maximum spacing

The maximum distance between bars can be based on a design check on the ability of the plain masonry to span between such bars acting compositely for the design condition. However, in members designed as fully rather than locally reinforced, a maximum spacing is recommended. The general recommendation for spacing main bars is the lesser of 500 mm or three times the effective depth. For distribution bars the general recommendation is to use the lesser of 500 mm or five times the effective depth (see Fig. 5.3).

Fig. 5.2. Minimum spacing of reinforcement

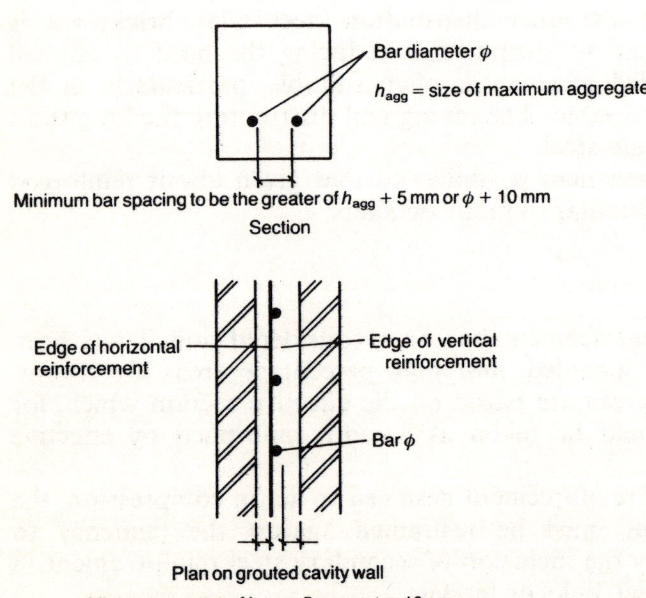

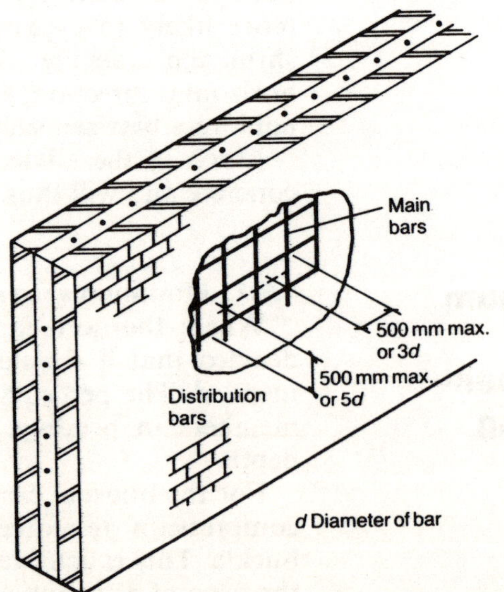

Fig. 5.3. Maximum spacing of reinforcement

5.3.4. Links

For linking reinforcement bars in compression (i.e. in columns and beams) the links should be spaced at a distance centre to centre, not greater than the least of the least lateral dimension of the section, the link diameter times 50 or the main bar diameter times 20 (see Fig. 5.1 (a)).

Each link must

- enclose each compression bar with an internal angle of not more than 135°
- anchor into the main body of the section
- be not less than 6 mm in diameter or 1/4 of the main bar diameter, whichever is the greater (see Fig. 5.1 (a)).

5.3.5. Nominal links

For nominal link reinforcement in beams (i.e. locations where the design shear stress approaches the design shear strength), nominal links should be spaced at centres not exceeding the lever arm, which is generally taken as 0·75 times the effective depth (see Fig. 3.24).

5.4. Maximum size of reinforcing bars

The maximum size of a reinforcing bar should be determined by design, taking into account local and average bond stresses, minimum bar spacing, location and method of grouting, but should not generally be greater than 25 mm.

5.5. Anchorage links, hooks and bends

For reinforcement to develop the design stress, it must be adequately bonded into the surrounding masonry and the local and average bond stresses must be within acceptable limits (see chapters 2 and 3).

When the calculated lengths for designed bond length are below a certain limit, however, it is generally recommended that minimum rules of thumb requirements should govern. These rules are based on the practical considerations of anchorage, local quality control and design error. These demand a minimum practical length beyond that based on normal calculations.

The rules are as follows.

(a) Except at end supports, all bars should extend beyond the point at which they are no longer required a distance of 12 times the bar diameter or the effective depth of the section, whichever is the greater.

(b) Bars should not be curtailed within a tensile zone unless one of three conditions is satisfied

    (i)   the design shear strength of the section must be at least twice the design shear stress

    (ii)   the remaining reinforcement should provide twice the area of reinforcement required to resist the bending moment to continue through past the curtailment point

    (iii)   the curtailed bars should have an anchorage length past the point at which they are no longer required to resist bending of at least the anchorage length appropriate to their design strength f_y/γ_{ms} from the point at which they are no longer required to resist bending, where f_y is the characteristic strength of the reinforcing steel and γ_{ms} is the partial safety factor for the strength of the steel.

(c) At a simply supported end of a member, each tension bar should be anchored using an effective anchorage of 12 bar diameters past the centre line of the support, where no bend or hook begins before the centre of the support or an effective anchorage equivalent to 12 times the bar size plus $d/2$ from the face of the support, where d is the effective depth of the member and no bend begins before $d/2$ inside the face of the support (see Fig. 5.4).

(d)   The effective anchorage of a hook should be taken as the lesser of 24 times the bar diameter and 8 times the internal radius (see Fig. 5.5).

(e)   For a bend, the effective anchorage should be taken as the lesser of 24 times the bar diameter and 4 times the internal radius (see Fig. 5.5).

(f)   The internal radius of bends and hooks should in no case be less than twice the radius of the test bend guaranteed by the manufacturer.

(g)   Full anchorage of a link may be considered if it passes around a bar of at least its own size through an angle of 90° and continues beyond for a minimum length of 8 times its own diameter, or if it passes through an angle of 180° and continues for a minimum length of 4 times its own diameter (see Fig. 5.6).

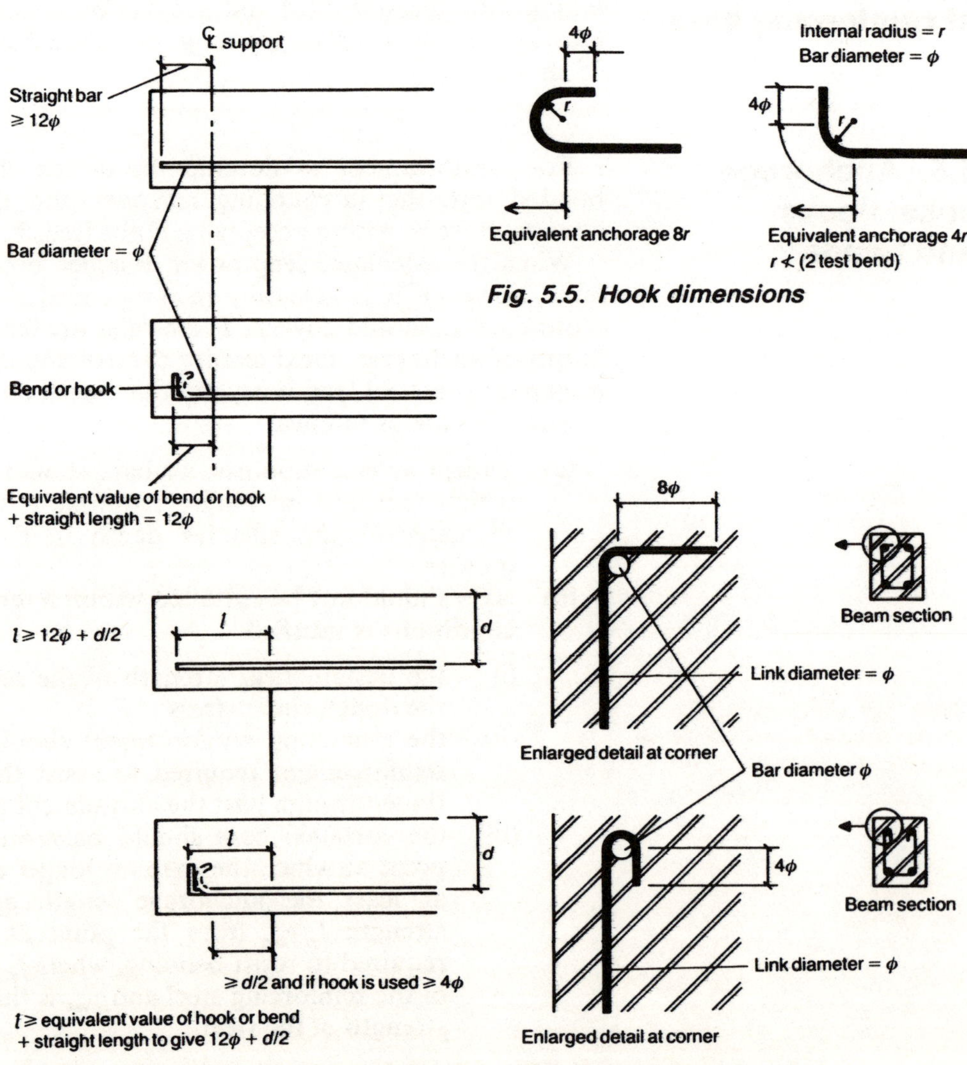

Fig. 5.5. Hook dimensions

Fig. 5.4. Anchorage lengths and hooks   **Fig. 5.6. Link anchorage**

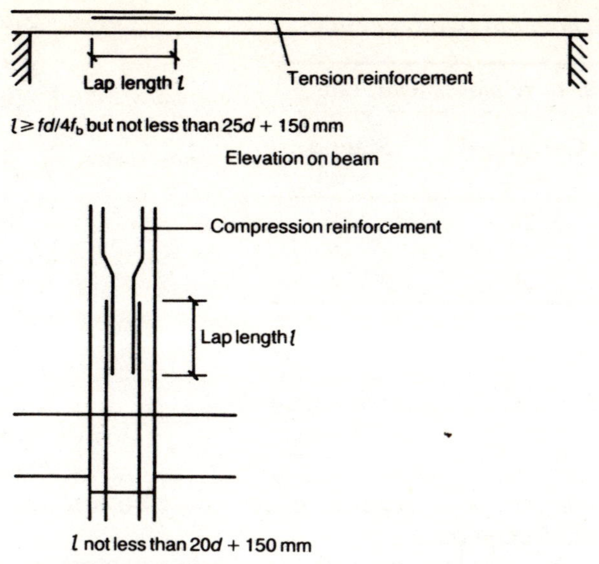

Fig. 5.7. Laps in reinforcement

5.6. Laps and joints in reinforcing bars

Laps and joints in reinforcement are necessary to assist construction, to reduce excessive lengths and so on. Generally such laps should be located away from areas of high stress and/or should be staggered. However, in all cases lap and joint locations should take into account the construction sequence and buildability as well as stress considerations.

It is also important to check the amount of steel in lap positions to make sure that the section does not become over-congested and prevent proper compaction of the mortar or grout. If this is found to be the case, then the section and/or reinforcement and pocket detail should be changed.

The lap length should generally be calculated from the stress consideration. However, it is recommended that for bars in tension it should not be less than the designed length or 25 times the bar diameter plus 150 mm. For bars in compression it should not be less than the design length or 20 times the bar diameter plus 150 mm (see Fig. 5.7).

5.7. Cover

5.7.1. Cover for fire protection

Masonry, particularly brick masonry, is very good as a fire protection cover to reinforcement. The entire thickness between the face of the reinforcement and the outer face of the section can be considered to provide fire resistance.

Guidance on fire protection for various requirements is given in Table 5.1, which indicates the notional period of fire resistance for minimum cover values.

5.7.2. Cover for bond and durability

In addition to fire protection, the cover to reinforcement performs other essential and important functions. Essentially the grout cover must be adequate to ensure composite behaviour, particularly with regard to bond and anchorage, as already discussed in chapters 2 and 3. In addition durability requirements dictate minimum covers which are often greater than those required for bond.

Consideration of the cover for durability, unlike fire protection, should generally assume that brickwork or other masonry units will allow the passage and absorption of water, and thus the masonry cover should be

Table 5.1. Minimum cover to reinforcement for fire resistance

Notional period of fire resistance, h	Minimum cover, mm
1	25
2	50
4	75

Table 5.2. Durability of nominal cover to reinforcement

Exposure conditions	Nominal cover to reinforcement, mm		
	Unprotected steel	Galvanised steel	Stainless steel
1	20	20	15
2	50	30	15
3	70	50	15

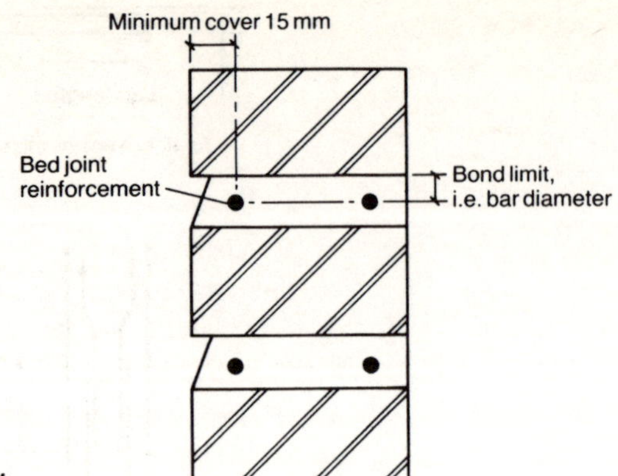

Fig. 5.8. Bed joint reinforcement

ignored with regard to its contribution to corrosion protection for the reinforcement.

In addition, since the corrosion protection provided by mortar is generally less than that for a similar thickness of dense concrete, the depth of the cover requirement should generally be greater than that for a concrete section under similar exposure conditions.

The exposure conditions can be divided into three categories: mild, moderate and severe. Mild conditions occur when masonry is completely protected against weather or aggressive conditions, except for a brief period of exposure to normal weather conditions during construction. Moderate conditions occur in masonry sheltered from severe rain and against freezing while it is saturated with water, in external work and in buried brickwork continually submerged in fresh water. Severe conditions occur when masonry is exposed to driving rain and freezing while it is wet, subjected to heavy condensation or corrosive fumes, exposed to alternate wetting and drying or continuously submerged in salt water or moorland water.

Table 5.2 gives recommendations for minimum cover requirements for these categories and applies to all reinforcement whether it is a main bar or secondary, such as distribution steel or links.

For bed joint reinforcement, used in external walls or other exposed conditions, the Authors recommend that stainless steel or other suitably protected steel should be used. For stainless steel, the application of the 15 mm cover requirement should apply to the face but can be waivered for the vertical cover between the brick horizontal faces and the mortar bed (see Fig. 5.8).

It is desirable, and in many exposed situations essential, that a dense well-graded mortar mix of designation (i) or (ii) be used in conjunction with reinforced masonry.

The cover requirements for stainless steel are those required for bond of the bars and therefore the limits for stainless steel in Table 5.2 relate to bond cover requirements for reinforcement.

5.8. Reinforcement: positioning to suit masonry bonding

In order to ensure that the details are practical, the reinforcement should be located to suit simple masonry bonding, keeping cut units to a minimum, and the bonding should maintain adequate voids for grouting. The spacing of reinforcement should therefore satisfy both the structural and buildability requirements. The designer should always consider the practicality of the bonding arrangement related to the chosen bar spacing. Figs 5.9–5.11 indicate typical bonding arrangements for walls, beams and columns.

5.9. Restraints

The detailing of restraints for both reinforced and post-tensioned masonry must not only take into account the local structural restraint requirements but also those for overall stability and accidental damage.

In addition, the details must be practical, buildable, economic, durable and robust. The duration, magnitude and direction of forces to be resisted vary greatly from element to element and location to location. In addition, the durability and design life requirements for one project can be very different from those needed on another, dependent on location and project requirements. It is important that the detailer obtains from the designer a clear understanding of the forces to be resisted and the implications of his details on the design assumptions as well as the implications on the sequence of construction.

Since design and detailing are interactive processes, it is also important that the designer makes sure that he is satisfied that the details produced are compatible with both the design needs of the complete structure and the construction sequences on site.

The subject of restraint is vast. Only the important points are covered here. Further information is given in references 2 and 13. Since reinforced masonry tends to be used where forces exceed the economic conditions for plain masonry, the restraining forces for reinforced masonry tend to be relatively larger than for plain masonry.

It is essential that if elements are assumed to be restrained in the design then the resistance provided by the restraint must be adequate for that condition. The detail of the restraint and its connection is given in BS 5628: Part 2[1] which states

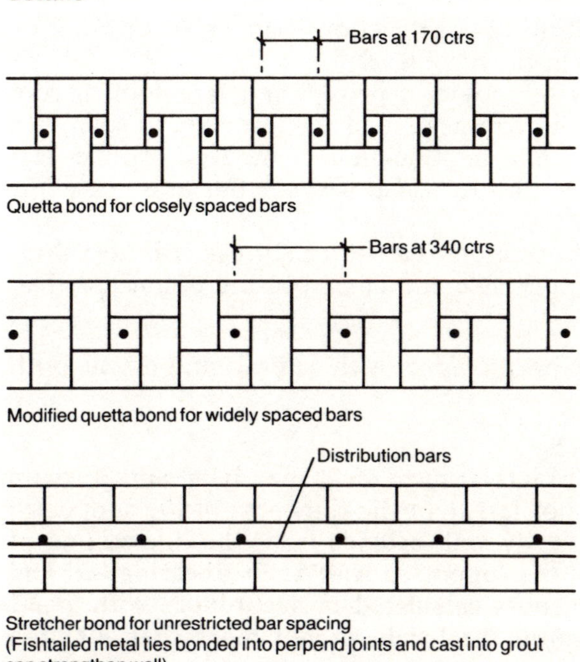

Fig. 5.9. *Typical reinforced brick wall bonding details*

Fig. 5.11 (right). *Plans on alternate courses of typical reinforced brickwork columns*

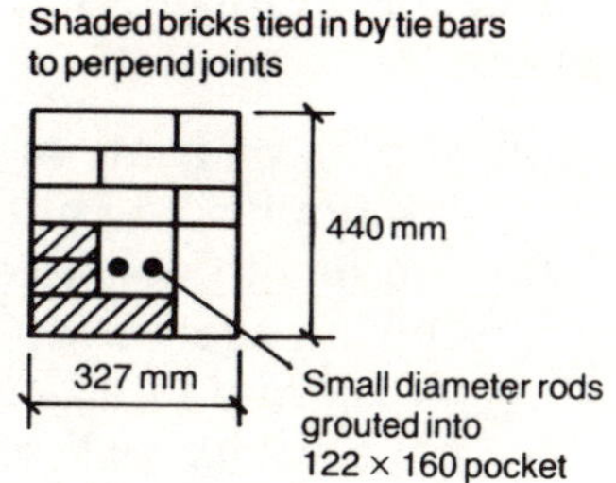

Fig. 5.10. *Typical section through shallow reinforced brickwork beam*

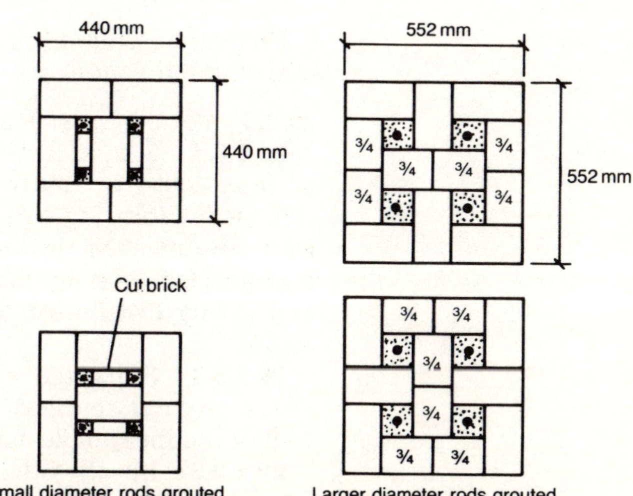

28.2   Lateral support.   A lateral support may be provided along either a horizontal or a vertical line, depending on whether the slenderness ratio is based on a vertical or horizontal dimension.

28.2.1   *Horizontal or vertical lateral supports.*   Horizontal or vertical lateral supports should be capable of transmitting to the elements of construction that provide lateral stability to the structure as a whole, the sum of the following design lateral forces:

(a)   the simple static reactions to the total applied design horizontal forces at the line of lateral support, and

(b)   2·5% of the total design vertical load that the wall or column is designed to carry at the line of lateral support; the elements of construction that provide lateral stability to the structure as a whole need not be designed to support this force.

However, the designer should satisfy himself that loads applied to lateral supports will be transmitted to the elements of construction providing stability, e.g. by the floors or roofs acting as horizontal girders.

28.2.2   *Horizontal lateral supports*
28.2.2.1   Simple resistance to lateral movement may be assumed in the case of houses of not more than three storeys where timber floor members, spaced apart at a distance of not more than 1·2 m, are connected by suitable joist hangers effectively fixed to the joist, as indicated in appendix C. In all other cases, including buildings of more than three storeys, a connection capable of providing simple resistance to lateral movement may be assumed where connections are of the form illustrated in appendix C.
28.2.2.2   Enhanced resistance to lateral movement may be assumed where:

(a)   floors or roofs of any form of construction span on to the wall or column from both sides at the same level;

(b)   an in-situ concrete floor or roof, or a precast concrete floor or roof giving equivalent restraint, irrespective of the direction of span, has a bearing of at least one-half the thickness of the wall or inner leaf of a cavity wall or column on to which it spans but in no case less than 90 mm;

(c)   in the case of houses of not more than three storeys, a timber floor spans on to a wall from one side and has a bearing of not less than 90 mm.

Preferably, columns should be provided with lateral support in both horizontal directions.

28.2.3   *Vertical lateral supports*
28.2.3.1   Simple resistance to lateral movement may be assumed where an intersecting or return wall not less than the thickness of the supported wall or loadbearing leaf of a cavity wall extends from the intersection at least ten times the thickness of the supported wall or loadbearing leaf and is connected to it by metal anchors calculated in accordance with **28.2.1** and evenly distributed throughout the height at not more than 300 mm centres.
28.2.3.2   Enhanced resistance to lateral movement may be assumed where an intersecting or return wall is properly bonded to the supported wall or loadbearing leaf of a cavity wall.
28.2.3.3   In all other cases of vertical lateral support, simple or enhanced resistance to lateral movement may be established by calculation.

Typical details of enhanced resistance are shown in Fig. 2.12. For most reinforced masonry the restraint forces should be analysed and the restraint designed. Often this will mean that the detail can be classed as a support detail and analysed as such.

5.10. Span support details

The main criteria for span support restraint details are that they must be capable of resisting the forces and moments assumed in the design of the elements which they support. This again demands that the detailer be fully aware of the designer's intentions and the implications of his details on the design.

For example, a reinforced masonry wall (retaining earth) which is assumed to be supported and restrained by a ground floor slab at the top of the wall is shown in Fig. 5.12.

The sequence of construction and the temporary supports for such conditions must be taken into account and any special requirement should be noted on the detail.

The temporary condition of the wall shown in Fig. 5.12 would be vulnerable to overturning during construction, prior to casting the slab and during the backfilling operation. Notes are therefore included relating to the temporary propping requirements for this condition.

In addition extra ties across the upper portion of the reinforced cavity are included to ensure adequate transfer of the restraint between the outer and inner leaf of the masonry, thus linking the force-path back into the ground floor slab.

The details incorporated may require extra calculations for the local condition where the wall reduces in section (i.e. becomes a normal cavity construction) in order to check the suitability of the adopted detail (see Fig. 5.12).

In other situations the secondary design conditions, not necessarily involving design calculations, may have to be catered for in the design of support details. These conditions include shrinkage and thermal movements (see Fig. 5.13) which indicate the effects of the detail at a beam seating, the need to control the movement and the damage that can result from an inadequate detail.

The shrinkage of the reinforced concrete beam, if not allowed to take place at the end bearing, could damage the local masonry around the bearing.

Fig. 5.12 (left). Reinforced masonry wall supported and restrained by ground floor slab at top

Fig. 5.13 (right). Beam bearing details

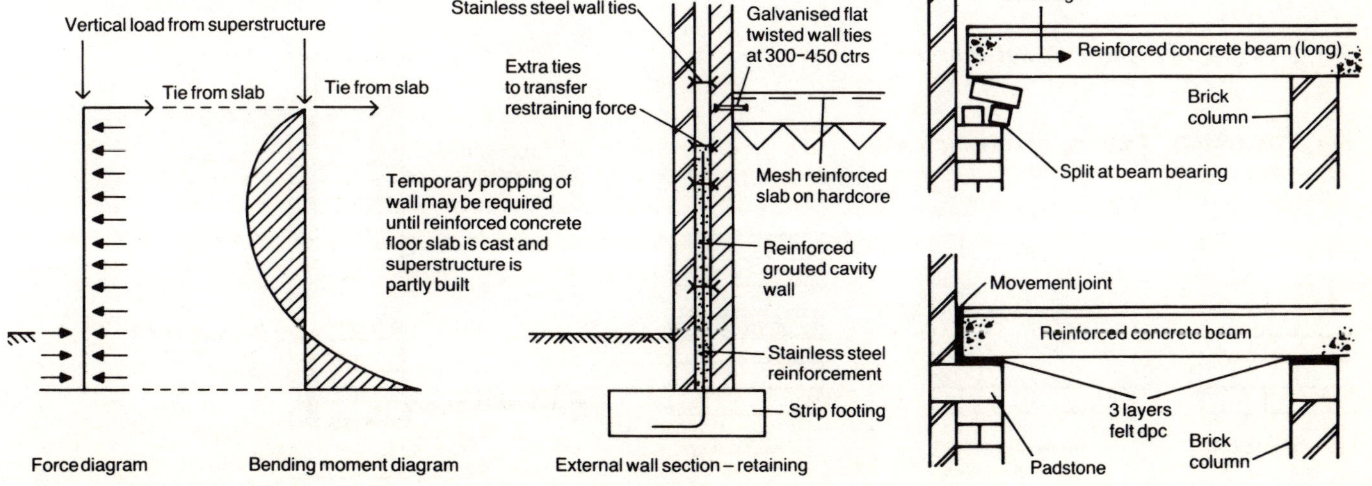

5.11. General arrangement drawings

General arrangement drawings are used as a means of communicating the overall impression of the construction, its mass and the basic materials to be used.

For reinforced masonry the information which should be provided on such drawings should include

- the scheme outline and basic dimensions
- the strengths and basic specification of the materials being used
- the thickness, location and size of all the structural elements and how they relate to each other
- the location of joint, restraints, straps, ties, damp-proof courses and reinforced elements
- information on special temporary works or other unusual requirements.

With reinforced masonry, the exact locations and spacing of the rods is generally detailed on additional detail drawings, whereas the location of more widely spaced post-tension rods is often shown on the general arrangement drawings (see section 8.10).

Additional information regarding general arrangement drawings is given in reference 2.

5.12. Damp-proof courses and damp-proof membranes

Damp-proof courses and damp-proof membranes are provided to prevent the ingress or rise of damp in masonry and, as such, are a building detailing requirement. However, their inclusion in certain instances and locations can have an effect on the structural performance of an element or of the total structure. This section highlights some of these situations.

Damp-proof courses and damp-proof membranes can be made from a wide variety of materials including bitumen felt, metals, slate, plastic and brick. The choice of the material to be used as it affects the structure must relate to the forces at that location. Thus it must neither squeeze out under high vertical loading nor slide under lateral loading. Three typical materials which prevent rising damp are shown in Fig. 5.14.

Bitumen felt damp-proof courses are usually the least expensive, but they have poor resistance to compressive forces and can therefore squeeze out under comparatively low compressive stress. They are also prone to damage through poor workmanship and are generally to be discouraged. Plastics are becoming increasingly popular in that they overcome some of the shortcomings of the felt materials but retain the advantage of flexibility to

Fig. 5.14 (below). Damp-proof courses

Fig. 5.15 (right). Tanking of retaining wall

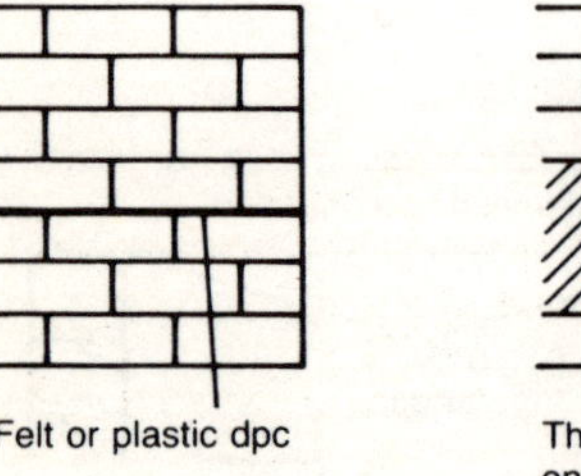
Felt or plastic dpc

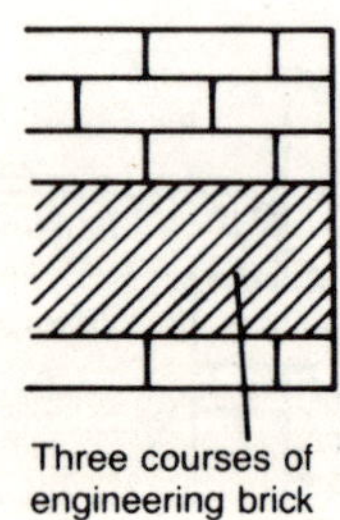
Three courses of engineering brick

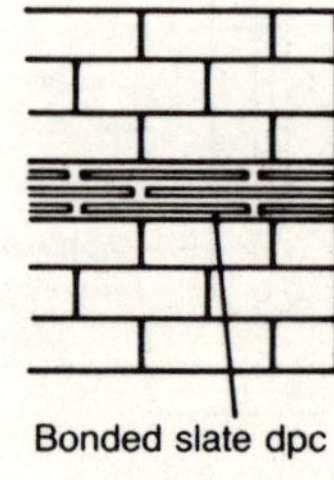
Bonded slate dpc

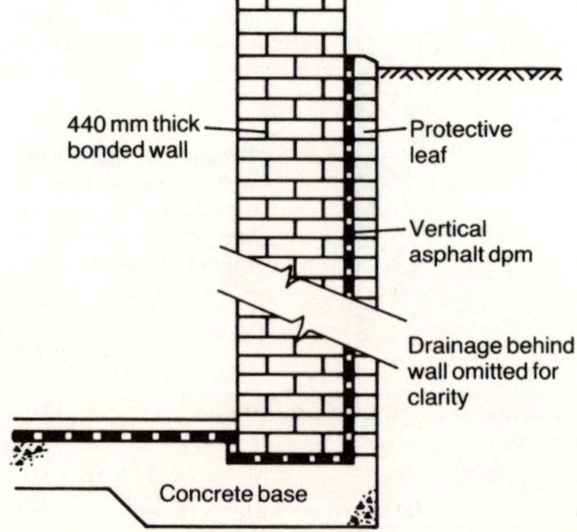

movement—a particularly useful quality where movements such as mining subsidence are expected. However, they are more likely to be critical where lateral sliding forces are significant and this must be given due consideration in their assessment.

Brick damp-proof courses have a high compressive resistance. They are formed by constructing two or three courses of specially selected engineering bricks to provide the necessary impermeable barrier.

One of the earliest materials used to form damp-proof courses was slate, usually provided in two or three layers and bonded to prevent a direct vertical route for the damp. Owing largely to the economy and convenience of more modern strip materials, slate is not now commonly used. Some slate damp-proof courses have suffered from flaking and, to avoid this, only good quality slate should be used.

In masonry retaining walls, a vertical damp-proof course is often included to prevent damp penetration through the wall from the retained earth. Fig. 5.15 shows a typical section through a reinforced masonry retaining wall.

A correct detail for such a wall is shown in which the designed thickness is maintained and fully bonded; to this is added a protective leaf of masonry between the earth and the damp-proof membrane. The purpose of this additional leaf is solely to protect the damp-proof membrane from damage during construction. The leaf can be of inferior quality to the remainder of the masonry as it is totally ignored in the design of the wall. To ensure the homogeneous action of the full thickness, the wall must be properly bonded. This tends to conflict with the request of some architects to display stretcher bond on the face brickwork. If this request is to be complied with, designed shear ties should be incorporated between the stretcher bond leaf and the remainder of the designed thickness. The horizontal forces from the retained earth are liable to cause the wall to slide across the damp-proof course. The stepped foundation shown in Fig. 5.15 shows a means of resisting this weakness in the damp-proof course. The sliding tendency is then transferred to the mortar bed joint, which is more likely to be able to provide the required resistance.

Further information on retaining wall detailing is given in reference 2.

Basis of prestressed masonry design

6.1. General

In prestressed masonry the improvement in resistance to applied bending (flexural) stresses within the masonry element is gained by applying a compressive stress to the element to counteract the applied flexural tensile stresses. Fig. 6.1 shows a typical prestressed wall with the elimination of flexural tensile stress.

The applied flexural tensile stress may be due to lateral loading or the result of eccentricity of axial loading. The effects and means of overcoming them are similar.

The structural masonry is prestressed (i.e. stressed before the applied flexural tensile stress is applied) with an additional compression. The additional compression applied to the section (i.e. the prestress force) may be applied at the centroid of the section. Alternatively, if the prestress force is applied at an eccentricity to the centroid, an additional countcracting moment may be introduced into the section.

This additional moment should be applied to counter that caused by the applied loadings in order to counterbalance the effect (see the prestressed diaphragm wall in Fig. 6.2). This means that the gain in flexural resistance of

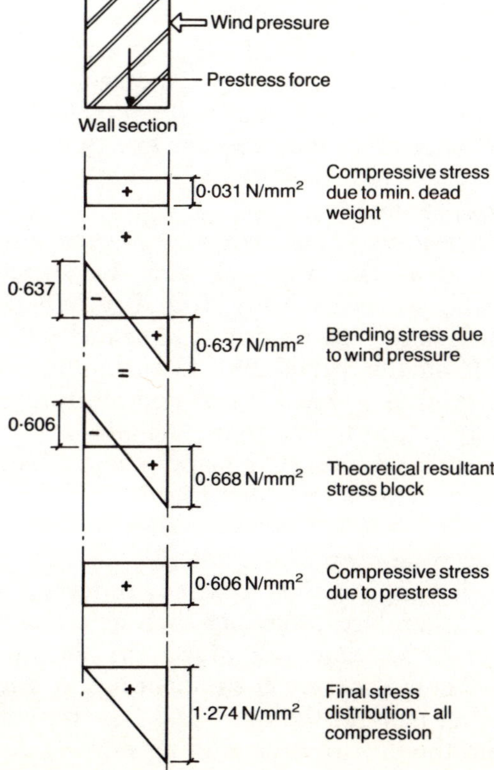

Fig. 6.1. Stress blocks in prestressing concept

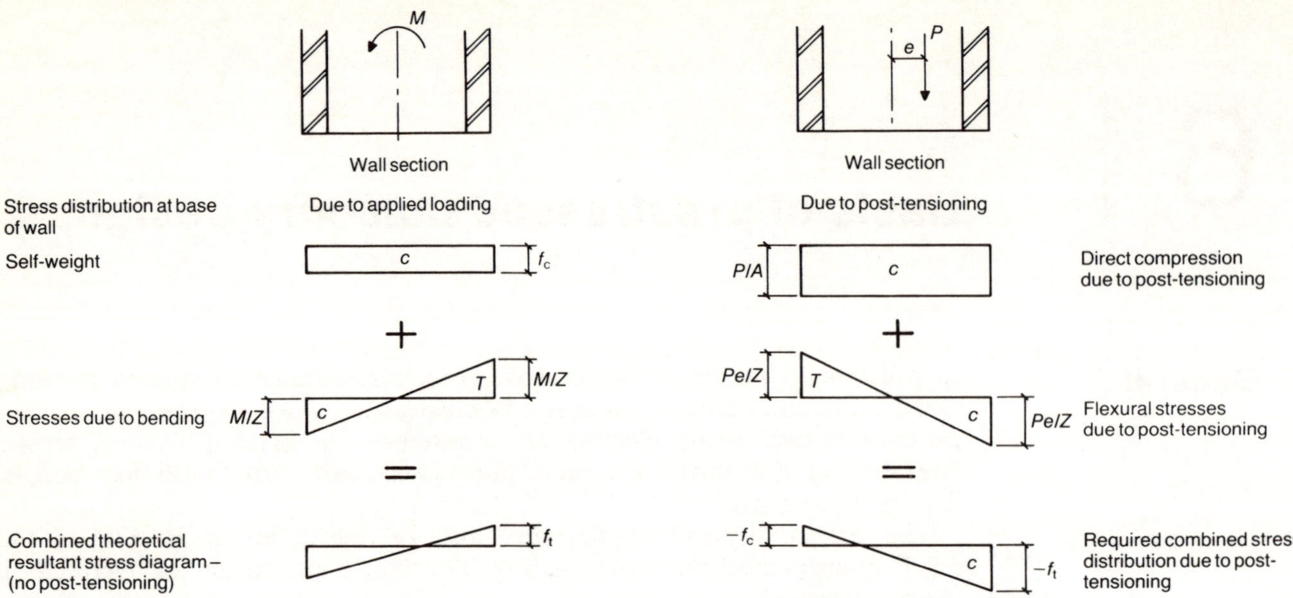

Fig. 6.2. Stress blocks in prestressed diaphragm wall

the section due to prestressing is not directly accompanied by a cumulative increase in compressive stress, which may otherwise limit the potential resistance (see chapter 2). The most suitable eccentricity for the prestress force may be readily determined to give the optimum levels of prestress for the given loading conditions.

In general the prestress is placed concentrically when reversal of applied moments is possible (e.g. wind loading) and eccentrically when the applied moments are predominantly in one direction (e.g. retaining walls). There are cases, however, where eccentrically placed prestress may be used to advantage even where there is reversal of applied moments (e.g. in asymmetric sections).

There are two main methods of prestressing: post-tensioning and pretensioning. This book is mainly concerned with post-tensioned masonry. In both techniques, as in prestressed concrete, high tensile steel is used and not mild steel because of the excessive loss of prestress in mild steel due to creep.

6.1.1. Post-tensioning

Post-tensioning is mainly a site process where one end of the prestressing tendon is fixed (see Fig. 6.3) and the structural masonry member (wall, column and so on) is built around it. When the masonry has reached its required strength, the tendon is stretched (i.e. tensioned), anchored and then released from the stretching or tensioning device. The structural masonry member is thus subject to a compressive force as the released tendon attempts to return to its original length.

The member is built first and after achieving its required strength the tendons are stressed (post-tensioned).

6.1.2. Pretensioning

Pretensioning is mainly a factory process for the long-line continuous production of standard members such as concrete block lintels. The masonry hollow units are laid out along the line and the stressing wires passed through them. They are then anchored at one end of the line and attached to a moving plate at the other end. The moving plate is hydraulically jacked back from the line and the wire tensioned. Concrete or grout is poured into

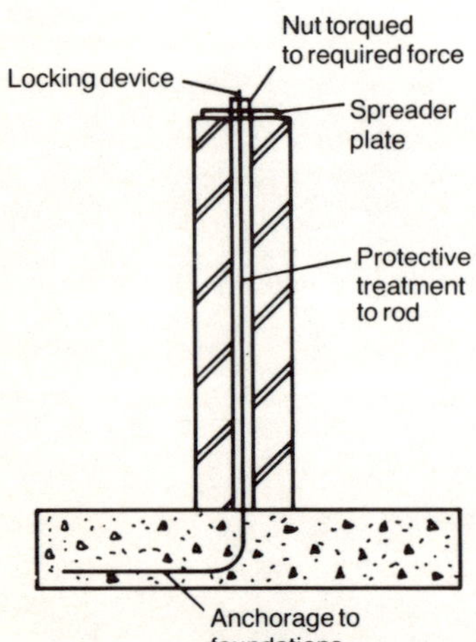

Fig. 6.3. Post-tensioned cantilever cavity wall

the units forming the standard member, to bond and fix the stressed units to the hollow unit members. When the concrete has reached its required strength the moving plate is released, and the prestress is thus transferred to the standard members. The wires are cut between the members and the now prestressed members are moved to the despatch area.

The wires are stressed first (pretensioned) and then the member is cast.

6.2. Methods of post-tensioning

The generic term for prestressing steel is 'tendon'. This includes separate wires, groups of wires (known as strands) and rods. Wires and strands are usually used in pretensioning and rods in post-tensioning.

Although strands could be used in post-tensioned masonry they are generally avoided because of the site difficulties in positioning and supporting them during the construction of the masonry.

There are two major types of post-tensioning technique: partial or low-level prestressing and full or high-level prestressing.

6.2.1. Low level prestressing

As its name implies, the low level prestressing technique is used when only relatively low levels of prestressing force are required. Typical examples are

- free-standing (i.e. cantilever) walls subject to lateral wind and crowd pressure
- normal height and tall single-storey structures (where walls act either as a free or propped cantilever) subject to wind pressure; this can convert the masonry from mere passive cladding to active structure thus eliminating the cost, construction delays and complexity of other structural framing
- hollow box columns supporting single-storey roof structures where the prestress is required to tail down the roof against wind suction forces and so on
- improved resistance to impact or blast.

There is, perhaps surprisingly, a wide and economic application for this technique which can be carried out by relatively unsophisticated builders. The technique is simple and does not require high technology.

Relatively small diameter—10–20 mm—rods can be bent at the base and cast in the in situ concrete foundation strip (see Fig. 6.3) or anchored by special plates.

The masonry member wall or column is built around the rods. When the mortar has achieved its required strength a spreader plate is placed over the rod and bedded on top of padstone or capping beam on top of the masonry. A nut is screwed on and tightened by the use of a torque spanner. Further tightening of the nut extends and hence tensions the rod. The torque spanner is provided with a gauge which can be preset to give the required torque and thus extension and prestressing of the rod. When the required level of prestress is reached, the nut is locked in position by a locking device and the torque spanner removed. Owing to inaccuracy in realistic calculations of the torque to prestress relationship which inaccuracy is itself due to errors in assessing the on-site friction and temperature effects, the system is not as accurate or reliable as jacking, and allowance is required to compensate for such inaccuracy.

6.2.2. High level prestressing

The high level prestressing technique is very similar to that used in post-tensioned concrete and is used when higher levels of prestress are required.

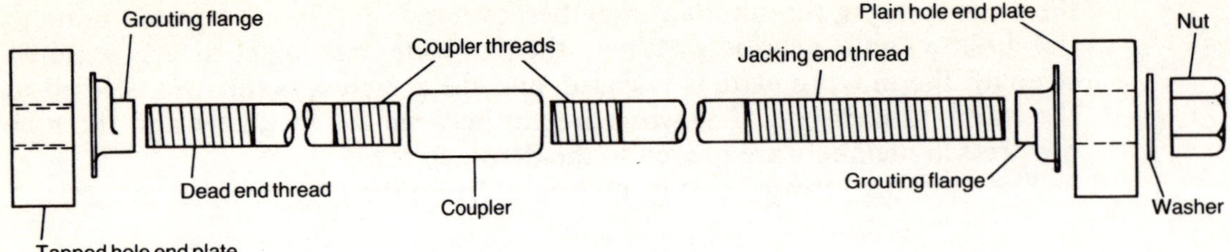

Fig. 6.4. Typical details of post-tensioning rod

Typical applications are retaining walls, tanks, culverts, heavily loaded beams and similar conditions when the loads, perpendicular to the member, and bending moments are relatively high.

Medium to large diameter rods—16–40 mm—are fixed at one end to an anchorage and spreader plate. The masonry member, topped by a capping beam or padstone, is built around the rods and when the mortar has gained its required strength (usually about 14 days after laying), a spreader plate and washer are placed over the rod and a nut is screwed on. A post-tensioning hydraulic jack extends (and thus tensions) the rod. The force is measured and read on the jack's pressure gauge. The pressure gauge must be checked and correlated at reasonably frequent intervals to ensure that the correct magnitude of prestressing is being applied to the rods. When the correct pressure has been reached the nut is screwed up tightly and the jack removed (see Fig. 6.4).

6.3. Geometric sections: symmetrical and asymmetric

Although solid square and rectangular sections can be prestressed, this does not make for economy of material or structural efficiency. Structurally efficient sections for prestressing (as is shown in section 6.4) should have a high section modulus to cross-sectional area ratio (i.e. Z/A), which in turn results in a high radius of gyration to cross-sectional area ratio (r/A), where A is cross-sectional area, r is radius of gyration and Z is section modulus. Typical sections used in prestressed masonry are diaphragms, fins and U or channel sections.

The diaphragm wall can be considered as a series of connected hollow boxes, a fin wall as a series of connected T sections and a fin wall broken by floor to roof glazing in alternate bays as a series of disconnected U sections. The ring or tubular section is useful in columns where the hollow area is used to house services. The U section can be used for prestressed walls and also for horizontal culverts.

6.4. Basic principles of prestressed masonry

Consider the column which is shown in Fig. 6.5. It is subject to a bending moment M inducing tensile and compressive stresses $f = M/Z$. The tensile stress M/Z can be eliminated by applying concentrically a prestressing force P. The compressive stress due to P is P/A.

If the magnitude of P is such that the prestress compressive stress P/A is equal to the bending stress M/Z, then under the combined action of bending and prestress there will be no tensile stress.

Although the tensile stress has been eliminated the compressive stress has increased.

If P is applied at an eccentricity e as shown in Fig. 6.6, the stresses due to prestress are $(P/A) \pm (Pe/Z)$. Although the stress $(P/A) + (Pe/Z)$ may eliminate a bending tensile stress, $(P/A) - (Pe/Z)$, which may still be compressive, would add to the compressive stress in the element due to the compressive stress caused by bending.

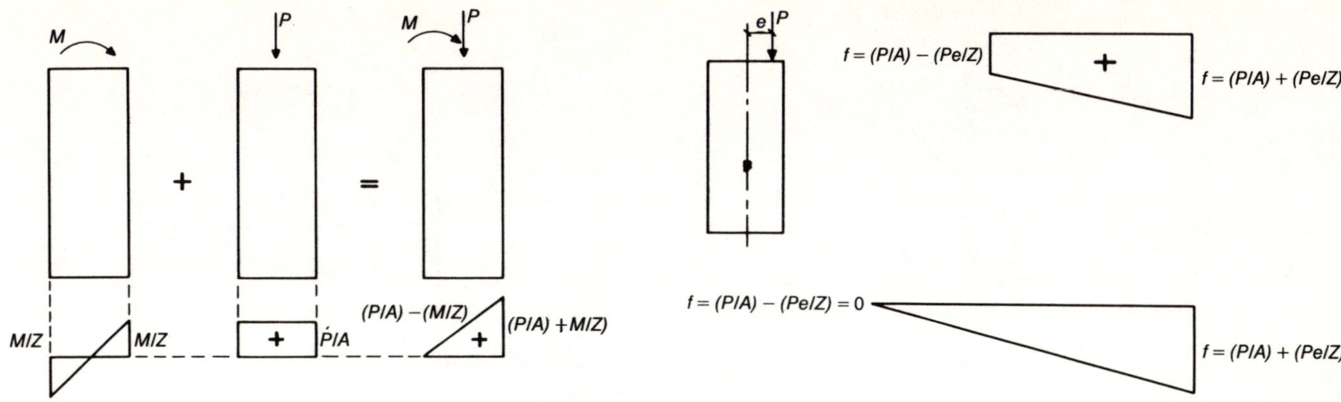

Fig. 6.5 (left). Column with applied concentric prestressing force

Fig. 6.6 (top right). Column with prestressing force applied at eccentricity

Fig. 6.7 (bottom right). Column with prestressing force applied at greater eccentricity

If e is increased further the stress distribution could be as shown in Fig. 6.7.

There is now no increase in the combined compressive stress and if $(P/A) + (Pe/Z) = M/Z$ then the bending tensile stress is eliminated.

When $f = 0 = (P/A) - (Pe/Z)$

$$P/A = Pe/Z \qquad (6.1)$$

$$e = Z/A \qquad (6.2)$$

When $f = P/A + Pe/Z$ from equation (6.1), then

$$f = 2P/A$$

and since this bending $f = M/Z$

$$M = 2PZ/A \qquad (6.3)$$

This shows the importance of a high Z/A ratio in prestress design. For a given value of P an increase in Z/A results in a greater resistance of the section to bending. It is for this reason that the Authors advocate the use of geometric sections in prestressed masonry design, as is done in prestressed concrete design.

Furthermore, if the section modulus is high then the second moment of area I is also high. Since the radius of gyration $r = \sqrt{(I/A)}$, increased Z gives increased I, which in turn increases r. If r is increased the slenderness ratio SR is decreased, resulting in a higher capacity reduction factor β and a higher design stress.

Although this labours the obvious points to design engineers, it shows that the advice in BS 5628[1] on solid rectangular sections is inadequate and that the use of such sections is structurally inefficient.

6.5. Design strengths: direct and flexural

As shown in chapter 2, in dealing with the limit state design of plain masonry, the strengths are expressed in terms of characteristic strength (direct compression) and characteristic flexural tensile strength (tensile strength in bending). As discussed in sections 6.1 and 6.4, in general post-tensioned masonry is designed to give zero net flexural tensile stresses and hence the design is based on compressive strengths. As is shown in Fig. 6.8, in post-tensioned masonry flexural compression as well as direct compression occurs, and both need to be considered when material strengths are being assessed.

As discussed in chapters 2 and 3, masonry, like concrete, has a higher flexural compressive strength than direct compressive strength, and for post-

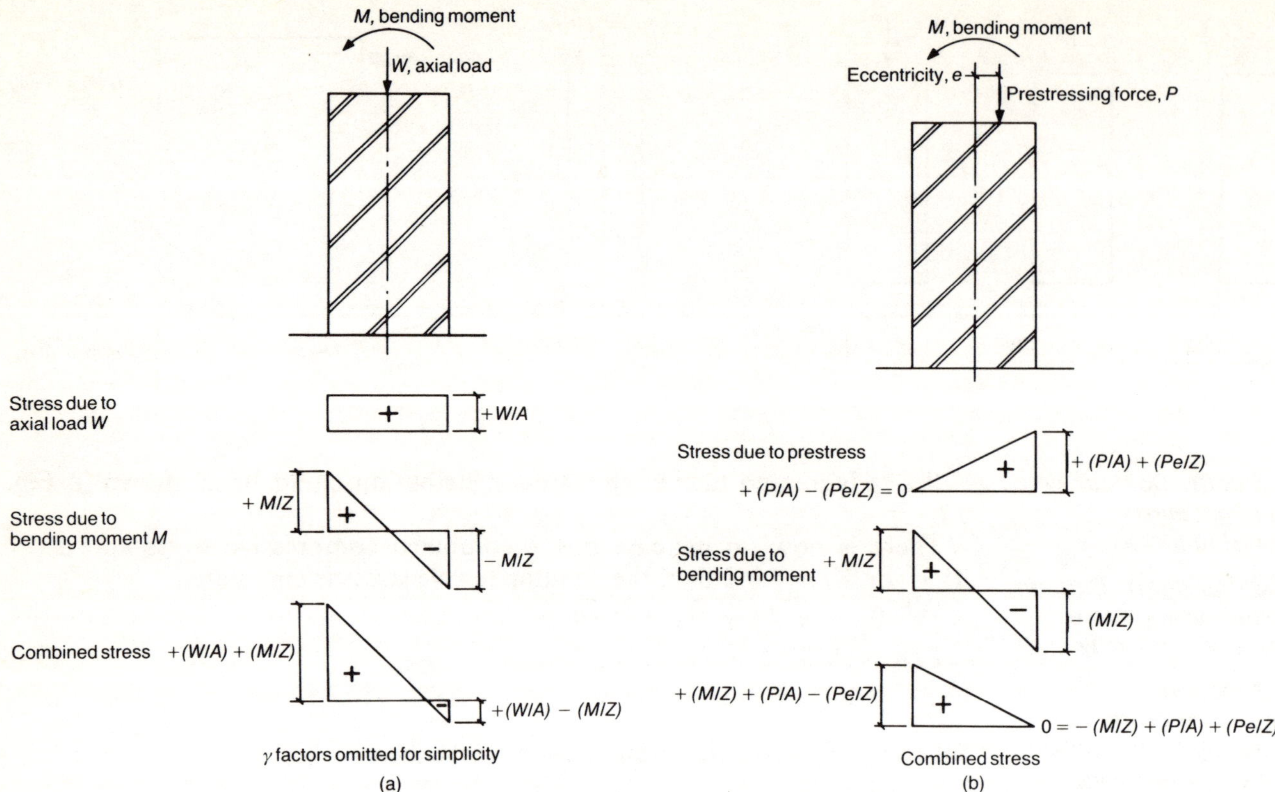

Fig. 6.8. Combined stress

tensioned masonry the same values as for reinforced masonry discussed in those chapters should be used for ultimate limit state.

For serviceability and transfer stresses the figures are adjusted. The basic assumptions made are similar to those for reinforced masonry but with additions to take into account the differences in the use of steel for prestressing as compared with reinforcing. The appropriate recommendations for both these adjustments as given in BS 5628: Part 2[1] are as follows.

28   Design for the ultimate limit state
28.1   Bending

When analysing a cross section to determine the design moment of resistance, M_d, the following assumptions should be made.

(a)   Plane sections remain plane when considering strain distribution in the masonry in compression.

(b)   The distribution of stress is uniform over the whole compression zone and does not exceed:

$$f_k/\gamma_{mm}$$

where f_k is the characteristic compressive strength of masonry; γ_{mm} is the partial safety factor for compressive strength of masonry.

(c)   The maximum strain at the outermost compression fibre is 0·0035.

(d)   The tensile strength of the masonry can be ignored.

(e)   Plane sections remain plane when considering the strains in bonded tendons and any other reinforcement, whether in tension or in compression.

(f)   Stresses in bonded tendons, whether initially tensioned or untensioned, and in any other reinforcement are derived from the appropriate stress–strain curve. A typical stress–strain curve is shown in figure 5.

(g) The stresses in unbonded tendons in post tensioned members do not exceed the values derived from figure 6.

(h) The effective depth, d, to unbonded tendons is determined by taking full account of the freedom of the tendons to move.

Note. Unbonded tendons may be restrained by cross ducts.

28.2 Axial loading

Prestressed masonry elements subjected to axial loading or vertical loading having a resultant eccentricity not exceeding 0·05 times the thickness of the wall should be designed in accordance with clause **24**. The design axial load resistance of tension members should be taken as equal to the design axial load of the prestressing tendons and any other reinforcement, making no allowance for any tensile strength of the masonry.

28.3 Shear

For prestressed sections, the shear stress, v, due to design loads at any cross section in a member may be calculated using the following equation:

$$v = V/bd_c$$

where V is the shear force due to design loads; b is the width of section, for rectangular sections, or width of web, for T and I sections; d_c is the depth of masonry in compression.

Where v exceeds f_v/γ_{mv}, where γ_{mv} is the partial safety factor for shear strength of masonry, shear reinforcement should be provided as described in **22.5.1**.

29   Design for the serviceability limit state

29.1   The compressive strength of the masonry at transfer should be at least 2·5 times the compressive stress induced by the prestressing forces, for approximately uniform distribution of prestress, or 2·0 times this stress for approximately triangular distribution of prestress.

29.2   The compressive stress in the masonry after all losses have occurred should not exceed:

(a) $0·33f_k$, for approximately uniform distribution of prestress;
(b) $0·4f_k$, for approximately triangular distribution of prestress;

where f_k is the characteristic compressive strength of masonry.

29.3   Where the area of concrete infill represents more than 10% of the section under consideration, elastic analysis should be undertaken using the transformed area calculated from the values of elastic modulus given in **19.1.7** (see also **16.1.2**).

29.4   The deflection of members should be calculated following the recommendations of **16.2.2.1**.

6.6. Design strength: initial and long-term

Values for the characteristic direct compressive strength f_k of masonry are given in BS 5628:[1] Part 2 for walls tested at an age of 28 days. As it is not generally practical from a construction point of view to wait for 28 days to elapse before post-tensioning, any design check for the initial loadings should be based on the compressive strength at the relevant age f_{ki}. The Authors consider that, other than in exceptional circumstances, post-tensioning should not be undertaken until 14 days have elapsed or until the mortar has reached a specified strength. The post-tensioning force to be applied is generally increased to allow for the effects of losses (see section 6.11). The initial post-tensioning force is thus greater than that applied over the long term.

The post-tensioning force before losses occur may, in the Authors' opinion, be considered a temporary load, and that remaining after losses as a permanent load. It is considered reasonable that, within the context of the design approach given in this book, the value of the design strength used could be modified when considering the short-term duration of this initial loading case, i.e. the post-tension force before losses. However, this varies from the recommendations given in BS 5628: Part 2.[1] At present an increase of 20% is considered appropriate, giving a design direct compressive strength of $1{\cdot}25 f_{ki}/\gamma_{mm}$ and a design flexural compressive strength of $1{\cdot}25(1{\cdot}20 f_{ki})/\gamma_{mm}$. If the loss in prestress is 20%, to allow for such loss the prestress must be increased by 25%, i.e. $1{\cdot}25 - 20\% = 1{\cdot}25 - 0{\cdot}25 = 1{\cdot}0$.

6.7. Design strength: capacity reduction factors

The capacity reduction factor β was introduced into the design of plain brickwork as a means of allowing for the reduction in compressive load carrying capacity due to buckling failure. As a result of buckling of the section, flexural stresses can be induced even if the applied loading is nominally centroidal.

The use of the capacity reduction factor discussed in chapter 2 tends to obviate the need to consider flexural stresses separately from direct stress when considering simple sections subject to basically axial loading. When, as in post-tensioned and reinforced masonry, flexural stresses are specially calculated, the Authors suggest that the relevance of the capacity reduction factor requires careful consideration. In walls and similar solid rectangular sections the possibility of buckling generally relates only to direct stress. The tendency of the section to buckle when subject to flexure is limited to some extent by the restraint offered to the compression edge by that part of the section which is in tension. Buckling is also limited along the length of the wall by adjoining or interconnecting walls (see Fig. 6.9).

However, in geometric profiles such as the fin, if the flexural compression occurs totally within the flange, then it may not benefit from the presence of any adjacent masonry. It is possible then for the whole flexural compression zone to buckle independently. In this condition it is considered relevant to apply the capacity reduction factor to both the flexural compressive strength and the direct compressive strength. Thus when comparing the applied design stresses with the design strengths, careful consideration must be given to the application of the capacity reduction factor β. Consider, for example, a rectangular wall where the sum of the compressive stresses due to flexure

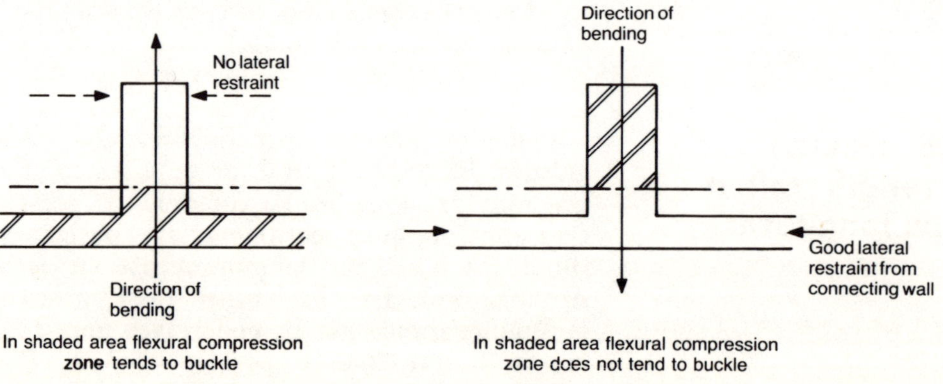

Fig. 6.9. Buckling characteristics of fin walls

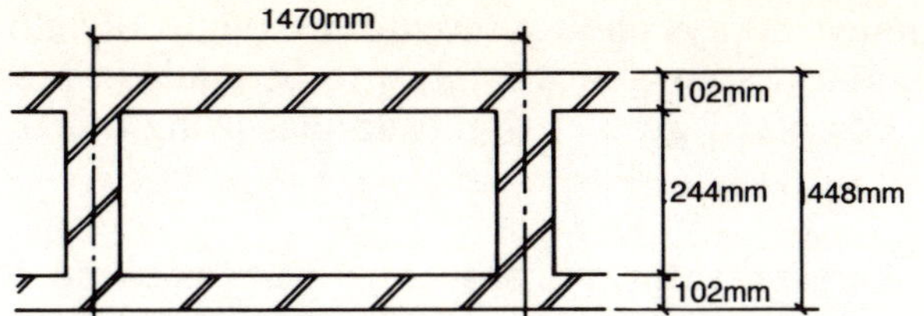

Fig. 6.10. Diaphragm wall section

f_{ubc} and direct compression f_{uac} must be less than the design strength. The compressive stress due to flexure plus the direct compressive stress is greater than or equal to the design compressive strength, i.e.

$$f_{ubc} + f_{uac}/\beta \geqslant 1{\cdot}2 f_k/\gamma_{mm}$$

The design strength quoted is for other than short-term loading and is the value of the flexural compressive strength. This is correct for the combined loading condition but for the direct compression alone f_{uac} should be compared with $\beta f_{ki}/\gamma_{mm}$.

The following equations summarise the design strength limitations.

For short-term loading, i.e. before losses

$$f_{uac} \leqslant 1{\cdot}25 f_{ki}/\gamma_{mm} \tag{6.4}$$

$$f_{ubc} \leqslant 1{\cdot}25(1{\cdot}2 f_{ki})/\gamma_{mm} \tag{6.5}$$

$$f_{ubc} + f_{uac}/\beta \leqslant 1{\cdot}25(1{\cdot}2 f_{ki})/\gamma_{mm} \tag{6.6}$$

For the general case, i.e. after losses

$$f_{uac} \leqslant f_k/\gamma_{mm} \tag{6.7}$$

$$f_{ubc} \leqslant 1{\cdot}2 f_k/\gamma_{mm} \tag{6.8}$$

$$f_{ubc} + f_{uac} \leqslant 1{\cdot}2 f_k/\gamma_{mm} \tag{6.9}$$

Equations (6.6) and (6.9) do not apply to fins, diaphragms and other special cases where β should also apply to f_{ubc}.

The capacity reduction factor β is based on the slenderness ratio SR of the structural element. BS 5628[1] bases SR on the effective height/effective thickness and not on the normal structural design method of effective height/radius of gyration.

While the concept of effective thickness is adequate for normal walls and solid rectangular or square columns, it is not applicable, in the Authors' opinion, to geometric sections such as the diaphragm and fin. For such sections the radius of gyration concept should be used. This point has important structural and economic consequences.

The diaphragm wall shown in Fig. 6.10 has a radius of gyration of 170 mm and an overall depth of 448 mm. A solid wall of the same radius of gyration would have to be 590 mm thick and the Authors suggest that this effective depth should be used in determining the slenderness ratio of the wall, and not its overall depth.

If the wall has an effective height of 7·62 m then its slenderness ratio would be given by

- overall depth: 7·62/448 = 17 and therefore $\beta = 0{\cdot}8$
- equivalent depth: 7·62/590 = 12·95 and therefore $\beta = 0{\cdot}91$

i.e. there is almost a 14% increase in design strength.

Since engineers familiar with prestressed concrete design rarely use solid rectangular sections (because of their relatively low Z/A and r/A values) but almost always choose the more structurally efficient I, T and box sections, a similar design choice is likely to be made in prestressed masonry.

Chapter 7 gives design examples using this theory.

6.8. Cracked and uncracked sections

Generally the purpose of post-tensioning brickwork is to overcome its relatively weak tensile resistance. Plain brickwork is usually designed to limit the tensile stresses due to applied loading to within acceptable limits. Alternatively tensile limits may be exceeded and elements designed on the

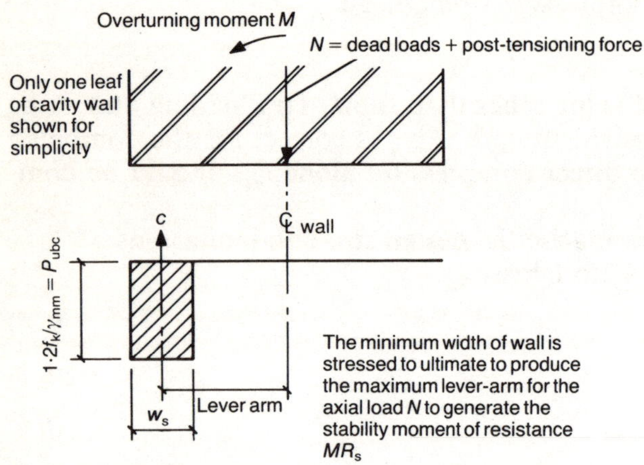

Fig. 6.11. Cracked section: ultimate resistance diagram

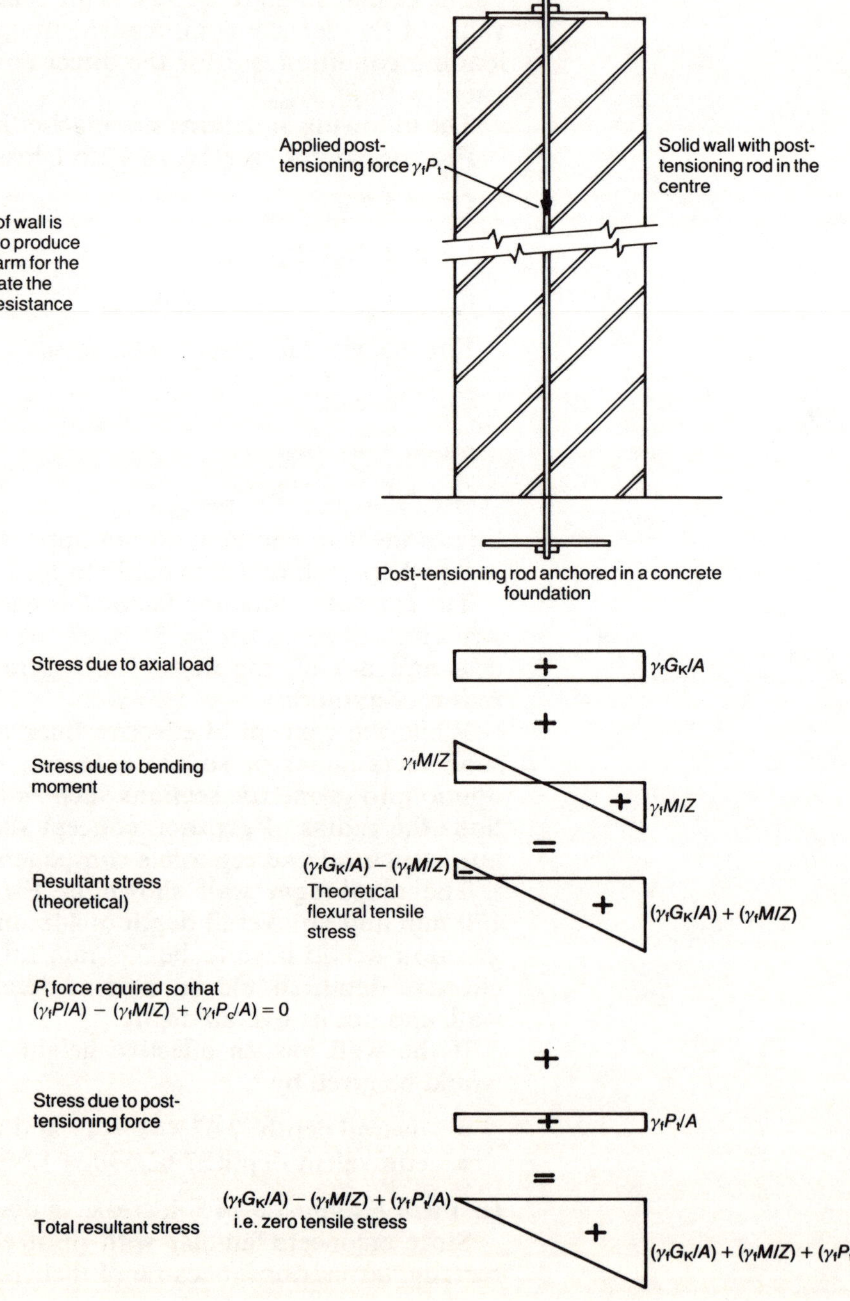

Fig. 6.12. Post-tensioned wall: stress diagrams

basis of a cracked section analysis. There appears to be no fundamental reason why a similar approach (i.e. cracked section analysis) should not be used in post-tensioned brickwork design for ultimate conditions. However, it is considered that in general tensile stresses should be limited to zero, at least for the working load condition, thus ensuring that the design section does not crack. The basis of design proposed by the Authors is therefore to design the wall initially for the serviceability limit state using an uncracked section with zero stress on the tensile face. Fig. 6.2 shows a wall section subjected to an axial load W from its own weight and a bending moment M, where A is the cross-sectional area of the section and Z the modulus of the section.

The post-tension force P is designed to cancel out to zero the stress on the tensile face. This is best achieved for this one direction bending by applying the post-tension force at an eccentricity e. This part of the analysis is carried out using working load conditions in the calculations.

The next step involves the checking of the designed section for the ultimate limit state. For this condition ultimate loads and moments are analysed and the section is checked against its ultimate design moment of resistance for the combined loading conditions. This ultimate resistance is based on a cracked section, as shown in Fig. 6.11.

Provided that the ultimate resistance is equal to or exceeds the ultimate design moments and forces, the section is considered to have adequate safety factors against ultimate failure. In addition, since the original sizing of the sections was based on the serviceability requirement for an uncracked section, this part of the analysis has also been satisfied.

The design checks outstanding at this stage then relate to the local stress conditions such as shear and bond, and the serviceability criteria of deflection.

Information on these aspects is given in chapters 2 and 3. Design examples are given in chapter 7.

6.9. Determination of post-tensioning force

The first stage in the design of post-tensioned brickwork is to establish the support system and the applied loads and moments. A trial section is adopted and then analysed to determine the theoretical flexural tensile stresses and any effective cracked section. Having established the limit to be applied then the applied compressive stress required to reduce the theoretical flexural tensile stress to the chosen limit may be determined. The post-tensioning force required is the force necessary to produce the requisite levels of applied compressive stress to eliminate the theoretical tensile stress or to reduce it to the chosen limit.

Consider the section shown in Fig. 6.12 subject to an applied moment $\gamma_f M$ and an axial load of G_K due only to its own weight. The combined stresses at the base of the wall are $(\gamma_f G_K/A) \pm (\gamma_f M/Z)$. The theoretical tensile stress is then $(\gamma_f G_K/A) - (\gamma_f M/Z)$.

Thus if the net tensile stress under this theoretical condition is to be limited to zero, a post-tensioning force is required which produces a compressive stress greater than, or equal to, the numerical values of the theoretical tensile stress.

The effect of the post-tensioning force is thus to eliminate the tensile stress, but it should be noted that if applied axially it also adds directly to the total compressive stress which now equals $(\gamma_f G_K/A) + (\gamma_f M/Z) + (\gamma_f P_t/A)$.

Although this is not always a problem, and the effect is unavoidable for bending which can be applied in either direction, there are cases where high

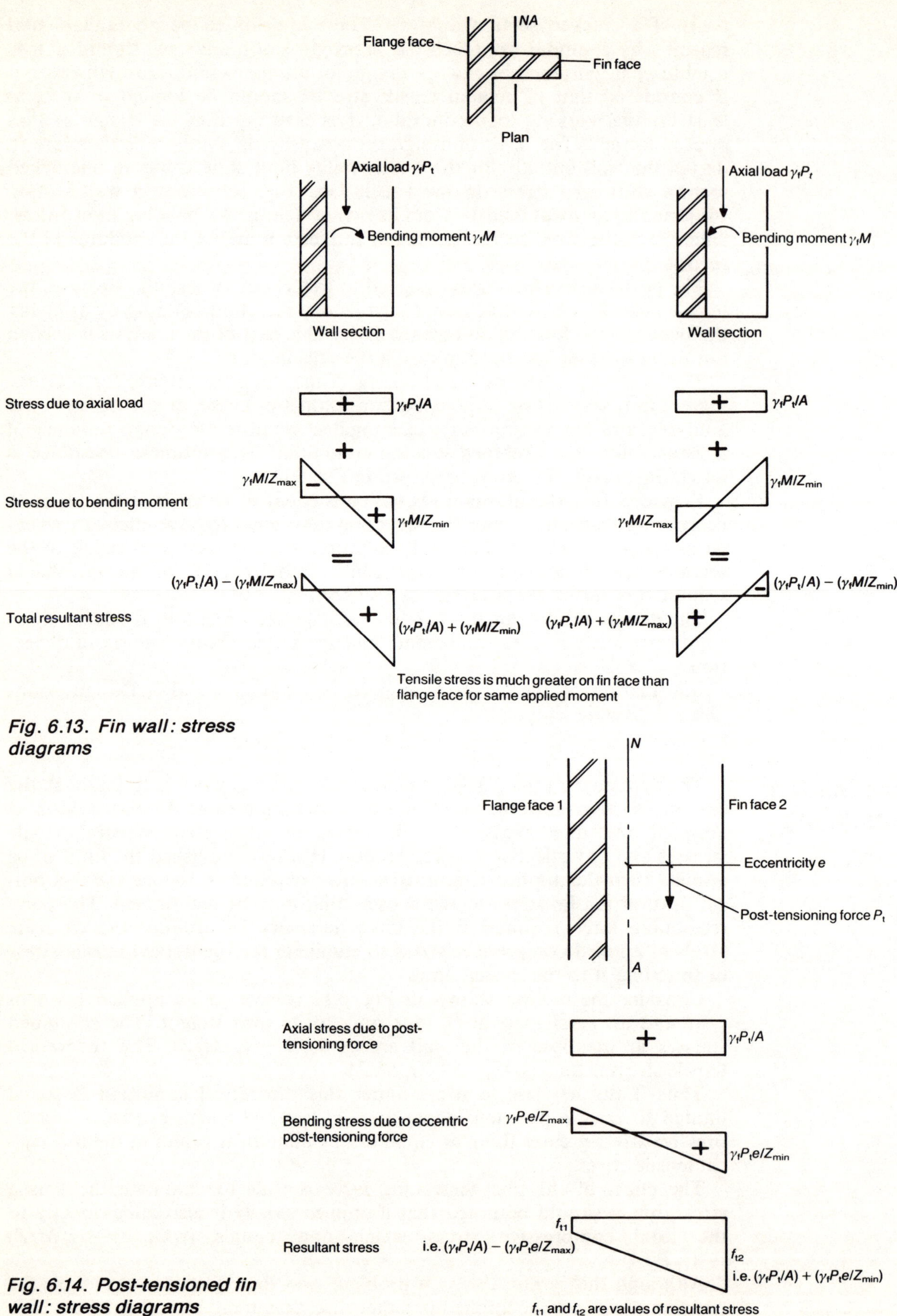

Fig. 6.13. Fin wall: stress diagrams

Fig. 6.14. Post-tensioned fin wall: stress diagrams

compressive stresses may not be acceptable and the section may be uneconomic. The solution then is to apply an eccentric post-tensioning force.

6.10. Eccentric post-tensioning

In some cases there are advantages to be obtained by applying the post-tensioning force at a suitable eccentricity. For retaining walls (and similar elements with the applied lateral loading acting predominantly in one direction only) it is generally most economical to provide the maximum design eccentricity to the post-tensioning force. The value of this eccentricity (and the design calculations) is then generally amended by the practicalities of the construction and detailing (see chapters 7 and 8).

A reversible loading, such as wind loading, generally requires an axial post-tensioning force applied at the centroid of the section. A uniform com-

Fig. 6.15. Post-tensioned fin wall: stress diagrams

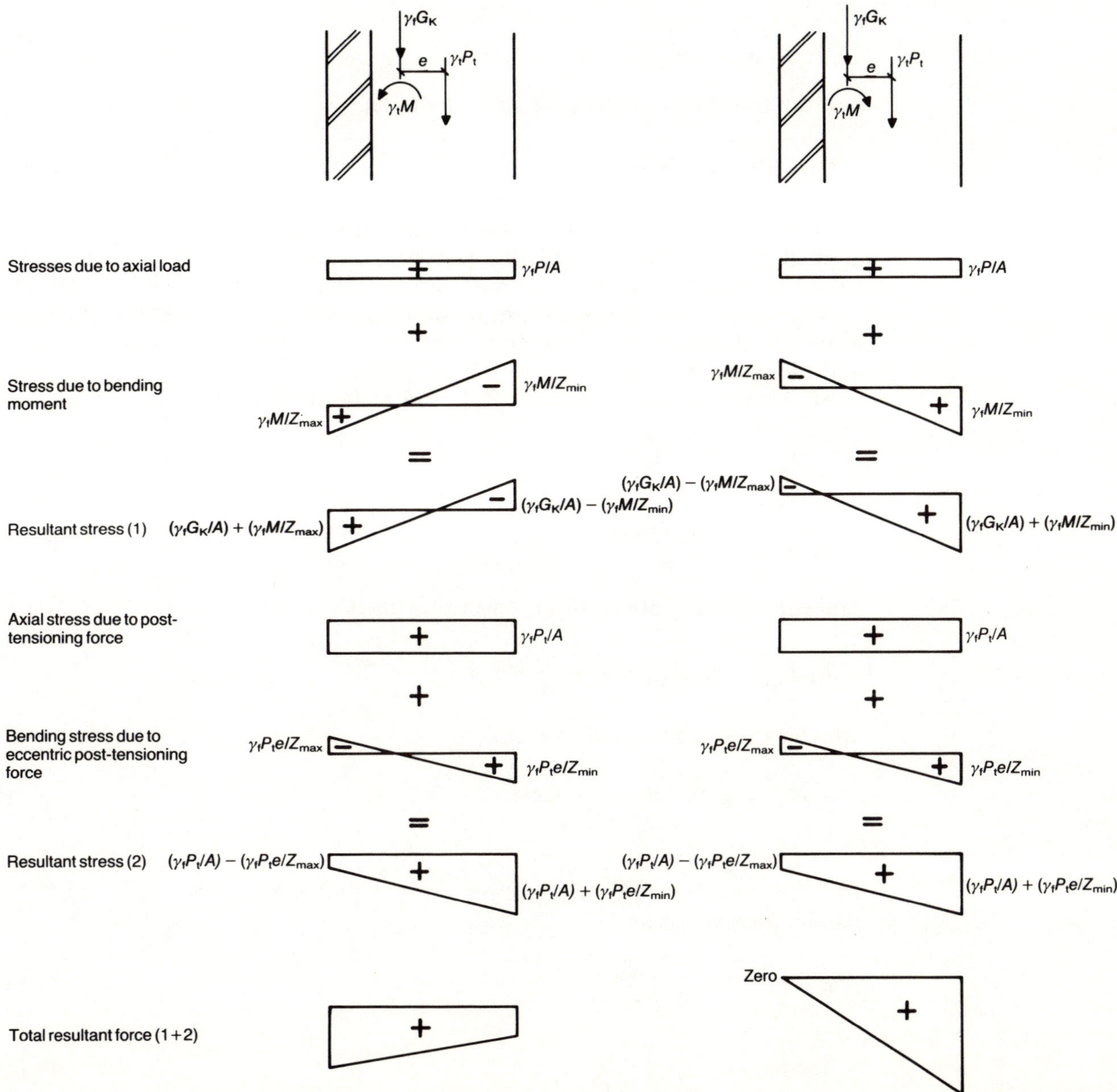

pression is thus applied to counteract the maximum flexural tensile stresses resulting from the bending in either face of the section, depending on the direction of loading being considered. The magnitudes of the flexural tensile stresses occurring at each face will, for symmetrical sections, generally be approximately equal.

However, for asymmetric sections, such as the fin wall, the respective maximum tensile and compressive stresses occurring in the outer fibres or face can vary considerably under the same value of applied loading when acting in reverse. This is due to the variation in the section modulus Z for each face (see Fig. 6.13). A post-tensioning force can be applied eccentrically to add a greater compression to the fin face than to the flange face (see Fig. 6.14).

This counteracts, proportionately, the different values of tensile stress (see Fig. 6.15). The post-tensioning force and eccentricity may be calculated as follows (with reference to Fig. 6.14).

The resulting compressive stress in face 1 is

$$f_{t1} = \frac{\gamma_f P_t}{A} - \frac{\gamma_f P_t e}{Z_{max}} \tag{6.9}$$

and the resulting compressive stress in fin face 2 is

$$f_{t2} = \frac{\gamma_f P_t}{A} + \frac{\gamma_f P_t e}{Z_{min}} \tag{6.10}$$

In order to balance out the theoretical tensile stresses resulting from the applied loading, the compressive stresses f_{t1} and f_{t2} should have the same values as those shown in Fig. 6.13. Thus f_{t1} and f_{t2} are known, as are the area of section A and the section moduli Z_{max} and Z_{min}. Equations (6.9) and (6.10) can therefore be solved to determine both the post-tensioning force P_t and the eccentricity e.

Multiplying equation (6.9) by Z_{max} and equation (6.10) by Z_{min} gives

$$f_{t1} Z_{max} = \frac{\gamma_f P_t e Z_{max}}{A} - \gamma_f P_t e \tag{6.11}$$

$$f_{t2} Z_{min} = \frac{\gamma_f P_t e Z_{min}}{A} + \gamma_f P_t e \tag{6.12}$$

and summing equations (6.11) and (6.12) gives

$$f_{t1} Z_{max} + f_{t2} Z_{min} = \frac{\gamma_f P_t e Z_{max}}{A} + \frac{\gamma_f P_t e Z_{min}}{A}$$

which can be rearranged to give

$$\gamma_f P_t e = A \frac{f_{t1} Z_{max} + f_{t2} Z_{min}}{Z_{max} + Z_{min}} \tag{6.13}$$

The value of P_t can thus be determined for equation (6.13) and substituted in either equation (6.9) or equation (6.10) to find e. Equations (6.9) and (6.10) can be rearranged as

$$e = \left(\frac{\gamma_f P_t}{A} - f_{t1} \right) \frac{Z_{max}}{\gamma_f P_t} \tag{6.14}$$

$$e = \left(f_{t2} - \frac{\gamma_f P_t}{A} \right) \frac{Z_{min}}{\gamma_f P_t} \tag{6.15}$$

6.11. Loss of prestress

The magnitude of the prestressing force, which is initially applied to a structural masonry element, will lessen over the service life of the element, as it does in prestressed concrete. The losses will occur during the application of the prestressing force, immediately after the application and transfer, and over the longer term.

At the various stages considered in the structural design it is therefore necessary, in calculating the force in the prestressing tendons and the stresses in the element, to make allowances for the appropriate loss of prestress. The losses in the initially applied force are due to any or all of the following causes

- relaxation of the tendons (see section 6.11.1)
- elastic deformation of the masonry (see section 6.11.2)
- moisture movement of the masonry (see section 6.11.3)
- creep of masonry (see section 6.11.4)
- draw-in of tendons during anchorage (see section 6.11.5)
- friction (see section 6.11.6)
- thermal movement (see section 6.11.7).

6.11.1. Relaxation of the tendons

The loss of prestress used in design should be (as in prestressed concrete) the maximum calculated relaxation of the tendon after a 1000 hours' duration (as given in BS 5896[1] or BS 4486[8], as appropriate) for stressing force applied equal to that at transfer. This loss of prestress is for a jacking force greater than, or equal to, 70% of the characteristic breaking load of the tendon, but it should be noted that the maximum initial prestress should not be allowed to exceed 70% of that load. If the jacking force is less than 70% then the relaxation loss is assumed to decrease linearly from 8% for an initial prestress of 70% to 0% for an initial prestress of 50% or less.

When a load equal to, or greater than, the relevant jacking force has been applied to a tendon for a short time (i.e. before it has been anchored and the jacking force has been released), no reduction in the loss due to relaxation should be assumed.

Stressing or over-stressing the rod before application does not reduce the relaxation value to be allowed for in design.

6.11.2. Elastic deformation of masonry

The masonry will contract elastically under the compression induced by the prestressing force by an approximate amount $c = PL/AE_\mathrm{m}$, where c is elastic contraction, P is the prestressing force, L is the length of the member, E_m is the elastic modulus of masonry and A is the cross-sectional area.

The bulk of the elastic contraction will take place during the application

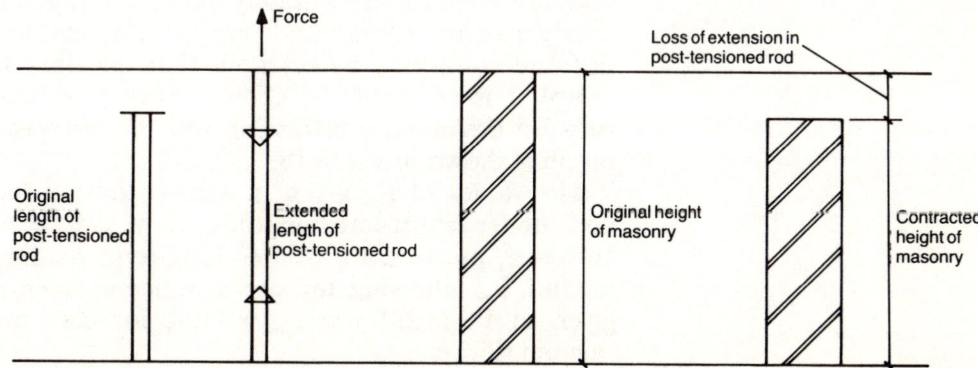

Fig. 6.16. Stress losses due to shrinkage of masonry

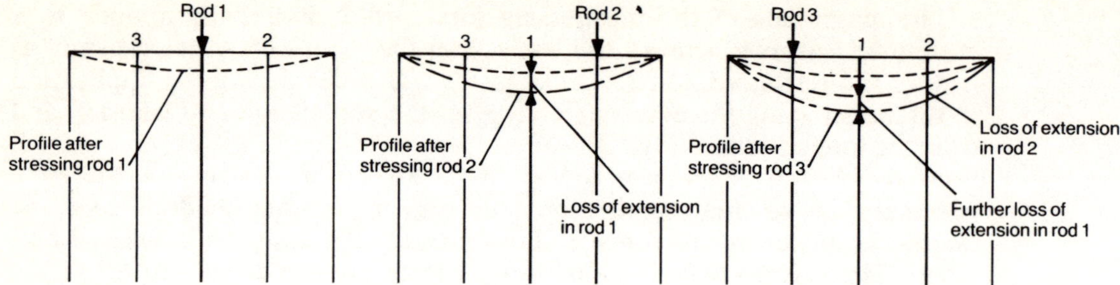

Fig. 6.17. Accumulative stressing of post-tensioned wall

of the prestress but masonry does not react immediately to compression. It takes time to bed down (and creep) and there can be residual contraction after the tendon has been anchored. This further contraction in the masonry will relieve some of the strain in the tendon and thus stress in it, resulting in a loss or prestress force in the tendon (see Fig. 6.16).

In pretensioned masonry (and sometimes in post-tensioned masonry) when the tendons are all stressed simultaneously, the loss of prestress in all the tendons is uniform. However, it is common in practice for the tendons to be stressed individually, when the loss of prestress is no longer uniform. Consider the post-tensioned wall shown in Fig. 6.17. Wall profile A shows the profile of the wall before any rods are stressed and the dotted profile is the profile of the wall when the stress in rod 1 is transferred to the wall. Profile B shows, dotted, the profile of the wall after rod 2 has been stressed and its force transferred to the wall. There is a loss of extension resulting in loss of prestress (and thus prestress force) in rod 1. When rod 3 is stressed (profile C) there is a loss of extension in rod 2 and a further loss of extension in rod 1.

It will be seen that there is thus a continuing loss of prestress in rods as other rods are stressed and contract the wall. This loss can be appreciable. BS 5628:[1] Part 2 suggests that the loss is

$$\frac{(E_s/E_m) \times \text{stress in masonry}}{2}$$

where E_s and E_m are the short-term elastic moduli of the rods and the masonry, respectively.

In practice this magnitude of loss is usually unacceptable and uneconomic and it is usual to restress the rods. (This is commonly known as topping up the prestress.) There will be further elastic contraction of the wall, but this is usually insignificant. By restressing the rods the loss of prestress due to elastic contraction of the masonry may be practically eliminated. Masonry does not contract immediately under a compressive force (because it is not a highly elastic material) and, since it has been found by the Authors in practice and confirmed by research that masonry takes time to bed down (or creep), it is advisable to allow a period of time (a minimum of 24 hours) between the initial prestressing and the restressing. A qualitative strain/time graph is shown in Fig. 6.18.

The values of E_m given in section 2.8 are the short-term elastic moduli, and under short-term loading they can be used for design purposes. However, prestressing creates long-term loading and the long-term elastic moduli E_m (allowing for the significant creep strains and also shrinkage) given in BS 5628[1] are $E_m = 450 f_k$ for clay masonry and $E_m = 300 f_k$ for calcium silicate.

The stress–strain relationship in clay brickwork is non-linear, tends to be parabolic and ends with a falling branch (see Fig. 6.19).[15]

Various researchers have related E_m to f_k with results varying from $E_m = 500f_k$ to $E_m = 1000f_k$. Differing combinations of brick and mortar strengths can give the same f_k and so it does not necessarily follow that they will have the same E_m.

Designers are advised to use the values for E_m suggested in BS 5628[1] and to check on site by measuring the contraction of brickwork under a known compressive force.

6.11.3. Moisture movement

Fired-clay brickwork expands with time due mainly to the take-up of moisture. This expansion is rapid when the bricks are leaving the firing kiln but slows down with time (for this reason bricks should not be laid until 14 days after leaving the kiln). Since brickwork should not be prestressed until the mortar has reached its required strength (often as long as 14 days) the total time lapse results in a reduced continuing expansion and its effect on prestressing may generally be ignored.

Concrete blocks and bricks and calcium silicate brickwork tend to shrink with time and it is suggested that designers may accept the assumption in BS 5628:[1] Part 2 that the maximum shrinkage strain is 500×10^{-6} in determining the loss of prestress.

6.11.4. Loss of prestress in masonry due to creep

Concrete blockwork and brickwork will creep similarly to in situ or precast concrete under compressive load. Fired-clay brickwork creeps but to

Fig. 6.18 (below). Strain plotted against time

Fig. 6.19 (right). Stress/strain curves for clay brickwork in compression: (a) four types of brick in 1:1/4:3 mortar; (b) dimensionless stress–strain curves

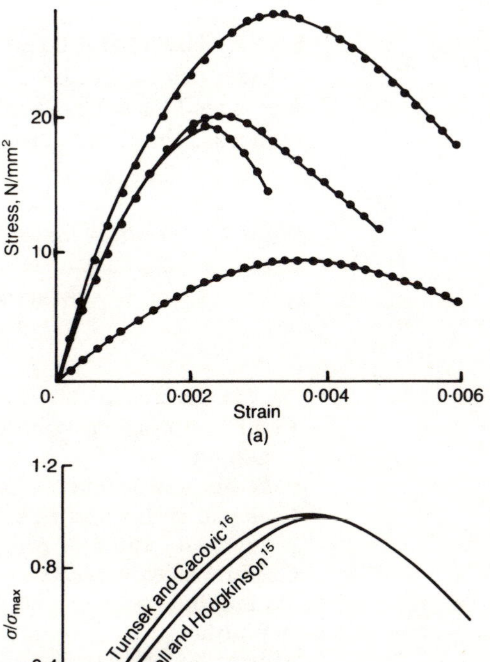

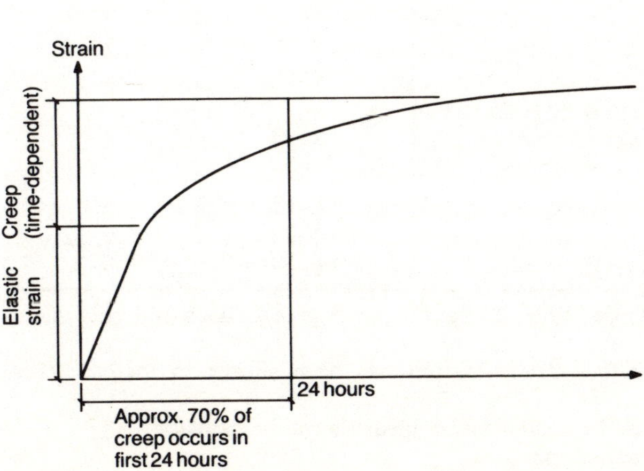

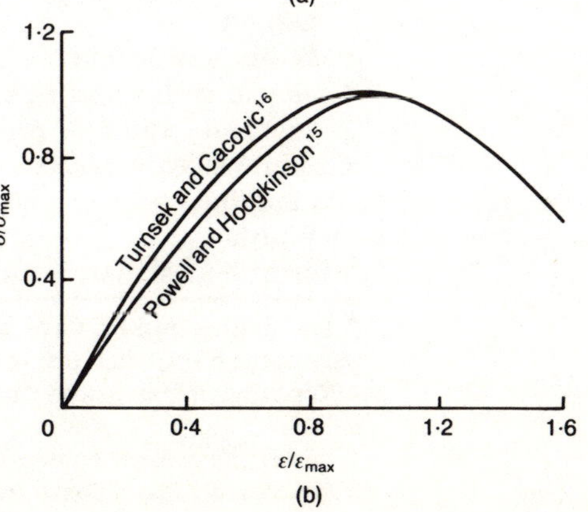

a considerably lesser extent (and it is thought that a major part of the creep of the brickwork as a whole is due to the mortar joints). The Authors have no experience of prestressing calcium silicate bricks or lightweight concrete blockwork. They suggest a value of creep loss in fired clay brickwork of 10–15% and in concrete blockwork of 25–30%.

6.11.5. Anchorage draw-in

Anchorage draw-in losses occur mainly in factory-made pretensioned units. At present this technique would appear to be limited to the manufacture of concrete block lintels by the long-line method. These draw-in losses are known to manufacturers by their own testing. If the designer should wish to use the pretensioning method on site then the recommendations for anchorage losses given in BS 8110[12] should be considered.

In post-tensioning work, particularly using end threaded rods stressed by jacking, the small anchorage losses may be assumed to be embraced within the global losses (see section 6.11.8).

6.11.6. Friction

All the work carried out by the Authors has been done with straight tendons. The use of curved tendons requires single wires or strands of wire and it has not been found practical to use them on site. The tendons need to be supported during construction (which is difficult) and to be maintained in position after construction.

Obviously in straight rods there are no friction losses due to curvature of tendons. Should the designer wish to use curved tendons, the advice given in BS 8110[12] for losses should be followed. Friction at any position on the straight rod is avoided by ensuring that there is adequate clearance of the rod through all holes, voids and ducts.

6.11.7. Thermal movement

Masonry expands with increase of temperature and contracts with decrease. The expansion could further extend the rods and thus increase the prestress; the contraction could decrease it. Since thermal movement is very

Table 6.1. Coefficients of linear thermal expansions

Material	Coefficient of linear thermal expansion $\times 10^{-6}$	
	Per °C	Per °F*
Fired-clay bricks and blocks†		
Length	4–8	2–4
Width and height	8–12	4–7
Concrete bricks and blocks‡ (depending on the		
aggregate and mix proportions)	7–14	4–8
Calcium silicate bricks		
Length	11–15	6–8
Width	14–22	8–12
Mortars§ (designations (iii) and (iv))	11–13	6–7

* The figures quoted have been rounded and the figures per degree Celsius and equivalents per degree Fahrenheit are not exact.
† Repeated freezing and thawing before determination has been shown to increase these values.
‡ Increasing cement content increases the coefficient of linear thermal expansion.
§ Appreciably less if based on calcareous sand.

small (see Table 6.1) it is unlikely to be of concern at usual levels of prestress and temperature conditions. However, if the prestressing is carried out in hot weather it is advisable to restress in the cool of early morning.

Where low levels of prestress are used (e.g. in a free-standing wall prestressed to counteract tensile bending stresses due to lateral wind pressure) it is advisable to check the prestress due to differential thermal movement between the masonry and the rod. (The rod also expands with increase in temperature, but the increase is not equal to that of the masonry nor does the rod temperature necessarily rise in proportion to the masonry, which may shelter the rod.)

Low levels of prestress can result in low levels of strain in the rods and this strain can be lost due to the cumulative losses caused by elastic contraction, relaxation of the rod and creep in the masonry. It is important to take these causes of loss of prestress into account in designing the initial prestress and to design for higher prestress at transfer.

6.11.8. Global losses

Provided that

- elastic contraction of the masonry (and hence loss of prestressing force) is counteracted by restressing
- the element is post-tensioned so that anchorage draw-in losses are minimal
- the tendons are straight (as they are in most applications) and therefore there is practically no loss due to friction.

the remaining major losses are the relaxation of the tendons and creep of the masonry.

The losses due to relaxation of the tendons have been thoroughly investigated and calculations are thus reasonably reliable, but there has been little research on the creep of masonry under prestress (although Lenczner has done excellent work on creep under constant compression[17] and on creep due to prestress[18]). It is necessary then to make a rough estimate of the global losses. From experience the Authors would suggest an allowance of 20% for clay brick masonry and about 35% for concrete masonry.

6.12. Anchorages and end blocks

Anchorages and end blocks for prestressed masonry are very similar to those used in prestressed concrete.

6.12.1. Design

All anchorages such as bearing plates and end blocks should be designed to resist shear and bending stresses.

6.12.2. Bearing stresses

The bursting tensile force F_{bst} in end blocks or anchorages at both ends of the rod should be designed (as in prestressed concrete) for the greater of the tendon jacking load and the load in the tendon at ultimate limit state.

The local bearing stress in the masonry, in contact with the anchorage, should not exceed $1 \cdot 5 f_k / \gamma_{mm}$. At a depth of $0 \cdot 4h$ below the bearing (where h is the clear height of the wall), allowing for a 45° dispersion of stress, the stress should not exceed $f_k \beta / \gamma_{mm}$ (see Fig. 6.20).

6.12.3. Transmission length in pretensioned members

When (as is uncommon in prestressed masonry) the structural member is pretensioned, the length of the member required to transmit the initial pre-

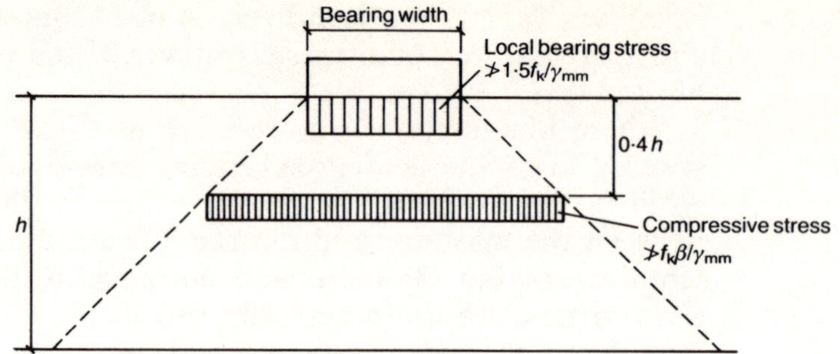

Fig. 6.20. Stresses at local bearings

stress force in the tendon to the concrete surrounding it is similar to that in pretensioned concrete. The required transmission length depends on such variables as the strength of the concrete or grout, and the diameter type and deformation (crimped, indented and so on) of the wire. It is better to check the transmission length by tests at the factory and, for a trial run, the transmission length should be 160 times the diameter of the wire. If small diameter strands are used the trial run transmission length could be obtained from BS 8110.[12]

6.13. Partial factors of safety on post-tensioning γ_f

The partial factors of safety γ_f are applicable to loadings, moments and post-tensioning forces in deriving the various design equations. The numerical values for the partial factors of safety for dead, imposed and wind loadings are all as used for reinforced brickwork and are given in BS 5628: Parts 1 and 2[1] and in chapter 2. The partial factor of safety γ_f on loads is applicable to some degree to the post-tensioning force. It appears that no research has yet been carried out to determine statistically the appropriate γ_f values and so the designer must assess the likely variables to determine a suitable figure. In the absence of other guidance the Authors suggest that the partial factors of safety applicable to dead loading should be used. The partial factors of safety for materials are also as for reinforced masonry but with variations relating to serviceability limit state (see section 6.5).

6.14. Design strength of post-tensioning rods

The post-tension wire strands and rods are generally of high yield steel to BS 4486[8] or BS 5896.[9] The characteristic strengths f_y and the relevant partial factor of safety γ_{ms} are as given in Tables 2.4 and 3.3. The steel is generally stressed to its design level, subject to slight variation due to changes, throughout its service life. (This is unlike most reinforcement which is only stressed to the maximum design value when the full dead and imposed loading is applied—a situation which in many cases occurs infrequently and for short periods, if at all.)

In view of this, and in order to limit the relaxation of the steel, the stress in the post-tensioning rods is limited to 70% of the general maximum design value. Thus the design strength of the post-tensioning rods is given as $0{\cdot}7f_y/\gamma_{ms}$.

6.15. Design of post-tensioning rods

Knowing the magnitude of the post-tensioning force to be applied and the design strength of the rods, the required area of the rods can be determined.

The design post-tensioned force is less than the area of the rods multiplied by the design strength of the rods, i.e.

$$1{\cdot}25\gamma_f P_t < A_s(0{\cdot}7f_y/\gamma_{ms})$$

Therefore

$$A_s \geqslant 1\cdot 25\gamma_{ms}\,\gamma_f\,P_t/0\cdot 7f_y \qquad\qquad (6.16)$$

assuming 20% losses. This area of steel required is the actual area, i.e. the net tensile area of the rod. As post-tensioning rods are generally threaded, the reduction in the area of the rod due to threading must be allowed for.

Worked examples of this theory are given in chapter 7.

It is possible to vary the number of rods or the area of the rods or both, in a given situation; the most suitable, practical and economic combination must be chosen by the designer/detailer. The spacing of the rods is often governed by such practical considerations as accommodating the required bearing plates or positioning the rods within voids in the construction. The size of rods is also governed by practical considerations (see chapter 8).

6.16. Application of post-tensioning force

The compression within the masonry as a result of the post-tensioning is induced by the action of a rod or strand in tension on anchorages at each end. One end is usually fixed and the other adjustable.

There are two main methods of applying the post-tensioning force to the brickwork section. Both involve the stretching of the steel rod or strand, the difference in the methods being the system used to apply the force. The main methods are

- by means of jacking (usually used for high levels of prestress)
- by means of a torque wrench (usually used for relatively low levels of prestress to counteract wind pressure stresses).

The jacking system allows a direct reading of the post-tensioning force on the dial gauge mounted on the hydraulic pump which applies pressure to the jack. The rod or strand is locked into the jaws of the jack by means of wedges. When the required predetermined tensile force has been applied and cross-checked by measurement of the extension of the rod or strand, the lock nut is tightened against the anchorage plate, which fixes the rod/strand at the required extension/force. Each jack and pump is calibrated to ensure accurate results in the application of the post-tension force.

The torque wrench system stretches the rod by a nut turned on the threaded adjustable end of the rod, by using a torque wrench, against an anchorage plate. This extends the rod which produces a tensile stress within it. The required tension within the rod is related to the amount of tightening on the nut, i.e. the torque which is the turning moment required to produce the requisite strain in the rod. The value of the torque required is dependent on the amount of prestress required and a number of other factors. The most significant of these are the type and pitch of the thread and the friction developed between the contact surfaces of the nut, bolt and spreader plates, and so on. The advice of the manufacturer of the equipment used to torque the rods should be sought on the relationship between bolt tension and torque. For the critical locations where accurate assessment of the applied force is necessary, the stress in the rod should be monitored during load applications, because the accuracy of the relationship between applied torque and individual force is dependent on many variables.

The method of determining torque that follows is based on a general engineering formula derived from test research. It is based on using lightly oiled metric threads, with self-finish nuts and bolts and a hardened washer between the nut and spreader plate. The torque required is

$$\frac{\text{rod tension} \times \text{rod diameter}}{5} = \frac{1\cdot 25\gamma_f\,P_t \times d}{5}$$

Torque values are usually expressed in kgf m units whereas the post-tensioning force is usually expressed in kN units $(1\,\text{kN} = 10^3/9\cdot81\,\text{kgf} = 102\,\text{kgf})$.

Since the actual force produced in the rod is dependent on many variables the engineer should satisfy himself from experience or testing that the force finally achieved is in compliance with the design requirements.

6.17. Checking of flexural and direct stress: all load cases

The magnitude of the post-tensioning force is based generally on the condition of no tensile stress in the section under the worst loading combination and after losses have occurred. The tensile stresses are therefore catered for. The combined flexural compressive stresses should then be checked for the various combinations of loading after losses for the limitations given in equations (6.7)–(6.9).

The direct compressive stresses should be checked for the various combinations of dead and imposed loads and the post-tensioning force, before losses.

6.18. Vertical and horizontal shear stress

Having calculated the post-tensioning force and checked the direct flexural stresses, the shear stresses should be considered. The design of post-tensioned masonry in shear is treated generally in the same way as plain masonry. The post-tensioning force generally adds to the applied axial loadings on the section and thus affects the design shear strength but not the applied shear load.

The general expression for shear stress vh is

$$vh = VA\bar{y}/Ib \qquad (6.17)$$

where vh is the shear stress, V is the horizontal or vertical shear force, A is the area of the cross-section to one side of the position where shear stress is being checked, $\bar{y}$ is the distance from the neutral axis to the centroid of the area, I is the second moment of the area and b is the width at the position being checked.

Details and dimensions on shear are given in sections 3.6 and 3.15.

The design shear strength is equal to the characteristic shear strength f_v divided by the partial factor of safety for the material strength in shear γ_{mv}. The characteristic horizontal shear strength is given in BS 5628: Part 2[1] for mortar designations (i) and (ii) as $f_v = (0\cdot35 + 0\cdot6g_B)$ N/mm^2 with a maximum value of $1\cdot75$ N/mm^2, where g_B is the design vertical load in the wall. Worked examples are given in chapter 7.

19.1.3.3 *Shear in prestressed sections.* For prestressed sections with bonded or unbonded tendons the characteristic shear strength of masonry, f_v, may be obtained from the following formula:

$$f_v = (0\cdot35 + 0\cdot6g_B)\ \text{N/mm}^2$$

where g_B is the design load per unit area due to the loads acting at right angles to the bed joints, including prestressing loads (in N/mm^2).

Note. In elements prestressed parallel to the bed joints $g_B = 0$, giving $f_v = 0\cdot35$ N/mm^2.

For simply supported prestressed beams or cantilever retaining walls where the ratio of the shear span, a, to the effective depth, d, is six or less, f_v may be increased by a factor $\{2\cdot5 - 0\cdot25(a/d)\}$.

In all cases f_v should not be taken to be greater than $1\cdot75$ N/mm^2.

For post-tensioned masonry, the Authors advise that only mortar designations (i) and (ii) should be used. A maximum value of 1·75 N/mm² may be reasonable for plain clay brickwork but not necessarily so for prestressed brickwork.

The post-tensioning force may be included in the design vertical load g_B and thus can increase the horizontal shear strength.

In masonry sections designed for flexure on the basis of an uncracked section, the horizontal shear resistance of the section may be taken as occurring over the whole plan cross-sectional area. When the cracked section analysis is applicable, only the uncracked section of masonry will provide resistance to horizontal shear. However, the vertical stress on this reduced area will be increased considerably and thus the horizontal shear resistance will be maintained at a comparably high level.

There is no guidance on the analysis of shear in cracked sections in BS 5628[1] nor on the maximum stresses for geometric shapes; only average horizontal shear is specified in that code. Generally only horizontal shear on rectangular uncracked sections is dealt with and designers working outside this restriction should assess the shear resistance in each case on the basis of sound engineering principles. For the present the Authors recommend the adoption of the values in BS 5628 for the average horizontal shear resistance as being also the maximum for geometric sections in any plane.

Horizontal and vertical shear stress can govern design in certain instances (e.g. heavily loaded retaining walls) when the post-tensioning force may need to be increased solely to increase the horizontal shear resistance.

6.19. Principal tensile stress

Masonry is a combination of units and mortar which are bonded together in traditional patterns. The methods of combining the elements, together with the shape of the units, means that masonry exhibits different structural characteristics when it is loaded in different directions. For example, brickwork flexural tensile strength, flexural compressive strength and shear strength differ for each direction. The analysis of sections is thus more complicated than that of a homogeneous material such as structural steel. Even when using structural steel in shapes such as I beams, the analysis can be more complex than for simple rectangular shapes.

In addition, when considering the analysis of more complex geometric shapes such as I sections, even in homogeneous materials, the analysis of the sections must be considered in greater detail because of the possibility of combined stress levels being greater than the individual values. Thus in masonry, because of its nature, and more particularly where geometric shapes other than simple rectangular sections are used, a more detailed analysis of stresses is often required.

Masonry design generally requires the applied maximum tensile and compressive stresses to be compared with the relevant respective material strengths as well as the applied maximum shear stress. In addition, particularly with geometric shapes, the principal stresses should be calculated and compared with the relevant material strength. The derivation of principal stresses is given in reference 19.

For a section in bending subject to a normal stress σ_x and a shear stress τ_{xy} acting together, a principal stress may be produced which is numerically larger than the stress at the extreme edges of the section. The principal stresses are given by the following.[19] The principal compressive stress is

$$\sigma_{max} = \frac{\sigma_x}{2} + \sqrt{\left(\frac{\sigma_x^2}{2} + \tau_{xy}^2\right)} \qquad (6.18)$$

the principal tensile stress

$$\sigma_{\min} = \frac{\sigma_x}{2} - \sqrt{\left(\frac{\sigma_x^2}{2} + \tau_{xy}^2\right)} \qquad (6.19)$$

and the maximum shearing stress is

$$\tau_{\max} = \frac{\sigma_{\max} - \sigma_{\min}}{2} = \sqrt{\left(\frac{\sigma_x^2}{2} + \tau_{xy}^2\right)} \qquad (6.20)$$

The values of σ_x will be the appropriate values as obtained for the point being analysed from consideration of the applied moments and axial forces with the section properties, e.g. $(P/A) \pm (M/Z) \pm (Pe/Z)$.

The shear stress τ_{xy} will be calculated from equation (6.17) for the corresponding section being considered.

6.20. Principal tensile strength

In many cases it will be the principal tensile stress that will need to be checked. However, the determination of the design principal tensile strength to compare with the design applied stress is not straightforward. There has been little research on the subject; few results are readily available on this phenomenon in brickwork, and there is no guidance in the codes of practice. Designers must therefore assess a suitable value on the basis of their engineering judgment. The Authors consider, admittedly from limited research experience, that the principal tensile strength is in general likely to be half the shear strength, i.e. $f_v/2$.

Designers are advised to check principal tensile stress in heavily prestressed geometric sections but, due to the high partial safety factors, there is rarely a serious design problem. Problems that may arise can be solved by thickening the section in the area of high principal stress.

6.21. Shear lag

With geometric sections, the tensile and compressive stresses in the flanges arising from flexure result in shear transfer between the ribs and the flanges. In design the stress is often assumed to be transferred from the rib across the full width of the flange at the rib/flange interface, so that the stress in the flange is uniform for the full width. In pure theory, however, the stress appears (as could be expected) to be at a maximum across the width of the rib, reducing to a minimum value within the section of the flange between the ribs (see Fig. 6.21).

This difference in stress level is termed 'shear lag'. It has been observed in steel box girders and other similar structural elements. The result of the shear lag in such sections can be an increase in the flexural stresses and deflection. As could be expected, the magnitude of the shear lag effect appears to be related to the ratio of span to depth and span to rib spacing. The shorter the span the greater the effect of this shear lag.

It seems that for relatively short spans the increase in the flexural compressive stress could be as much as 25%.

Shear lag is, however, a localised effect and away from the rib/flange interface the stresses will redistribute to more equal levels. As increases of up to 50% are considered appropriate for localised compressive stress effects, in most cases it is considered that the shear lag effect may be discounted in any analysis of stresses. In particular instances of very high stresses (i.e. approaching ultimate stresses and/or where the span to depth and span to rib spacing ratios are comparatively low, and/or where tensile stresses may be critically affected) the effect of shear lag should be considered and stress

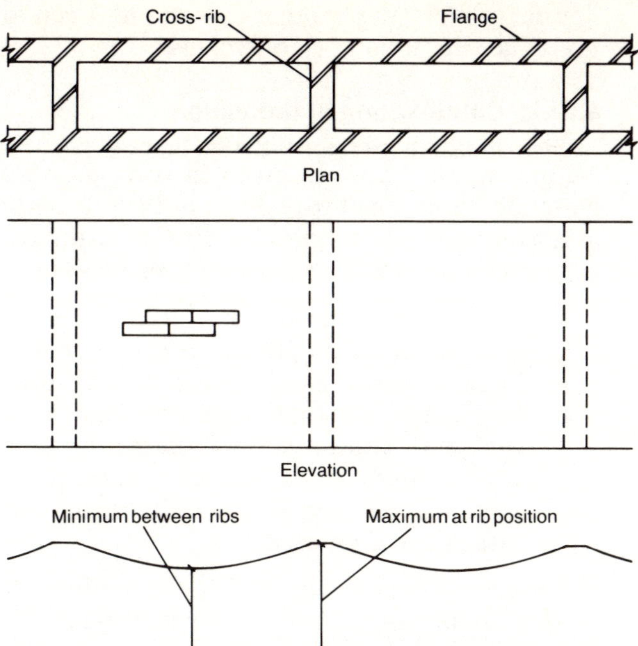

Fig. 6.21. Diaphragm wall: stress distribution in flanges

levels adjusted appropriately. It should be noted that the critical effect of shear lag is restricted by the normal considerations of maximum flange widths embraced within the design guidance given for geometric shapes and additional analysis will rarely be required.

6.22. Deflection

6.22.1. General

The deflection should not be excessive (i.e. adversely affect the performance of the structure, applied finishes, partitions and so on) and should not exceed the values given in the following extract from BS 5628: Part 2.[1]

16.2.2.1 *Deflection*. The deflection of the structure or any part of it should not adversely affect the performance of the structure or any applied finishes, particularly in respect of weather resistance.

The design should be such that deflections are not excessive, with regard to the requirements of the particular structure, taking account of the following recommendations.

(*a*) The final deflection (including the effects of temperature, creep and shrinkage) of all elements should not, in general, exceed length/125 for cantilevers or span/250 for all other elements.

(*b*) Consideration should be given to the effect on partitions and finishes of that part of the deflection of the structure taking place after their construction. A limiting deflection of span/500 or 20 mm, whichever is the lesser, is suggested.

(*c*) If finishes are to be applied to prestressed masonry members, the total upward deflection, before the application of finishes, should not exceed span/300 unless uniformity of camber between adjacent units can be ensured.

In any calculation of deflections (see appendix C) the design loads and the design properties of materials should be those recommended for the serviceability limit state in clauses **18** to **20**. For reinforcement, stresses lower than the characteristic strengths given in table 4 may need to be used to reduce deflection or control cracking.

6.22.2. Calculations of deflection

The design loads should be those given in section 2.4, the partial safety factors should be those given in sections 2.8 and 2.9 and the properties of materials should be those given in the relevant tables.

Elastic analysis should be used to estimate (not necessarily to calculate accurately) the deflection. The following assumptions may be made.

- The gross cross-sectional area of the masonry, ignoring the steel is used in calculating the stiffness of the member.
- Plane sections remain plane after bending.
- The steel (whether reinforcement or prestressing and whether in tension or compression) is elastic.
- The masonry is also elastic (for short-term loading the short-term E_m value can be used; conversely for long-term loading the long-term E_m value should be used).

Deflection is given by $K(WL^3/EI)$ where K is a constant depending on type of loading (e.g. uniform, concentrated), W is the load, L is the span, E is the elastic modulus and I is the second moment of the area.

The estimate of deflection is unlikely to be accurate because

- it is difficult to measure precisely the live loading, and this affects W
- the true span depends on the end fixity of the member and since this can only be estimated (or postulated) L is only an estimate
- E_m can vary depending on the stress, load, duration, time and so on
- clause 19.2 of BS 5628:[1] Part 2 gives three methods of assuming I values of reinforced and post-tensioned masonry (see the following extract) and none of these is 100% accurate; if the section has cracks, then the I value should be for the cracked section and not necessarily the gross cross-section.

19.2   Analysis of structure

When analysing any cross section within the structure, the properties of the materials should be assumed to be those associated with their design strengths appropriate to the limit state being considered. Due allowance should be made when materials with different properties are used in combination. Where the member to be designed forms part of an indeterminate structure, the method of analysis employed to determine the forces in the member should be based on as accurate a representation of the behaviour of the structure as is practicable.

When elastic analysis is used to determine the force distribution throughout the structure, the relative stiffnesses of the members may be based throughout on any one of the following cross sections:

(a)   the entire masonry section, ignoring the reinforcement;
(b)   the entire masonry section including the reinforcement on the basis of the modular ratio derived from the appropriate values of modulus of elasticity given in **19.1.7**;
(c)   the compression area of the masonry cross section combined with the reinforcement on the basis of the modular ratio as derived in (b).

If W, L, E and I are variable, or are estimates, then it is unlikely that the predicted deflection will equal the actual. It is common in practice for the predicted 'calculated' deflection to well exceed the actual deflection.

7

Design examples: prestressed masonry

7.1. Cavity wall subject to lateral wind and axial loads

The cavity wall shown in Fig. 7.1 comprises two leaves of brickwork using bricks with a minimum compressive strength of 27·5 N/mm² set in a designation (ii) mortar. The bricks may be taken to have a density of 20 kN/m³ and normal manufacturing control is assumed, giving a partial safety factor γ_{mm} of 2·3. This brickwork has an f_k value of 7·9 N/mm².

A design check would demonstrate that a plain masonry wall of this height would fail in bending for the loading conditions shown and hence post-tensioning rods are introduced to provide the additional bending tensile strength required.

The reinforced concrete roof slab is assumed to be capable of providing simple lateral support to the head of the wall and these lateral forces are assumed to be transferred through the roof slab to stiff gable shear walls capable of providing overall stability to the structure. Hence the wall is designed as a propped cantilever with the design bending moment diagram as shown in Fig. 7.2.

The upper anchorage of the post-tensioning rods is formed by steel spreader plates recessed into the top of the reinforced concrete slab with sleeves cast into the slab through which the rods pass, as shown in Fig. 7.3. This detail provides for the post-tensioning force to be shared equally between the two leaves of masonry, and the stiffness of the reinforced concrete slab is such that zero eccentricity of the post-tensioning forced on each

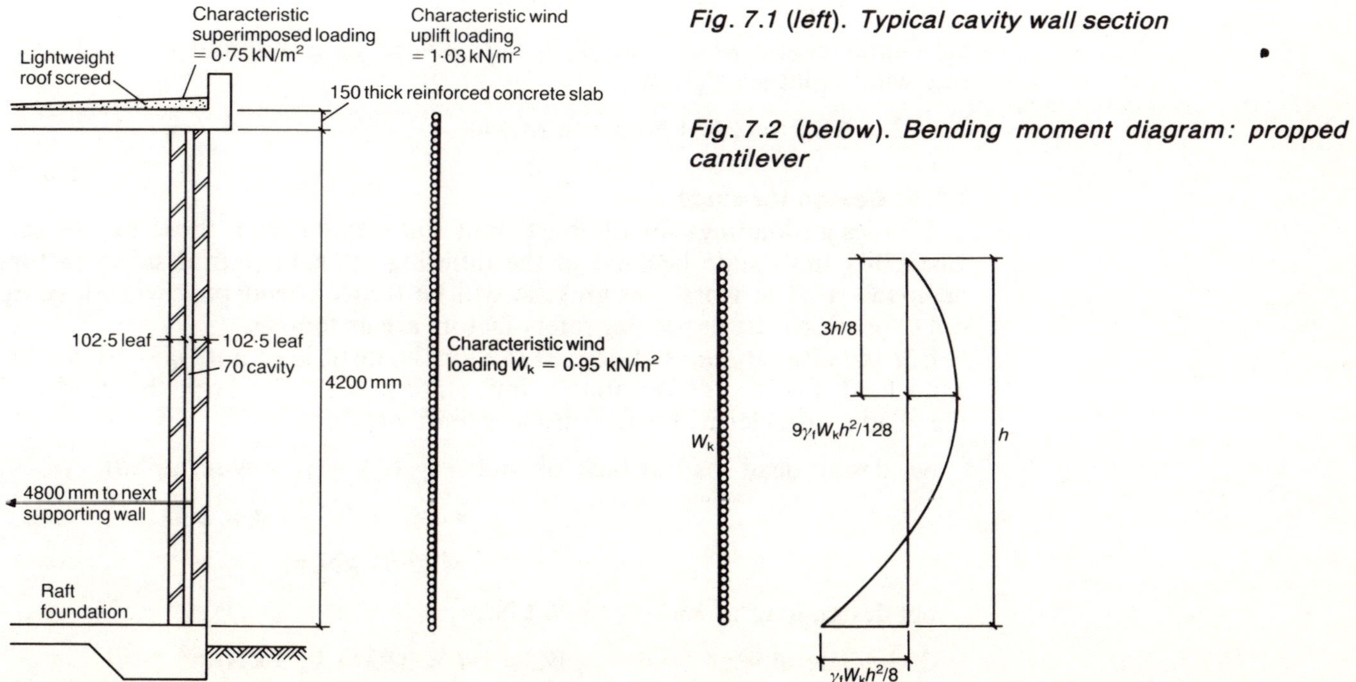

Fig. 7.1 (left). Typical cavity wall section

Fig. 7.2 (below). Bending moment diagram: propped cantilever

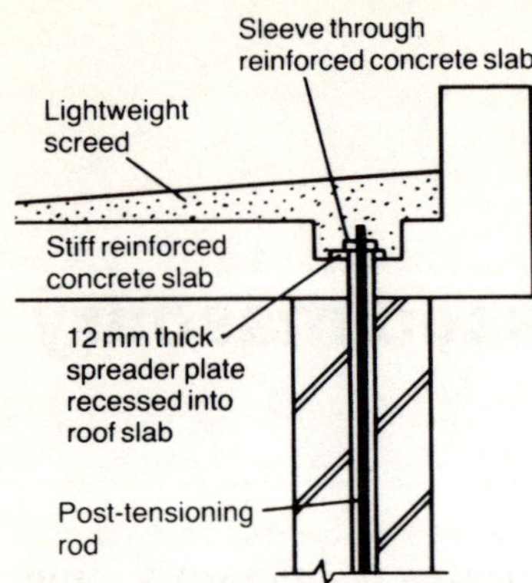

Fig. 7.3. Upper anchorage detail

leaf width may be assumed. The recess in the slab should not be too deep and it may be better to provide the cover protection to the steel using the screed, thus minimising or eliminating the slab recess.

7.1.1. Basic design method

The design method adopted for this example is to calculate the theoretical flexural tensile stress which would be likely to develop as a result of the lateral loading for the serviceability limit state, in the absence of any post-tensioning force. An axial compressive stress is then applied by way of the post-tensioning force (in addition to the existing axial stress from the minimum vertical loading) to eliminate the theoretical flexural tensile stress previously calculated. Checks are then carried out to establish that the wall specified can support the applied vertical and lateral loading as well as the post-tensioning force for the various design conditions and limit states

7.1.2. Characteristic loadings

The characteristic dead load G_k (at the base of the wall) comprises

the roof slab: $0.15 \times 24 \times 4.8/2 = 8.64$ kN/m

plus the finishes: say, 3.14 kN/m

plus the self-weight of the wall: $2 \times 0.1025 \times 20 \times 4.2 = 17.22$ kN/m

i.e. total $G_k = 8.64 + 3.14 + 17.22 = 29.00$ kN/m run of wall.

The characteristic dead load G_k (at $3h/8$ from the top of the wall) comprises

the roof slab: 8.64 kN/m

plus the finishes: say, 3.14 kN/m

plus the self-weight of the wall $(3h/8)$: $3/8 \times 17.22 = 6.46$ kN/m

i.e. total $G_k = 8.64 + 3.14 + 6.46 = 18.24$ kN/m run of wall.

The characteristic superimposed load Q_k is as given

$0.75 \times 4.8/2 = 1.80$ kN/m run of wall

the characteristic wind load W_k is 0.95 kN/m² as given, and the characteristic wind uplift is as given

$1.03 \times 4.8/2 = 2.47$ kN/m run of wall

7.1.3. Design loadings

The design loadings for ultimate limit states differ from those for the serviceability limit state because of the differing relevant partial safety factors on loads γ_f. The worst design case will be that of dead plus wind loading only, for which the particular safety factors are as follows.

For the ultimate limit state $\gamma_f = 0.9$ for the dead load and $\gamma_f = 1.4$ for the wind load. For the serviceability limit state $\gamma_f = 1.0$ for both the dead and the wind loads. Hence, for the ultimate limit state

net design dead load at base of wall $= (\gamma_f\, G_k) - (\gamma_f \times \text{wind uplift})$

$$= 0.9 \times 29 - 1.4 \times 2.47$$

$$= 22.64 \text{ kN/m}$$

net design load at $3h/8 = 12.96$ kN/m

design lateral wind load $= \gamma_f\, W_k = 1.4 \times 0.95 = 1.33$ kN/m²

For the serviceability limit state

net design dead load at base of wall $= 1{\cdot}0 \times 29 - 1{\cdot}0 \times 2{\cdot}47$

$$= 26{\cdot}53 \text{ kN/m}$$

net design load at $3h/8 = 15{\cdot}77$ kN/m

design lateral wind load $= 1{\cdot}0 \times 0{\cdot}95 = 0{\cdot}95$ kN/m^2

7.1.4. Design bending moments

For the ultimate limit state, at the base of the wall

$$M_b = 1{\cdot}33 \times 4{\cdot}2^2/8$$

$$= 2{\cdot}93 \text{ kNm}$$

and at $3h/8$ level

$$M_w = 9 \times 1{\cdot}33 \times 4{\cdot}2^2/128$$

$$= 1{\cdot}65 \text{ kNm}$$

For the serviceability limit state, at the base of the wall

$$M_b = 0{\cdot}95 \times 4{\cdot}2^2/8$$

$$= 2{\cdot}09 \text{ kNm}$$

and at $3h/8$ level

$$M_w = 9 \times 0{\cdot}95 \times 4{\cdot}2^2/128$$

$$= 1{\cdot}18 \text{ kNm}$$

7.1.5. Serviceability limit state: theoretical flexural tensile stress

Referring back to the basic design method in section 7.1.1 to calculate the theoretical flexural tensile stress it is assumed that the design loads (lateral and vertical) are shared equally between the two leaves of the cavity wall.

Hence, considering the serviceability limit state, the axial compressive stress is

$$g_d = \frac{26{\cdot}53 \times 10^3}{2 \times 102{\cdot}5 \times 1000}$$

$$= 0{\cdot}13 \text{ N/mm}^2$$

The theoretical flexural tensile stress is $g_d - M_b/Z$ but M_b per wall leaf is

$$2{\cdot}09/2 = 1{\cdot}045 \text{ kNm}$$

and Z per wall leaf is

$$1000 \times 102{\cdot}5^2/6 = 1{\cdot}751 \times 10^6 \text{ mm}^3$$

Therefore the theoretical flexural tensile stress is

$$0{\cdot}13 - \frac{1{\cdot}045 \times 10^6}{1{\cdot}751 \times 10^6} = 0{\cdot}13 - 0{\cdot}6$$

$$= -0{\cdot}47 \text{ N/mm}^2$$

To eliminate this theoretical flexural tensile stress a post-tensioning stress of $+0{\cdot}47$ N/mm^2 is required. The stress relationships for this process are shown in Fig. 7.4.

The post-tensioning force required is not calculated until the ultimate limit state condition has been checked.

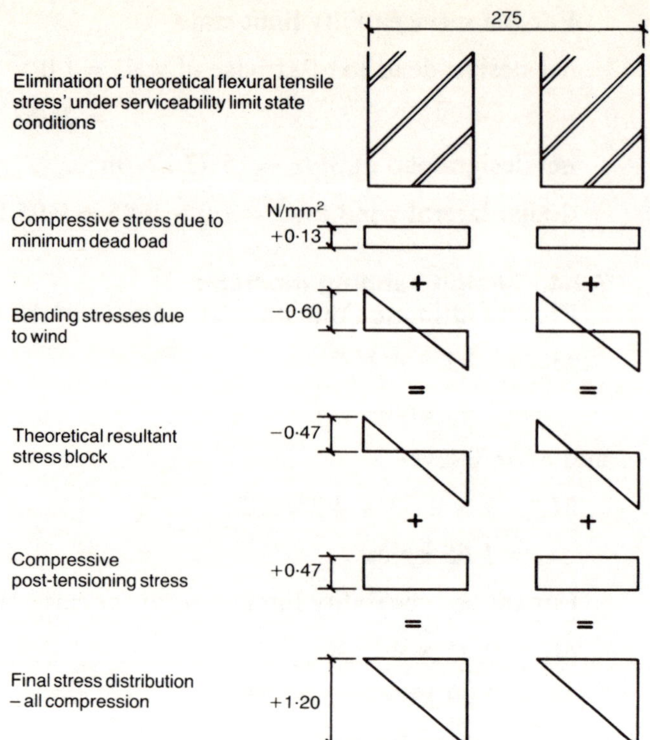

Fig. 7.4. Elimination of theoretical flexural tensile stresses

7.1.6. Ultimate limit state: cracked section

For the ultimate limit state, the Authors consider that the wall may be designed as a cracked section at its base and that the post-tensioning force required for this condition should be based on the moment of resistance of the cracked section.

The ultimate stress condition for a cracked section design is shown in Fig. 6.11.

The minimum width of stress block is stressed to ultimate conditions. Prior to the application of the post-tensioning force, an extremely narrow width of wall will be stressed to ultimate, which relates to the design dead loads from the roof and the self-weight of the masonry. For this example, this will not be adequate to generate the required moment of resistance and therefore the additional load required, in the form of the post-tensioning force, can be calculated. This calculation involves two unknowns: the force and the lever arm. The larger the vertical load or force (dead loads plus post-tensioning force) the greater will be the width of the ultimate stress block but this, in turn, reduces the lever arm.

The ultimate stress on the rectangular stress block P_{ubc} is shown in Fig. 6.11 as $1\cdot2f_k/\gamma_{mm}$, which for the masonry specification given for this example is

$$1\cdot2 \times 7\cdot9/2\cdot3 = 4\cdot12 \text{ N/mm}^2$$

The applied moment at the base for the ultimate limit state condition has been calculated in section 7.14 as 2·93 kNm and the moment of resistance MR_s produced by the post-tensioning force must be equal to or greater than this applied moment.

Let C be the compressive force due to dead loads plus the post-tensioning force. Then $MR_s = C \times$ lever arm for each leaf of masonry.

In some cases the leaf thicknesses will differ and in others the compressive force C will differ in each leaf, or a combination of both circumstances may

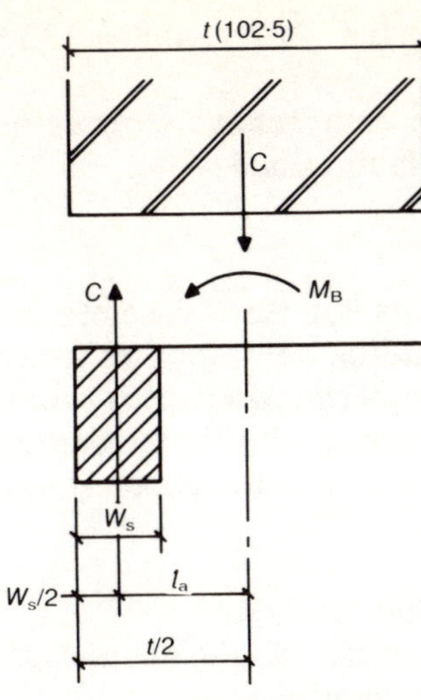

Fig. 7.5. Stress block at base of wall

occur. For this example the dead and superimposed loads as well as the post-tensioning force are considered to be shared equally between the two leaves, which are of the same thickness.

From Fig. 7.5

$$MR_s = Cl_a \times 2$$

Therefore

$$MR_s/2 = Cl_a$$

but

$$l_a = \frac{t}{2} - \frac{W_s}{2}$$

$$w_s = C/P_{ubc}$$

$$= \frac{C}{4 \cdot 12 \times 10^3}$$

and hence

$$l_a = \frac{0 \cdot 1025}{2} - \frac{C}{2 \times 4 \cdot 12 \times 10^3}$$

giving

$$\frac{MR_s}{2} = C\left(\frac{0 \cdot 1025}{2} - \frac{C}{2 \times 4 \cdot 12 \times 10^3}\right)$$

$$\frac{2 \cdot 93}{2} = C(0 \cdot 05125 - 0 \cdot 00012C)$$

Therefore

$$C = 30 \cdot 94 \text{ kN/m run of wall}$$

of which $22 \cdot 64/2 = 11 \cdot 32$ kN/m is provided by the net design dead load (the roof plus the self-weight of the masonry minus the wind uplift) per leaf.

Hence the design post-tensioning force is

$$(30 \cdot 94 - 11 \cdot 32)2 = 39 \cdot 24 \text{ kN/m}$$

which is the total for both leaves of the cavity wall.

The ultimate stress block for this condition is shown in Fig. 7.6.

7.1.7. Design post-tensioning force

For the serviceability limit state, the design post-tensioning force required was $0 \cdot 47 \times 2 \times 102 \cdot 5 \times 1000 \times 10^{-3} = 96 \cdot 4$ kN/m, which exceeds the equivalent value of $39 \cdot 24$ kN/m calculated for the ultimate limit state.

In order to achieve a minimum design post-tensioning force of $96 \cdot 40$ kN/m, a characteristic post-tensioning force after losses of

$$96 \cdot 4/\gamma_f = 96 \cdot 4/0 \cdot 9$$

$$= 107 \cdot 11 \text{ kN/m}$$

$$= \frac{107 \cdot 11 \times 10^3}{2 \times 102 \cdot 5 \times 1000} = 0 \cdot 522 \text{ N/mm}^2$$

must be provided.

The design of the rods to produce the required force and the bearing stresses in the local masonry are considered in sections 7.1.11–7.1.13 and

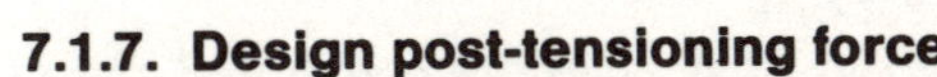

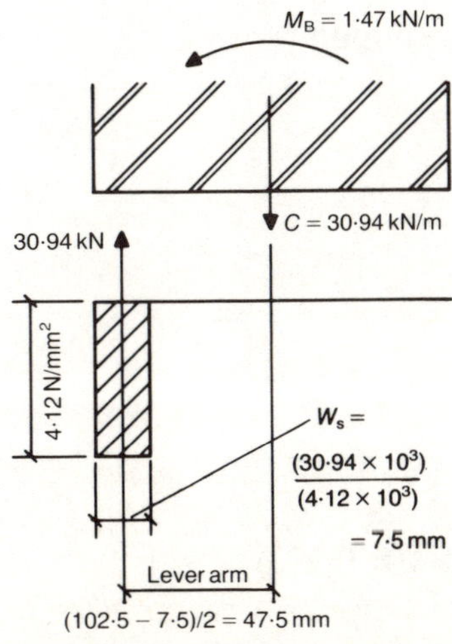

Fig. 7.6. Ultimate stress block at base of wall

should include an allowance of 20% for losses (i.e. $107 \cdot 11/0 \cdot 8 = 133 \cdot 9$ kN/m) where appropriate.

Consideration is now given to the effect of these compressive stresses on the wall section provided, to ensure that stability is maintained.

7.1.8. Compressive stresses after losses

Because of the numerous combinations of factors for partial safety on loading, it cannot be immediately obvious which loading combination (dead plus superimposed, dead plus wind or dead plus superimposed plus wind) will provide the most onerous design condition. Each of the three loading combinations is therefore considered individually, taking into account the capacity reduction factor β where applicable.

7.1.8.1. Case (a): dead plus superimposed (axial load only)

This check is carried out within the central $0 \cdot 4h$ of the height of the wall as buckling/compressive stresses only are being considered.

The design axial stress is

$$f_{uac} = \frac{\gamma_f G_k + \gamma_f Q_k}{\text{area}} + \text{post-tensioning stress}$$

For the case where the characteristic dead load G_k is $29 \cdot 0$ kN/m, the characteristic superimposed load Q_k is $1 \cdot 8$ kN/m and γ_f is the partial safety factor on loads which, for this loading combination is $1 \cdot 4$ and $1 \cdot 6$ for G_k and Q_k, respectively

$$f_{uac} = \frac{(1 \cdot 4 \times 29 \times 10^3) + (1 \cdot 6 \times 1 \cdot 8 \times 10^3)}{2 \times 102 \cdot 5 \times 1000}$$

$$+ \gamma_f \times \text{characteristic post-tensioning stress}$$

$$= \frac{40\,600 + 2880}{205\,000} + (1 \cdot 4 \times 0 \cdot 522)$$

$$= 0 \cdot 944 \text{ N/mm}^2$$

in which the design post-tensioning stress is derived from the product of the characteristic post-tensioning stress and the partial safety factor on loads which is taken to be $1 \cdot 4$—equivalent to a dead loading.

7.1.8.2. Case (b): dead plus wind (lateral loads, propped cantilever action)

The design axial stresses are calculated as follows.

At base level the design axial stress is

$$f_{uac} = \frac{\text{design dead load} - \text{design wind uplift}}{\text{area}}$$

$$+ \text{design post-tensioned stress}$$

$$= \frac{\gamma_f G_k - \gamma_f \times \text{wind uplift}}{\text{area}} + (\gamma_f \times 0 \cdot 522)$$

$$= \frac{(1 \cdot 4 \times 29 \times 10^3) - (1 \cdot 4 \times 2 \cdot 47 \times 10^3)}{2 \times 102 \cdot 5 \times 1000} + (1 \cdot 4 \times 0 \cdot 522)$$

$$= 0 \cdot 181 + 0 \cdot 731$$

$$= 0 \cdot 912 \text{ N/mm}^2$$

At $3h/8$ level

$$f_{\text{uac}} = \frac{(1\cdot4 \times 18\cdot24 \times 10^3) - (1\cdot4 \times 2\cdot47 \times 10^3)}{2 \times 102\cdot5 \times 1000} + (1\cdot4 \times 0\cdot522)$$

$$= 0\cdot108 + 0\cdot731$$

$$= 0\cdot839 \text{ N/mm}^2$$

and the design flexural stress is

$$f_{\text{ubc}} = \pm M_{\text{A}}/Z$$

i.e. both compressive and tensile. However, the combined design stresses are limited so that there is zero tensile stress.

Section modulus is

$$Z = 2 \times 1000 \times 102\cdot5^2/6$$

$$= 3\cdot5 \times 10^6 \text{ mm}^3 \text{ total for both leaves}$$

The design flexural stresses are calculated as follows.
At the base level the design flexural stress is

$$f_{\text{ubc}} = \pm M_{\text{A}}/Z$$

$$= \frac{2\cdot93 \times 10^6}{3\cdot5 \times 10^6}$$

$$= \pm 0\cdot84 \text{ N/mm}^2$$

and at $3h/8$ level

$$f_{\text{ubc}} = \pm \frac{1\cdot65 \times 10^6}{3\cdot5 \times 10^6}$$

$$= \pm 0\cdot47 \text{ N/mm}^2$$

Hence the combined stresses are calculated as follows.

At base level the combined stresses are $f_{\text{uac}} \pm f_{\text{ubc}}$. Slenderness does not affect the stresses at this level, as full restraint may be assumed to be provided by the raft foundation. Slenderness considerations would enter the calculations at a height of $0\cdot3h$ from the base of the wall (see appendix B of BS 5628: Part 1 and chapter 5 of reference 13) but at this level, for a propped cantilever design, the bending moment has reduced significantly and the experienced designer would choose to eliminate the design check. For a simply supported or a free cantilever design condition, this would not necessarily be the case.

The combined stresses are therefore

$$0\cdot912 \pm 0\cdot84 = +1\cdot752 \text{ or } +0\cdot072 \text{ N/mm}^2$$

At $3h/8$ level the combined stresses are $f_{\text{uac}} \pm f_{\text{ubc}}$ where the design axial strength is affected by the slenderness of the wall at this level.

The capacity reduction factor β is calculated in the normal way from the slenderness ratio and the eccentricity of the applied loading. The stiffness of the slab at the upper anchorage of the post-tensioning rods can be assumed to transmit the post-tensioning force into each leaf of the wall with zero eccentricity. Hence the slenderness ratio is

$$\frac{h_{\text{ef}}}{t_{\text{ef}}} = \frac{0\cdot75 \times 4200}{2/3(2 \times 102\cdot5)}$$

$$= 23$$

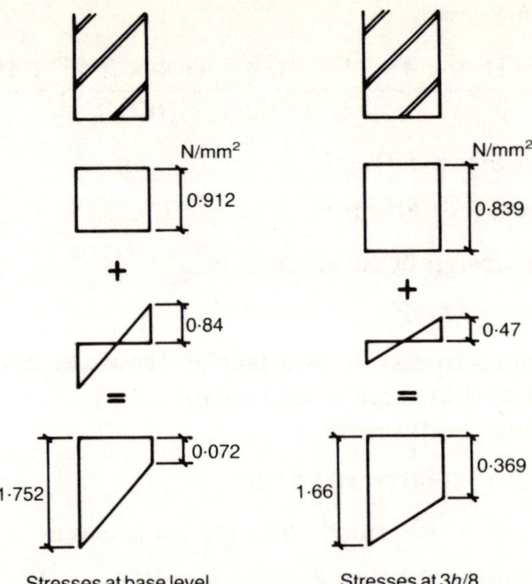

Fig. 7.7. Stress diagrams: dead plus wind condition

Therefore, for zero eccentricity

$$\beta = 0.58$$

and the design strength is $0.58f_k$. The combined stresses are therefore

$$0.839 \pm 0.47 = +1.66 \text{ or } +0.369 \text{ N/mm}^2$$

The stress diagrams for this loading condition are shown in Fig. 7.7.

7.1.8.3. Case (c): dead plus superimposed plus wind

The design axial stresses are calculated as follows.
At the base level the design axial stress is

$$f_{uac} = \frac{(1.2 \times 29 \times 10^3) - (1.2 \times 2.47 \times 10^3) + (1.2 \times 1.8 \times 10^3)}{2 \times 102.5 \times 1000}$$

$$+ (1.2 \times 0.522)$$

$$= 0.166 + 0.626$$

$$= 0.792 \text{ N/mm}^2$$

and at $3h/8$ level

$$f_{uac} = \frac{(1.2 \times 18.24 \times 10^3) - (1.2 \times 2.47 \times 10^3) + (1.2 \times 1.8 \times 10^3)}{2 \times 102.5 \times 1000}$$

$$+ (1.2 \times 0.522)$$

$$= 0.103 + 0.626$$

$$= 0.729 \text{ N/mm}^2$$

The design flexural stresses are calculated as follows.
At base level the design flexural stress is

$$f_{ubc} = \pm 0.84 \times 1.2/1.4$$

$$= \pm 0.72 \text{ N/mm}^2$$

The design flexural stresses are as for case (b) loading but the adjustment factor $1.2/1.4$ takes account of the different γ_f factor for this loading combination.

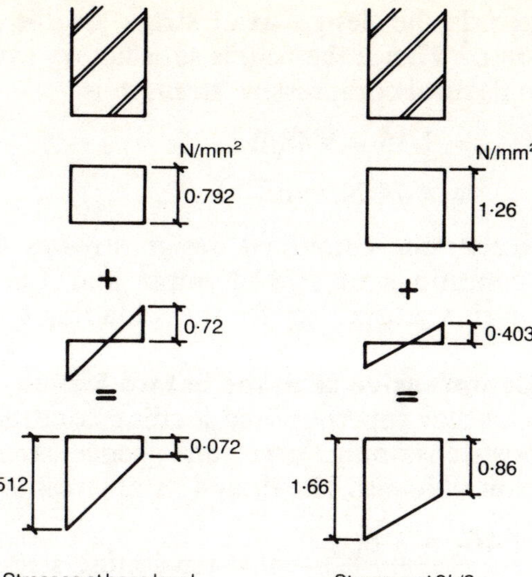

Fig. 7.8. Stress diagrams: dead plus super plus wind condition

At $3h/8$ level

$$f_{ubc} = \pm 0{\cdot}47 \times 1{\cdot}2/1{\cdot}4$$

$$= \pm 0{\cdot}403 \text{ N/mm}^2$$

Hence the combined stresses are given as follows.
At base level

$$f_{uac} \pm f_{ubc} = 0{\cdot}792 \pm 0{\cdot}72$$

$$= +1{\cdot}512 \text{ or } +0{\cdot}072 \text{ N/mm}^2$$

and at $3h/8$ level

$$\frac{f_{uac}}{\beta} \pm f_{ubc} = \frac{0{\cdot}729}{0{\cdot}58} \pm 0{\cdot}403$$

$$= +1{\cdot}66 \text{ or } +0{\cdot}86 \text{ N/mm}^2$$

The stress diagrams for this loading condition are shown in Fig. 7.8.

7.1.9. Design strengths

The characteristic compressive strength is $f_k = 7{\cdot}9$ N/mm², to which the β factor is applied for slenderness.

The characteristic flexural compressive strength is

$$f_f = 1{\cdot}2f_k = 9{\cdot}48 \text{ N/mm}^2$$

The characteristic flexural tensile strength is limited to zero.
The design strengths (after losses) are calculated as follows.
At the base level the design compressive strength is

$$1{\cdot}15f_k/\gamma_{mm} = 1{\cdot}15 \times 7{\cdot}9/2{\cdot}3$$

$$= 3{\cdot}95 \text{ N/mm}^2$$

(1·15 allows for the narrow wall factor.)
At $3h/8$ level the design compressive strength is

$$1{\cdot}15\beta f_k/\gamma_{mm} = 1{\cdot}15 \times 0{\cdot}58 \times 7{\cdot}9/2{\cdot}3$$

$$= 2{\cdot}29 \text{ N/mm}^2$$

which exceeds the design axial stress for the case (*a*) loading condition of
0·944 N/mm². Hence the wall is satisfactory for this loading condition.

Design flexural compressive strength is

$$1·15 f_f / \gamma_{mm} = 1·15 \times 9·48/2·3$$

$$= 4·74 \text{ N/mm}^2$$

which exceeds the combined design stresses for both case (*b*) and case (*c*)
loading conditions of 1·752 N/mm² and 1·66 N/mm², respectively. Hence
the wall is also satisfactory for these loading conditions.

7.1.10. Compressive stresses before losses

The dead plus superimposed loading condition only is considered here.

The post-tensioning force, and hence stress, are increased by 20% in
anticipation of losses. The design axial stress is

$$f_{uac} = \frac{1·4 G_k + 1·6 Q_k}{\text{area}} + \text{post-tensioning stress before losses}$$

$$= \frac{(1·4 \times 29 \times 10^3) + (1·6 \times 1·8 \times 10^3)}{2 \times 102·5 \times 1000} + \frac{1·4 \times 0·522}{0·8}$$

$$= 1·126 \text{ N/mm}^2$$

and the design compressive strength of the wall before losses is

$$1·2 \times 1·15 \beta f_{ki} / \gamma_{mm} = 1·2 \times 1·15 \times 0·58 \times 7·9/2·3$$

$$= 2·75 \text{ N/mm}^2$$

where f_{ki} is the characteristic compressive strength of the masonry at the age
at which the post-tensioning is applied, and the 1·2 factor accounts for the
short-term duration of the loading while the post-tensioning losses are
taking place. For the purposes of this design example, it has been assumed
that the masonry achieves its full 28 day characteristic strength before the
post-tensioning is applied (hence $f_{ki} = f_k$).

The design compressive strength (before losses) exceeds the design axial
stress (before losses) and hence the wall is satisfactory for this loading condition.

7.1.11. Design of post-tensioning rods

The characteristic post-tensioning stress required in the masonry is
0·522 N/mm², which is equivalent to a characteristic post-tensioning force of
107·11 kN/m run of wall. To allow for the possibility of 20% losses in the
post-tensioning force, the equivalent characteristic post-tensioning force
must be initially increased to 107·11/0·8 = 133·9 kN/m run of wall.

To limit relaxation of the steel and hence minimise some of the losses, the
stress in the post-tensioning rods is restricted to $0·7 f_y / \gamma_{ms}$.

Large diameter high yield bars are used for the post-tensioning rods.
Hence

$$0·7 f_y / \gamma_{ms} = 0·7 \times 460/1·15$$

$$= 280·0 \text{ N/mm}^2$$

Therefore the area of steel required is

$$133·9 \times 10^3/280·0 = 478·0 \text{ mm}^2 \text{ per m run of wall.}$$

Use high yield bars of 25 mm diameter placed at 900 mm centres (A_s provided is 546 mm²).

7.1.12. Torque to produce rod tension

There is a considerable variation in the recommendations given by manufacturers of torqueing equipment for the calculation of torque requirements.

The amount of tension induced by a particular torque is dependent on many factors, the two most significant being the pitch and type of the thread and the coefficients of friction between the contact surfaces of nuts, bolts, spreader plates and so on. This latter aspect is largely dependent on the type and quality of the original finish to these components and the degree of lubrication and general protection during the construction of the works prior to the post-tensioning.

The calculation that follows is based on a general engineering formula derived from test research and assumes that lightly oiled, metric threads with self-finish nuts and a hardened washer between the nut and the spreader plate are used

$$\text{torque required} = \text{bar tension} \times \text{bar diameter}/5$$

but

$$\text{bar tension} = 133 \cdot 9 \times 10^3 \times 0 \cdot 9/9 \cdot 81$$

$$= 12\,284 \text{ kgf per rod placed at 900 mm centres}$$

Hence

$$\text{torque required} = 12\,284 \times 0 \cdot 025/5$$

$$= 61 \cdot 42 \text{ kgf m}$$

7.1.13. Design of spreader plate

The following calculation applies in the design of the spreader plate

$$\text{maximum design force/rod} = 133 \cdot 9\gamma_f$$

$$= 133 \cdot 9 \times 1 \cdot 4$$

$$= 187 \cdot 5 \text{ kN}$$

$$\text{design compressive strength} = 1 \cdot 5 \times 1 \cdot 15 \times 7 \cdot 9/2 \cdot 3$$

$$= 5 \cdot 925 \text{ N/mm}^2$$

in which a 1·5 strength increase factor has been applied to take account of the local bearing condition of the spreader plate passing its load through the concrete roof slab into the brickwork.

The area of spread required is

$$187 \cdot 5 \times 10^3/5 \cdot 925 = 31\,646 \text{ mm}^2$$

Fig. 7.9. Capping plate stress dispersal

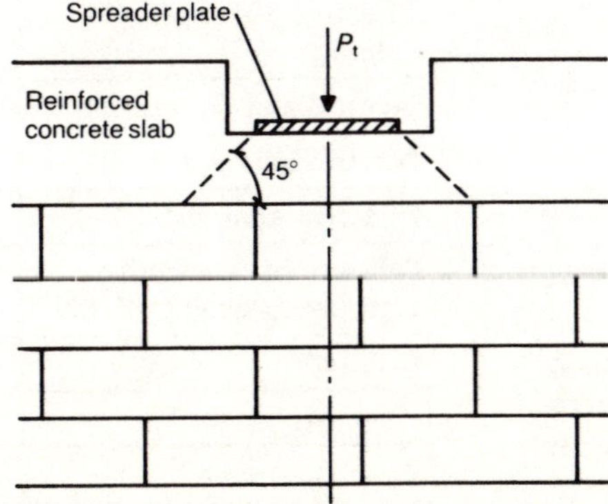

and the length of spread required is

$$31\,646/2 \times 102 \cdot 5 = 155 \text{ mm}$$

Allowing for a 45° dispersion of load through the reinforced concrete slab, as shown in Fig. 7.9, a 100 mm × 100 mm × 12 mm thick spreader plate would be suitable.

A shear check may be carried out using the principles given in the design example in section 7.3 but shear is unlikely to be a critical design factor in the case of a wall subject to wind loading and nominal axial loading.

7.2. Fin wall subject to lateral wind and axial loads

Take, for example, a sports hall of dimensions 30 m × 50 m × 9 m which is to be constructed in loadbearing brickwork using fin wall construction for its main vertical structure. The planning requirements are for the fins to project as little as possible, and hence post-tensioning of the fins is proposed to provide for this requirement. The wall panels between the fins will be of normal 255 cavity construction; there are no internal walls within the building. The roof construction and detailing are assumed to provide an adequate prop to the head of the wall and similarly ensure stability of the structure by transferring this propping force to the gable shear walls. The masonry to be used throughout is bricks with a compressive strength of 30 N/mm^2 set in a designation (ii) mortar. The partial safety factor for materials may be taken as 2·3. The fin profile to be used and its various dimensions and properties are shown in Fig. 7.10.

Typical fin wall sections and their properties are shown in Table 7.1. To suit the spacing of the roof beams the fins are placed at 3 m centres.

7.2.1. Design procedure

The design procedure is similar to that used for the previous example but consideration must be given to the fin profile being asymmetrical. The post-tensioning force was positioned concentrically in the previous example because the section was symmetrical and the wind loading (suction and pressure) in each direction could be assumed to be of a similar magnitude. For the asymmetrical fin wall profile, with similar wind loading, the stresses at the extreme edges of the fin and flange would differ considerably owing to

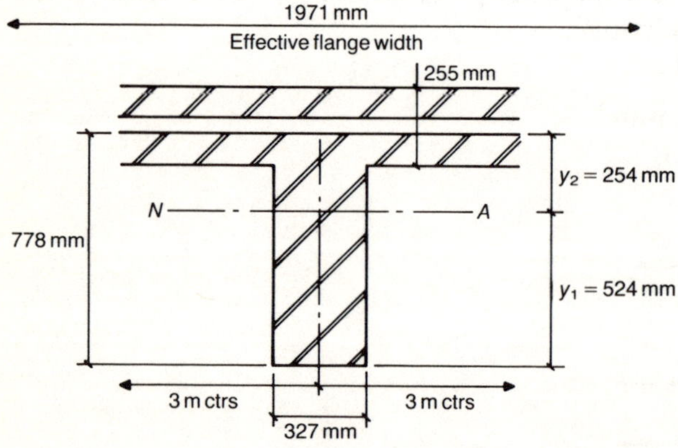

Fig. 7.10. Fin wall profile

* The effective area is the summation of the areas of the projecting fin and the flange over its effective length. The I and Z values are calculated in relation to the effective area.

Table 7.1. *Typical section properties for fin walls**

Fin reference letter	A	B	C
Fin size, mm	665 × 327	665 × 440	788 × 327
Effective width of flange, m	1·971	2·084	1·971
Neutral axis y_1, m	0·455	0·435	0·524
Neutral axis y_2, m	0·210	0·230	0·254
Effective area, m^2	0·386	0·4611	0·4262
Self-weight of effective area per m height W, kN	7·720	9·222	8·458
I_{NA}, m^4	0·01567	0·01939	0·02454
$Z_1 = I_{NA}/y_1$, m^3	0·03441	0·0445	0·04684
$Z_2 = I_{NA}/y_2$, m^3	0·07462	0·0843	0·09663

the two values of section modulus Z_1 and Z_2. Hence, the post-tensioning force is applied eccentrically in this situation, to make the maximum use of it in eliminating the tensile stress at each of the extreme faces of the section.

Figures 7.11 and 7.12 show the theoretical stresses which may occur in the wall, in the absence of the post-tensioning force, for wind suction and wind pressure loadings, respectively. For each direction of wind loading, a theoretical flexural tensile stress may be calculated at the extreme edges of either fin or flange and at either base level or at maximum span moment level, as applicable.

Having established the maximum theoretical flexural tensile stress at each of the extreme edges, an eccentric post-tensioning force may be calculated to eliminate these stresses. The basic theory of this process is indicated in Fig. 7.13 in which f_{t1} and f_{t2} are the maximum theoretical flexural tensile stresses at the extreme edges of the fin and flange, respectively. Then

$$f_{t1} = \frac{P}{A} + \frac{Pe}{Z_1} \tag{7.1}$$

$$f_{t2} = \frac{P}{A} - \frac{Pe}{Z_2} \tag{7.2}$$

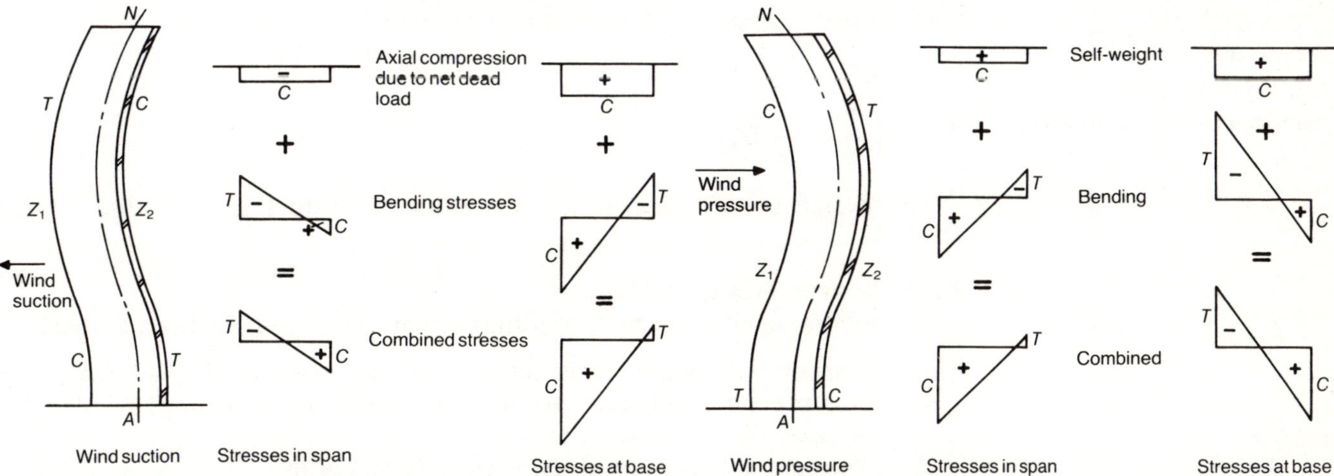

Fig. 7.11. Theoretical fin wall stresses: wind suction   Fig. 7.12. Theoretical fin wall stresses: wind pressure

D	E	F	G	H	J	K	L	M	N	P	Q	R
788 × 440	890 × 327	890 × 440	1003 × 327	1003 × 440	1115 × 327	1115 × 440	1227 × 327	1227 × 440	1339 × 327	1339 × 440	1451 × 327	1451 × 440
2·084	1·971	2·084	1·971	2·084	1·971	2·084	1·971	2·084	1·971	2·084	1·971	2·084
0·500	0·589	0·563	0·654	0·626	0·718	0·687	0·780	0·747	0·841	0·807	0·902	0·866
0·278	0·301	0·327	0·349	0·377	0·397	0·428	0·447	0·480	0·498	0·532	0·549	0·585
0·5152	0·4595	0·5601	0·4965	0·6098	0·5331	0·6591	0·5697	0·7084	0·6064	0·7577	0·6430	0·807
10·216	9·190	11·202	9·930	12·196	10·662	13·182	11·394	14·168	12·128	15·154	12·860	16·140
0·0303	0·0359	0·04426	0·05021	0·06187	0·06746	0·08312	0·088	0·10848	0·11208	0·13826	0·13992	0·17277
0·06059	0·06096	0·07862	0·07677	0·09883	0·09395	0·12099	0·11282	0·14522	0·13327	0·17132	0·15513	0·1995
0·10898	0·11928	0·13536	0·14387	0·16410	0·16992	0·19421	0·19687	0·226	0·22506	0·26039	0·25487	0·2953

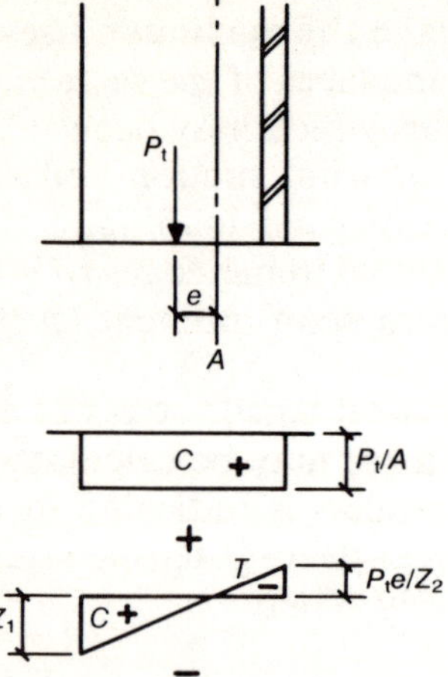

P_t = design post-tensioning force
A = area of effective fin section
e = eccentricity of P_t about neutral axis
Z_1 = minimum section modulus
Z_2 = maximum section modulus

Fig. 7.13. Eccentric post-tensioning of fin wall

A pair of simultaneous equations may be written from equations (7.1) and (7.2) to solve for P and e as follows. Cross-multiply by Z_1 and Z_2 respectively. Hence

$$f_{t1} Z_1 = \frac{PZ_1}{A} + Pe \tag{7.3}$$

$$f_{t2} Z_2 = \frac{PZ_2}{A} - Pe \tag{7.4}$$

Adding equations (7.3) and (7.4) gives

$$f_{t1}Z_1 + f_{t2}Z_2 = \frac{P}{A}(Z_1 + Z_2)$$

from which

$$P = \frac{(f_{t1}Z_1) + (f_{t2}Z_2)}{Z_1 + Z_2} A \tag{7.5}$$

The value of P calculated from equation (7.5) can then be substituted into equation (7.2) to calculate e.

Transposed, equation (7.2) gives

$$e = \left(\frac{P}{A} - f_{t2}\right)\frac{Z_2}{P}$$

$$e = \left(\frac{1}{A} - \frac{f_{t2}}{P}\right)Z_2 \tag{7.6}$$

The design should then follow the same procedure as was used for the example in section 7.1.

7.2.2. Characteristic loadings

The following characteristic loadings are assumed to have been derived

(a) characteristic wind loads: suction on leeward face $W_{k1} = 0.6$ kN/m², pressure on windward face $W_{k2} = 0.8$ kN/m² and gross wind uplift on roof $W_{k3} = 0.3$ kN/m²
(b) characteristic roof dead load: $G_k = 0.50$ kN/m²
(c) characteristic roof superimposed load: $Q_k = 0.75$ kN/m²

7.2.3. Design loadings

The wall panel between the fins is assumed to span horizontally between them and all wind loading on the wall panel is assumed to be transferred to the fins in this way. In reality the third line of support at ground slab level will reduce the amount of wind load which the fins are required to support, but in view of the minimal effect of this third line of support and for simplicity of analysis, this is ignored in this design example. The design of the wall panel is outside the scope of this example but could be shown to be perfectly adequate (see reference 13).

7.2.4. Design wind loads per fin: ultimate limit state

For the loading combination dead plus wind, the partial safety factors for loads are taken as 0.9 and 1.4 for dead and wind loads, respectively.

For the suction case the design wind load is

$3.0 \times 1.4 \times 0.6 = 2.52$ kN/m height of fin

for the pressure case the design wind load is

$3.0 \times 1.4 \times 0.8 = 3.36$ kN/m height of fin

and for the uplift case the design wind load is

$1.4 \times 0.3 = 0.42$ kN/m^2

The net roof uplift is the design dead load minus the design roof uplift, i.e.

$(0.9 \times 0.5) - 0.42 = 0$, say

Hence only the own weight of the masonry can be considered to assist under the wind loading condition, in addition to the post-tensioning force, in providing resistance to the theoretical flexural tensile stresses.

7.2.5. Design wind loads per fin: serviceability limit state

For the serviceability limit state the partial safety factor for loads is taken as 1·0 for dead and wind loads. For the suction case the design wind load is

$3.0 \times 1.0 \times 0.6 = 1.8$ kN/m height of fin

for the pressure case the design wind load is

$3.0 \times 1.0 \times 0.8 = 2.4$ kN/m height of fin

and for the uplift case the design wind load is

$1.0 \times 0.3 = 0.3$ kN/m^2

The net roof uplift is the design dead load minus the design roof uplift, i.e.

$(1.0 \times 0.5) - 0.3 = 0.2$ kN/m^2

For simplicity, such a small load may be ignored.

7.2.6. Design dead loadings: ultimate limit state

For the ultimate limit state, at base level the design dead loading is

$$8.458 \times 9 \times 0.9 = 68.51 \text{ kN}$$
$$= 0.161 \text{ N/mm}^2$$

and at $3h/8$ level it is

$$8.458 \times 9 \times 0.9 \times 3/8 = 25.69 \text{ kN}$$
$$= 0.060 \text{ N/mm}^2$$

7.2.7. Design dead loading: serviceability limit state

For the serviceability limit state, at base level the design dead loading is

$$8.458 \times 9 \times 1.0 = 76.122 \text{ kN}$$
$$= 0.179 \text{ N/mm}^2$$

and at $3h/8$ level it is

$$8.458 \times 9 \times 1.0 \times 3/8 = 28.54 \text{ kN}$$
$$= 0.067 \text{ N/mm}^2$$

7.2.8. Design bending moment

The wall is designed, taking advantage of the prop provided by the roof plate, as a propped cantilever. The design bending moment diagram is as shown in Fig. 7.14.

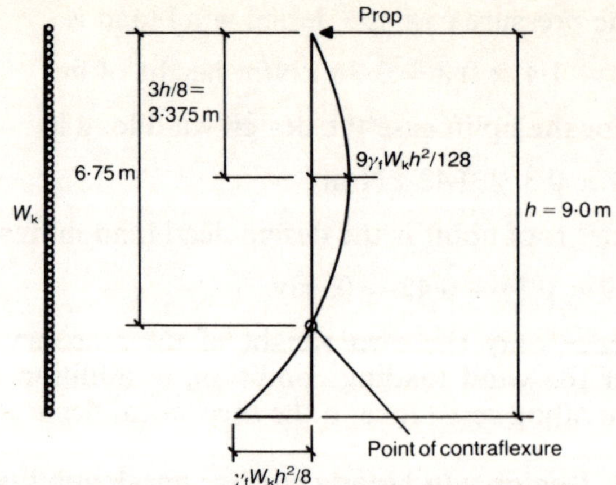

*Fig. 7.14. Propped
cantilever bending moments*

7.2.9. Design bending moments: ultimate limit state

For the suction case, at base level

$$M_{\mathrm{b}} = 2{\cdot}52 \times 9^2/8$$

$$= 25{\cdot}52 \text{ kNm}$$

and at $3h/8$ level

$$M_{\mathrm{w}} = 9 \times 2{\cdot}52 \times 9^2/128$$

$$= 14{\cdot}35 \text{ kNm}$$

For the pressure case, at base level

$$M_{\mathrm{b}} = 3{\cdot}36 \times 9^2/8$$

$$= 34{\cdot}02 \text{ kNm}$$

and at $3h/8$ level

$$M_{\mathrm{w}} = 9 \times 3{\cdot}36 \times 9^2/128$$

$$= 19{\cdot}14 \text{ kNm}$$

7.2.10. Design bending moments: serviceability limit state

For the suction case, at base level

$$M_{\mathrm{b}} = 1{\cdot}8 \times 9^2/8$$

$$= 18{\cdot}23 \text{ kNm}$$

and at $3h/8$ level

$$M_{\mathrm{w}} = 9 \times 1{\cdot}8 \times 9^2/128$$

$$= 10{\cdot}25 \text{ kNm}$$

For the pressure case, at base level

$$M_{\mathrm{b}} = 2{\cdot}4 \times 9^2/8$$

$$= 24{\cdot}30 \text{ kNm}$$

and at $3h/8$ level

$$M_{\mathrm{w}} = 9 \times 2{\cdot}4 \times 9^2/128$$

$$= 13{\cdot}67 \text{ kNm}$$

7.2.11. Serviceability limit state

7.2.11.1. Theoretical flexural tensile stress

For the suction case, at base level, the axial compressive stress due to self-weight is $+0\cdot179$ N/mm^2. The theoretical flexural stresses are

$$\pm \frac{M_b}{Z_2} = \pm \frac{18\cdot23 \times 10^6}{0\cdot097 \times 10^9}$$

$$= \pm 0\cdot188 \text{ N/mm}^2$$

Therefore the theoretical flexural tensile stress is $+0\cdot179 - 0\cdot188$ and

$$f_{t2} = -0\cdot009 \text{ N/mm}^2$$

For the suction case at $3h/8$ level, the axial compressive stress due to self-weight is $+0\cdot067$ N/mm^2 and the theoretical flexural stresses are

$$\pm \frac{M_w}{Z_1} = \pm \frac{10\cdot25 \times 10^6}{0\cdot047 \times 10^9}$$

$$= \pm 0\cdot218 \text{ N/mm}^2$$

Therefore the theoretical flexural tensile stress is $+0\cdot067 - 0\cdot218$ and

$$f_{t1} = -0\cdot151 \text{ N/mm}^2$$

For the pressure case the axial compressive stresses are as for the suction case. At base level, the theoretical flexural stresses are

$$\pm \frac{M_b}{Z_1} = \pm \frac{24\cdot3 \times 10^6}{0\cdot047 \times 10^9}$$

$$= \pm 0\cdot517 \text{ N/mm}^2$$

Therefore the theoretical flexural tensile stress is $+0\cdot179 - 0\cdot517$ and

$$f_{t1} = -0\cdot338 \text{ N/mm}^2$$

For the pressure case at $3h/8$ level, the theoretical flexural stresses are

$$\pm \frac{M_w}{Z_2} = \pm \frac{13\cdot67 \times 10^6}{0\cdot097 \times 10^9}$$

$$= \pm 0\cdot141 \text{ N/mm}^2$$

Therefore the theoretical flexural tensile stress is $+0\cdot067 - 0\cdot141$ and

$$f_{t2} = -0\cdot074 \text{ N/mm}^2$$

By inspection of these theoretical flexural tensile stresses, the pressure cases at base level and at $3h/8$ level give the most onerous values of f_{t1} and f_{t2}, respectively, for the calculation of the required post-tensioning force and its eccentricity.

7.2.11.2. Calculation of P and e, after losses

From equation (7.5)

$$P = \frac{(f_{t1}Z_1 + f_{t2}Z_2)A}{Z_1 + Z_2}$$

From equation (7.6)

$$e = \left(\frac{1}{A} - \frac{f_{t2}}{P}\right)Z_2$$

Therefore

$$P = \frac{(0{\cdot}338 \times 0{\cdot}047 \times 10^9) + (0{\cdot}074 \times 0{\cdot}097 \times 10^9)]0{\cdot}426 \times 10^6}{(0{\cdot}047 + 0{\cdot}097)10^9}$$

$$= 68{\cdot}25 \text{ kN}$$

where P relates to a γ_f of $1{\cdot}0$ for the serviceability limit state, and

$$e = \frac{1}{0{\cdot}426 \times 10^9} - \frac{0{\cdot}074}{68{\cdot}25 \times 10^3}\,0{\cdot}097 \times 10^9$$

$$= 126 \text{ mm}$$

7.2.12. Ultimate limit state: cracked sections

As for the example in section 7.2.11.2, for the ultimate limit state the wall may be designed as a cracked section at its base. The post-tensioning force required for this condition should be based on the moment of resistance of the cracked section. The ultimate stress condition for a cracked section is shown in Fig. 7.15.

For the masonry specification given

$$P_{\text{ubc}} = 1{\cdot}2 \times 8{\cdot}5/2{\cdot}3$$

$$= 4{\cdot}43 \text{ N/mm}^2$$

The design moment at base level has been previously calculated as $34{\cdot}02$ kNm and the moment of resistance produced by the combination of the dead loads and the post-tensioning force applied at their respective lever arms must be equal to or greater than the applied design moment.

$$MR_s = (P \times \text{lever arm}) + (\text{design dead loads} \times \text{lever arm})$$

The post-tensioning force required for this case cannot be directly calculated, as was the case for the previous example, because there is a third unknown resulting from the post-tensioning force being eccentric about the centroid of the section. Hence the characteristic post-tensioning force calculated for the serviceability limit state will be used initially to check if the MR_s produced exceeds the design moment

Fig. 7.15 (left). Cracked section: ultimate resistance diagram

Fig. 7.16 (right). Centroid of loads from flange face

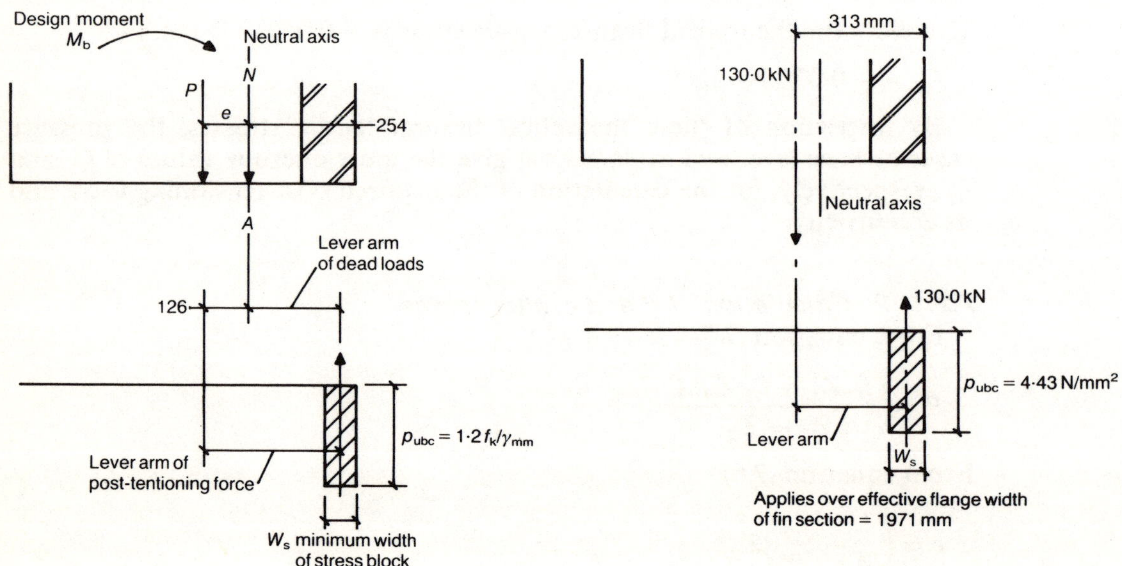

$$\text{design dead load} + \text{design post-tensioning force} = 68\cdot51 + (68\cdot25 \times \gamma_f)$$
$$= 68\cdot51 + (68\cdot25 \times 0\cdot9)$$
$$= 130\cdot0 \text{ kN}$$

The centroid of these loads is

$$\frac{(68\cdot51 \times 0\cdot254) + (68\cdot25 \times 0\cdot9 \times 0\cdot38)}{130\cdot0} = 0\cdot313 \text{ m}$$

from the flange face (see Fig. 7.16)

$$MR_s = 130\cdot0 \times 0\cdot313 - W_s/2$$

$$W_s = \frac{130\cdot0 \times 10^3}{4\cdot43 \times 1971}$$

$$= 15 \text{ mm}$$

$$MR_s = 130\cdot0(0\cdot313 - 0\cdot015/2)$$

$$MR_s = 39\cdot72 \text{ kNm}$$

which exceeds the design bending moment calculated earlier as 34·02 kNm and is therefore acceptable for the ultimate limit state.

7.2.13. Spread of post-tensioning force

The area A used in the equations for the calculations of P and e in section 7.2.11.2 is the effective area of the effective fin section which is based on a flange width of 1971 mm. The spacing of the fins for the example is 3000 mm, leaving a central length of wall 1029 mm long, which was not taken into account in calculating P and e. The effective fin section is shown in Fig. 7.17.

It may be argued that the effect of the post-tensioning force will spread into this central, 1029 mm long area, and therefore account of it should be included in the calculation of P and e. However, whatever effect P may have on this central area would be compensated for by a larger effective section giving higher section moduli Z_1 and Z_2. In the absence of any research work to investigate this phenomenon, it is considered that a reasonable and safe solution would be provided by considering the post-tensioning force to be effective over the area of the effective section only.

7.2.14. Characteristic post-tensioning force P_k

In order to achieve a minimum design post-tensioning force of 68·25 kN, a characteristic post-tensioning force of $68\cdot25/\gamma_f$ should be provided in the rods, where γ_f is taken as 0·9 as discussed in section 7.1.7.

Hence, the characteristic post-tensioning force P_k is

$$68\cdot25/0\cdot9 = 75\cdot83 \text{ kN}$$

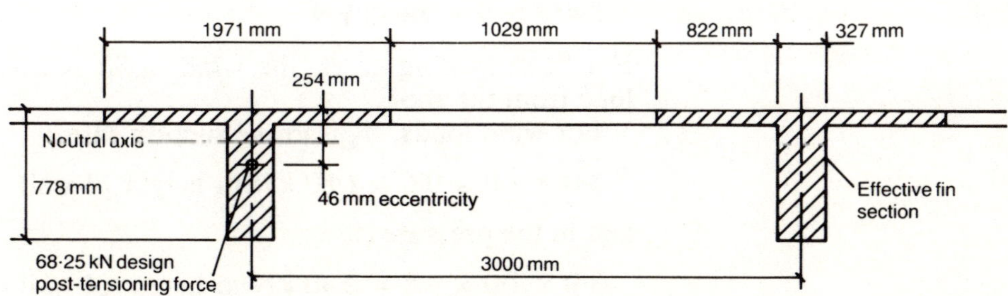

Fig. 7.17. Spread of post-tensioning force into flange

This force is now used to check the design compressive stresses in the wall and to establish the size of the post-tensioning rods.

7.2.15. Capacity reduction factors β

It is assumed that at the base of the wall a raft foundation has been provided which is able to provide full buckling restraint to both the fin and the flange depending on the particular direction of wind loading. A more robust raft foundation may be necessary for the post-tensioning fin wall than for the same section of plain fin wall.

Buckling will be a significant consideration at the $3h/8$ level; this is discussed fully in reference 13.

For the suction case, the maximum combined compressive stress is in the flange of the fin section.

The slenderness ratio SR is

$$\frac{2 \times \text{outstanding length of effective flange}}{\text{effective cavity wall thickness}} = \frac{2 \times 822}{2/3(102{\cdot}5 + 102{\cdot}5)}$$

$$= 12$$

The eccentricity of the compressive stress in the flange of the fin wall may be taken to be $0\text{–}0{\cdot}05t$. Hence $\beta = 0{\cdot}93$.

For the pressure case, the maximum combined compressive stress is at the end of the fin section. The slenderness ratio SR is

$$\frac{\text{distance between points of contraflexure}}{\text{actual thickness of fin}} = \frac{6750}{327}$$

$$= 20{\cdot}64$$

The eccentricity of this compressive stress may again be taken as between 0 and $0{\cdot}05t$. Hence $\beta = 0{\cdot}674$.

7.2.16. Check combined compressive stresses

The critical loading condition will be either loading case (*a*)—dead load plus wind (where the partial factors of safety for loads are $1{\cdot}4G_k$ and $1{\cdot}4W_k$)—or loading case (*b*)—dead load plus superimposed load plus wind (where the partial safety factors for loads are $1{\cdot}2G_k$, $1{\cdot}2Q_k$ and $1{\cdot}2W_k$).

7.2.16.1. Characteristic load per fin

For vertical loads, Q_k from the roof is

$$0{\cdot}75 \times 3{\cdot}0 \times 30/2 = 33{\cdot}75 \text{ kN}$$

and G_k at $3h/8$ level is

$$8{\cdot}458 \times 3{\cdot}375 = 28{\cdot}54 \text{ kN}$$

and G_k at base level is

$$8{\cdot}458 \times 9 = 76{\cdot}12 \text{ kN}$$

For both loading cases, the wind uplift is assumed to cancel out the dead load from the roof.

For wind loads, W_k is in the suction case

$$3{\cdot}0 \times 1{\cdot}0 \times 0{\cdot}6 = 1{\cdot}80 \text{ kN/m height of wall}$$

and in the pressure case

$$3{\cdot}0 \times 1{\cdot}0 \times 0{\cdot}8 = 2{\cdot}40 \text{ kN/m height of wall}$$

7.2.16.2. Design bending moments

For case (a)—dead load plus wind $1\cdot4W_{k}$—suction at $3h/8$ level

$$M_{w} = 9\gamma_{f}\,W_{k}\,h^{2}/128$$
$$= 9 \times 1\cdot4 \times 1\cdot80 \times 9^{2}/128$$
$$= 14\cdot35 \text{ kNm}$$

and suction at base level

$$M_{b} = \gamma_{f}\,W_{k}\,h^{2}/8$$
$$= 1\cdot4 \times 1\cdot80 \times 9^{2}/8$$
$$= 25\cdot52 \text{ kNm}$$

For case (a)—pressure at $3h/8$ level

$$M_{w} = 9\gamma_{f}\,W_{k}\,h^{2}/128$$
$$= 9 \times 1\cdot4 \times 2\cdot4 \times 9^{2}/128$$
$$= 19\cdot14 \text{ kNm}$$

and pressure at base level

$$M_{b} = \gamma_{f}\,W_{k}\,h^{2}/8$$
$$= 1\cdot4 \times 2\cdot4 \times 9^{2}/8$$
$$= 34\cdot02 \text{ kNm}$$

For case (b)—dead load plus superimposed load plus wind $1\cdot2W_{k}$—suction at $3h/8$ level

$$M_{w} = 14\cdot35 \times 1\cdot2/1\cdot4$$
$$= 12\cdot30 \text{ kNm}$$

and suction at base level

$$M_{b} = 25\cdot52 \times 1\cdot2/1\cdot4$$
$$= 21\cdot87 \text{ kNm}$$

For case (b)—pressure at $3h/8$ level

$$M_{w} = 19\cdot14 \times 1\cdot2/1\cdot4$$
$$= 16\cdot41 \text{ kNm}$$

and pressure at base level

$$M_{b} = 34\cdot02 \times 1\cdot2/1\cdot4$$
$$= 29\cdot16 \text{ kNm}$$

7.2.16.3. Combined compressive stresses for suction case

For case (a)—dead load plus wind loading—at base level the design axial stress due to self-weight G_{k} is

$$+\gamma_{f}\,G_{k}/A = +1\cdot4 \times 76\cdot12 \times 10^{3}/0\cdot43 \times 10^{6}$$
$$= +0\cdot248 \text{ N/mm}^{2}$$

The design axial stress due to the post-tensioning force P_{k} is

$$+\gamma_{f}\,P_{k}/A = +1\cdot4 \times 75\cdot83 \times 10^{3}/0\cdot43 \times 10^{6}$$
$$= +0\cdot247 \text{ N/mm}^{2}$$

The design flexural stress due to the post-tensioning force P_k is

$$+\gamma_f P_k e/Z_1 = 1\cdot4 \times 75\cdot83 \times 10^3 \times 126/0\cdot047 \times 10^9$$
$$= +0\cdot285 \text{ N/mm}^2$$

or

$$-\gamma_f P_k e/Z_2 = -1\cdot4 \times 75\cdot83 \times 10^3 \times 126/0\cdot097 \times 10^9$$
$$= -0\cdot138 \text{ N/mm}^2$$

The design flexural stress due to the applied moment is

$$+M_b/Z_1 = +25\cdot52 \times 10^6/0\cdot047 \times 10^9$$
$$= +0\cdot543 \text{ N/mm}^2$$

or

$$-M_b/Z_2 = -25\cdot52 \times 10^6/0\cdot097 \times 10^9$$
$$= -0\cdot263 \text{ N/mm}^2$$

Hence the maximum combined compressive stress is

$$(0\cdot248 + 0\cdot247 + 0\cdot285 + 0\cdot543) = +1\cdot323 \text{ N/mm}^2$$

and the minimum combined compressive stress

$$(0\cdot248 + 0\cdot247 - 0\cdot138 - 0\cdot263) = +0\cdot094 \text{ N/mm}^2$$

For case (*b*)—dead load plus superimposed load plus wind—at base level the design axial stress due to G_k and Q_k is

$$+\gamma_f G_k + \alpha_f Q_k/A = +(1\cdot2 \times 76\cdot12) + (1\cdot2 \times 33\cdot75)/0\cdot43 \times 10^3$$
$$= +0\cdot307 \text{ N/mm}^2$$

The design axial stress due to the post-tensioning force P_k is

$$+\gamma_f P_k/A = +1\cdot2 \times 75\cdot83 \times 10^3/0\cdot43 \times 10^6$$
$$= +0\cdot212 \text{ N/mm}^2$$

The design flexural stress due to the post-tensioning force P_k is

$$+\gamma_f P_k e/Z_1 = +1\cdot2 \times 75\cdot3 \times 10^3 \times 126/0\cdot047 \times 10^9$$
$$= +0\cdot244 \text{ N/mm}^2$$

or

$$-\gamma_f P_k e/Z_2 = -1\cdot2 \times 75\cdot83 \times 10^3 \times 126/0\cdot097 \times 10^9$$
$$= -0\cdot118 \text{ N/mm}^2$$

The design flexural stress due to the applied moment is

$$+M_b/Z_1 = +21\cdot87 \times 10^6/0\cdot047 \times 10^9$$
$$= +0\cdot465 \text{ N/mm}^2$$

or

$$-M_b/Z_2 = -21\cdot87 \times 10^6/0\cdot097 \times 10^9$$
$$= -0\cdot225 \text{ N/mm}^2$$

Hence the maximum combined compressive stress is

$$(0\cdot307 + 0\cdot212 + 0\cdot244 + 0\cdot465) = +1\cdot228 \text{ N/mm}^2$$

and the minimum combined compressive stress is

$$(0.307 + 0.212 - 0.118 - 0.225) = +0.176 \text{ N/mm}^2$$

For case (a)—dead load plus wind loading—at $3h/8$ level the design axial stress due to self-weight G_k is

$$+\gamma_f G_k/A = +1.4 \times 28.54 \times 10^3/0.43 \times 10^6$$
$$= +0.093 \text{ N/mm}^2$$

The design axial stress due to the post-tensioning force P_k is

$$+\gamma_f P_k/A = +1.4 \times 75.83 \times 10^3/0.43 \times 10^6$$
$$= +0.247 \text{ N/mm}^2$$

The design flexural stress due to the post-tensioning force P_k is

$$+\gamma_f P_k e/Z_1 = +1.4 \times 75.83 \times 10^3 \times 126/0.047 \times 10^9$$
$$= +0.285 \text{ N/mm}^2$$

or

$$-\gamma_f P_k e/Z_2 = -1.4 \times 75.83 \times 10^3 \times 126/0.097 \times 10^9$$
$$= -0.138 \text{ N/mm}^2$$

The design flexural stress due to the applied moment is

$$-M_w/Z_1 = -14.35 \times 10^6/0.047 \times 10^9$$
$$= -0.427 \text{ N/mm}^2$$

or

$$+M_w/Z_2 = +14.35 \times 10^6/0.097 \times 10^9$$
$$= +0.148 \text{ N/mm}^2$$

Hence the minimum combined compressive stress is

$$(0.093 + 0.247 + 0.285 - 0.427) = +0.198 \text{ N/mm}^2$$

and the maximum combined compressive stress is

$$(0.093 + 0.247 - 0.138 + 0.148) = +0.350 \text{ N/mm}^2$$

For case (b)—dead load plus superimposed load plus wind loading—at $3h/8$ level the design axial stress due to G_k and Q_k is

$$+(\gamma_f G_k + \gamma_f Q_k)/A = +[(1.2 \times 28.54) + (1.2 \times 33.75)]/0.43 \times 10^3$$
$$= 0.174 \text{ N/mm}^2$$

The design axial stress due to the post-tensioning force P_k is

$$+\gamma_f P_k/A = +1.2 \times 75.83 \times 10^3/0.43 \times 10^6$$
$$= +0.212 \text{ N/mm}^2$$

The design flexural stress due to the post-tensioning force P_k is

$$+\gamma_f P_k e/Z_1 = +1.2 \times 75.83 \times 10^3 \times 126/0.047 \times 10^9$$
$$= +0.244 \text{ N/mm}^2$$

or

$$-\gamma_f P_k e/Z_2 = -1.2 \times 75.83 \times 10^3 \times 126/0.097 \times 10^9$$
$$= -0.118 \text{ N/mm}^2$$

The design flexural stress due to the applied moment is

$$-M_\mathrm{w}/Z_1 = -12\cdot30 \times 10^6/0\cdot047 \times 10^9$$

$$= -0\cdot262 \ \mathrm{N/mm^2}$$

or

$$+M_\mathrm{w}/Z_2 = +12\cdot30 \times 10^6/0\cdot097 \times 10^9$$

$$= +0\cdot127 \ \mathrm{N/mm^2}$$

Hence the minimum combined compressive stress is

$$(0\cdot174 + 0\cdot212 + 0\cdot244 - 0\cdot262) = +0\cdot368 \ \mathrm{N/mm^2}$$

and the maximum combined compressive stress is

$$(0\cdot174 + 0\cdot212 - 0\cdot118 + 0\cdot127) = +0\cdot395 \ \mathrm{N/mm^2}$$

7.2.16.4. Combined compressive stresses for pressure case

For case (*a*)—dead load plus wind loading—at base level the design axial stress due to self-weight G_k is $+0\cdot248 \ \mathrm{N/mm^2}$, the design axial stress due to the post-tensioning force P_k is $+0\cdot247 \ \mathrm{N/mm^2}$ and the design flexural stress due to the post-tensioning force P_k is $+0\cdot285 \ \mathrm{N/mm^2}$ or $-0\cdot138 \ \mathrm{N/mm^2}$. These values are the same as those for the suction case in section 7.2.16.3.

The design flexural stress due to the applied moment is

$$-M_\mathrm{b}/Z_1 = -34\cdot02 \times 10^6/0\cdot047 \times 10^9$$

$$= -0\cdot724 \ \mathrm{N/mm^2}$$

or

$$+M_\mathrm{b}/Z_2 = +34\cdot02 \times 10^6/0\cdot097 \times 10^9$$

$$= +0\cdot351 \ \mathrm{N/mm^2}$$

Hence the minimum combined compressive stress is

$$(0\cdot248 + 0\cdot247 + 0\cdot285 - 0\cdot724) = +0\cdot056 \ \mathrm{N/mm^2}$$

and the maximum combined compressive stress

$$(0\cdot248 + 0\cdot247 - 0\cdot138 + 0\cdot351) = +0\cdot708 \ \mathrm{N/mm^2}$$

For case (*b*)—dead load plus superimposed load plus wind loading—at base level the design axial stress due to G_k and Q_k is $+0\cdot307 \ \mathrm{N/mm^2}$, the design axial stress due to the post-tensioning force P_k is $+0\cdot212 \ \mathrm{N/mm^2}$ and the design flexural stress due to the post-tensioning force P_k is $+0\cdot244 \ \mathrm{N/mm^2}$ or $-0\cdot118 \ \mathrm{N/mm^2}$. These values are the same as those for the suction case in section 7.2.16.3.

The design flexural stress due to the applied moment is

$$-M_\mathrm{b}/Z_1 = -29\cdot16 \times 10^6/0\cdot047 \times 10^9$$

$$= -0\cdot620 \ \mathrm{N/mm^2}$$

or

$$+M_\mathrm{b}/Z_2 = +29\cdot16 \times 10^6/0\cdot097 \times 10^9$$

$$= +0\cdot301 \ \mathrm{N/mm^2}$$

Hence the minimum combined compressive stress is

$$(0\cdot307 + 0\cdot212 + 0\cdot244 - 0\cdot620) = +0\cdot143 \ \mathrm{N/mm^2}$$

and the maximum combined compressive stress is

$$(0\cdot307 + 0\cdot212 - 0\cdot118 + 0\cdot301) = +0\cdot702 \ \mathrm{N/mm^2}$$

For case (*a*)—dead load plus wind loading—at $3h/8$ level the design axial stress due to self-weight G_k is $+0.093$ N/mm², the design axial stress due to the post-tensioning force P_k is $+0.247$ N/mm² and the design flexural stresses due to the post-tensioning force P_k is $+0.285$ N/mm² or -0.138 N/mm². These values are the same as those for the suction case in section 7.2.16.3.

The design flexural stress due to the applied moment is

$$+M_w/Z_1 = +19.14 \times 10^6/0.047 \times 10^9$$

$$= +0.407 \text{ N/mm}^2$$

or

$$-M_w/Z_2 = -19.14 \times 10^6/0.097 \times 10^9$$

$$= -0.197 \text{ N/mm}^2$$

Hence the maximum combined compressive stress is

$$(0.093 + 0.247 + 0.285 + 0.407) = +1.032 \text{ N/mm}^2$$

and the minimum combined compressive stress is

$$(0.093 + 0.247 - 0.138 - 0.197) = +0.005 \text{ N/mm}^2$$

For case (*b*)—dead load plus superimposed load plus wind loading—at $3h/8$ level the design axial stress due to G_k and Q_k is $+0.174$ N/mm², the design axial stress due to the post-tensioning force P_k is $+0.212$ N/mm² and the design flexural stress due to the post-tensioning force P_k is $+0.244$ N/mm² or -0.118 N/mm². These values are the same as those for the suction case in section 7.2.16.3.

The design flexural stress due to the applied moment is

$$+M_w/Z_1 = +16.41 \times 10^6/0.047 \times 10^9$$

$$= +0.349 \text{ N/mm}^2$$

or

$$-M_w/Z_2 = -16.41 \times 10^6/0.097 \times 10^9$$

$$= -0.169 \text{ N/mm}^2$$

Fig. 7.18. Critical design cases: wind suction and pressure

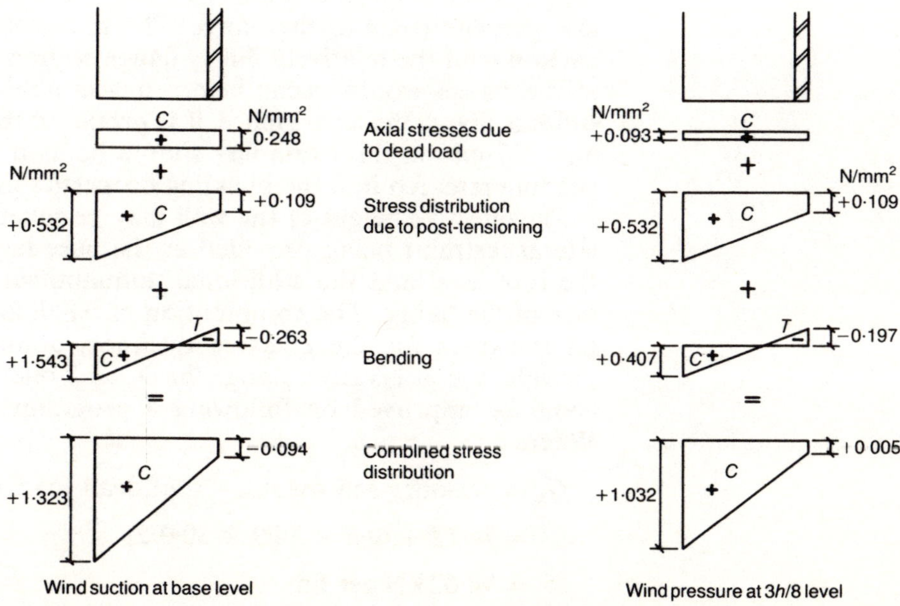

Hence the maximum combined compressive stress is

$$(0 \cdot 174 + 0 \cdot 212 + 0 \cdot 244 + 0 \cdot 349) = +0 \cdot 979 \text{ N/mm}^2$$

and the minimum combined compressive stress is

$$(0 \cdot 174 + 0 \cdot 212 - 0 \cdot 118 - 0 \cdot 169) = +0 \cdot 099 \text{ N/mm}^2$$

The two most critical design cases—for wind suction and wind pressure—given in Fig. 7.18 show the method of stress calculations. A number of the design cases presented did not require calculation to arrive at the two critical conditions stated. With experience, designers will acquire judgment and quickly decide on the critical condition.

7.2.16.5. Design flexural compressive strengths of wall, after losses

The calculations for base level and $3h/8$ level are as follows.

At base level, where $\beta = 1 \cdot 0$ by inspection, the design flexural compressive strength is

$$1 \cdot 2 f_k / \gamma_{mm} = 1 \cdot 2 \times 8 \cdot 4 / 2 \cdot 3 = 4 \cdot 38 \text{ N/mm}^2$$

and at $3h/8$ level, where $\beta = 0 \cdot 674$ by calculation, the design flexural compressive strength is

$$1 \cdot 2 \beta f_k / \gamma_{mm} = 1 \cdot 2 \times 0 \cdot 674 \times 8 \cdot 4 / 2 \cdot 3 = 2 \cdot 95 \text{ N/mm}^2$$

Comparison of the design flexural compressive strengths with the combined compressive stresses calculated in section 7.2.16 shows that the wall is acceptable for all loading cases already considered.

7.2.16.6. Check overall stability of wall

The wall is checked for overall stability, both before and after losses in the post-tensioning force, for the loading combination of dead load plus superimposed load plus post-tensioning force. Consideration has already been given to the local stability of the fin and flange in the design of flexural compressive strengths. In this design check consideration is given to buckling of the section as a whole.

As with the design of plain fin walls, it is questionable whether the narrow flange of the T profile is able to offer full resistance to buckling about the axis perpendicular to the flange. The robustness of the fin section in comparison with the relatively flimsy flange section suggests that local instability of the flange would occur before it was able to develop its full apparent stiffness about the axis. Hence, it is proposed that the actual thickness of the fin—327 mm in this example—should be used to give a conservative design pending research into the buckling properties of such composite sections.

The effective height of the wall may be taken as $0 \cdot 85h$ on the basis of full lateral restraint being provided at the base level, partial lateral restraint at the roof level and the additional unquantified restraint from the contribution of the flange. The combination of $0 \cdot 85h$ for the effective height and the fin thickness for the effective thickness about that axis is considered to provide a conservative design basis and one which it would be expected could be improved on following a programme of suitable research work. Where

$$G_k = \text{masonry self-weight} + \text{roof dead load}$$

$$= 76 \cdot 12 + (0 \cdot 5 \times 3 \cdot 09 \times 30 \cdot 0/2)$$

$$= 98 \cdot 62 \text{ kN per fin}$$

design compressive strength before losses is

$$\frac{(1\cdot4G_k) + (1\cdot6Q_k) + (1\cdot4P_k/0\cdot8)}{A}$$

$$= \frac{(1\cdot4 \times 98\cdot62) + (1\cdot6 \times 33\cdot75) + (1\cdot4 \times 94\cdot79)}{0\cdot42 \times 10^3}$$

$$= 0\cdot755 \text{ N/mm}^2$$

where P_k is increased by 20% in anticipation of the amount of loss in the force. The design strength before losses exceeds the design stress and is therefore acceptable for this loading condition.

Although the stiffness of the fin about the neutral axis is considerably greater, the post-tensioning force is eccentric.

The net eccentricity of G_k, Q_k and P_k may be calculated by taking moments of these loads and forces about the flange face

$$\frac{(98\cdot62 \times 254) + (33\cdot75 \times 254) + (75\cdot83 \times 380)}{98\cdot62 + 33\cdot75 + 75\cdot83}$$

$$= \frac{62\cdot43 \times 10^3}{208\cdot2}$$

$$= 300 \text{ mm from the flange face}$$

This assumes that Q_k is applied at the centroid. Hence the net eccentricity about the neutral axis is

$$300 - 254 = 46 \text{ mm}$$

which, in terms of the fin length, is $e = 0\cdot06 \times 778$.

This extremely small net eccentricity, in relation to the stiffness of the fin about its major axis, indicates that the section will not be critical about this axis from the point of view of overall stability.

7.2.17. Design of post-tensioning rods

The design stress in the high yield steel rods is limited to

$$0\cdot7f_y/\gamma_{ms} = 0\cdot7 \times 460/1\cdot15$$

$$= 370\cdot0 \text{ N/mm}^2$$

as discussed in section 7.1.11. The post-tensioning force before losses is

$$75\cdot83/0\cdot8 = 94\cdot79 \text{ kN}$$

and therefore the steel area required is

$$94\cdot79 \times 10^3/370 = 256 \text{ m}^2$$

This will be provided by one high-yield rod of 25 mm diameter positioned within the fin section to give an eccentricity about the neutral axis of 46 mm (see Fig. 7.17).

Torque required in the rods is

$$\text{bar tension} \times \text{bar diameter}/5 = 94\cdot79 \times 10^3 \times 0\cdot025/9\cdot81 \times 5$$

$$= 48\cdot3 \text{ kgf m}$$

7.2.18. Spreader plate design

The following calculations apply in spreader plate design.

The maximum design force in the rods is

$$94{\cdot}79\gamma_f = 94{\cdot}79 \times 1{\cdot}4 = 132{\cdot}71 \text{ kN}$$

and the design compressive strength is

$$1{\cdot}5 \times 8{\cdot}4/2{\cdot}3 = 5{\cdot}48 \text{ N/mm}^2$$

in which a 1·5 strength increase factor has been applied to take account of the local bearing condition of the spreader plate.

The area of the spread required is

$$132{\cdot}71 \times 10^3/5{\cdot}48 = 24\,217 \text{ mm}^2$$

which requires a spreader plate 102·5 mm wide by 240 mm long. To prevent excessive local bending of the spreader plate, 12 mm would be an appropriate thickness.

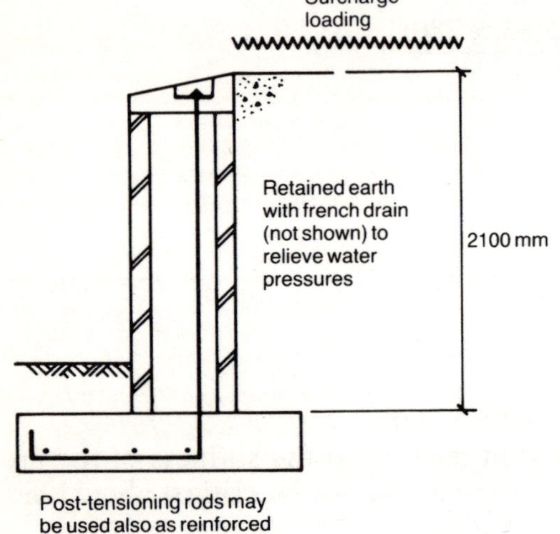

Fig. 7.19 (left). *Post-tensioned diaphragm retaining wall*

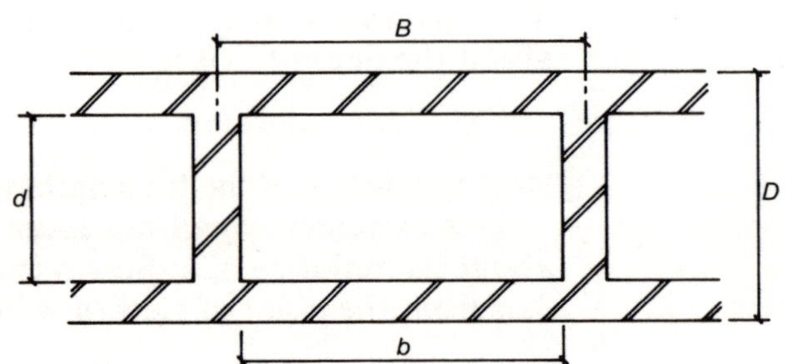

Fig. 7.20 (below). *Diaphragm walls: typical section properties (see Table 7.2)*

Table 7.2. *Typical section properties for diaphragm walls (to be read in conjunction with Fig. 7.20)*

Section	Dimensions, m				Section properties/diaphragm			Section properties/m		
	D	d	B	b	$I \times 10^{-3}$ m^4	$Z \times 10^{-3}$ m^3	A m^2	$I \times 10^{-3}$ m^4	$Z \times 10^{-3}$ m^3	A m^2
1	0·44	0·235	1·4625	1·36	8·91	40·49	0·324	6·09	27·69	0·222
2	0·44	0·235	1·2375	1·135	7·55	34·32	0·278	6·10	27·73	0·225
3	0·44	0·235	1·0125	0·91	6·21	28·83	0·232	6·13	27·88	0·229
4	0·5575	0·352	1·4625	1·36	16·18	58·04	0·337	11·06	39·69	0·230
5	0·5575	0·352	1·2375	1·135	13·74	49·29	0·290	11·10	39·83	0·234
6	0·5575	0·352	1·0125	0·91	11·31	40·57	0·244	11·17	40·07	0·241
7	0·665	0·46	1·4625	1·36	24·81	74·62	0·347	16·96	51·02	0·237
8	0·665	0·46	1·2375	1·135	21·12	63·52	0·301	17·07	51·33	0·243
9	0·665	0·46	1·0125	0·91	17·43	52·43	0·254	17·21	51·77	0·251
10	0·7825	0·5775	1·4625	1·36	36·56	93·45	0·359	24·99	63·90	0·245
11	0·7825	0·5775	1·2375	1·135	31·18	79·69	0·313	25·19	64·40	0·253
12	0·7825	0·5775	1·0125	0·91	25·82	66·01	0·267	25·50	65·20	0·264
13	0·89	0·685	1·4625	1·36	49·46	111·14	0·37	33·82	76·00	0·253
14	0·89	0·685	1·2375	1·135	42·4	95·3	0·324	34·26	77·01	0·262
15	0·89	0·685	1·0125	0·91	34·86	78·34	0·278	34·43	77·37	0·274

A shear check should be carried out using the principles given in the following design example.

7.3. Diaphragm retaining wall

The retaining wall shown in Fig. 7.19 is to be designed as a post-tensioned diaphragm wall constructed of facing bricks with a crushing strength of 35 N/mm² and a water absorption of 7% set in a designation (i) mortar. The partial safety factor γ_{mm} may be taken as 2·3.

For retaining walls, bending from the retained earth occurs in one direction only, and therefore an eccentric post-tensioning force can be applied to counteract it.

Table 7.2 gives typical diaphragm wall sections and their properties, and should be read in conjunction with Fig. 7.20; the data given are all for a constant thickness of cross-rib and flange.

7.3.1. Design procedure

The design procedure is broadly similar to that used for the design of the post-tensioned fin wall from the example in section 7.2.18. The stress diagrams showing the design process are given in Fig. 7.21.

The design equations for the required post-tensioning force and its eccentricity may then be derived as follows

$$(P_t/A) + (P_t e/Z_1) = f_t$$

$$(P_t/A) + (P_t e/Z_2) = -f_c$$

but for a diaphragm wall $Z_1 = Z_2$ and hence

$$(P_t Z/A) + P_t e = f_t Z$$

$$(P_t Z/A) - P_t e = -f_c Z$$

Adding these gives

$$2P_t Z/A = (f_t - f_c)Z$$

which may be rearranged as

$$P_t = (f_t - f_c)A/2$$

Fig. 7.21. Diaphragm wall: stress diagrams

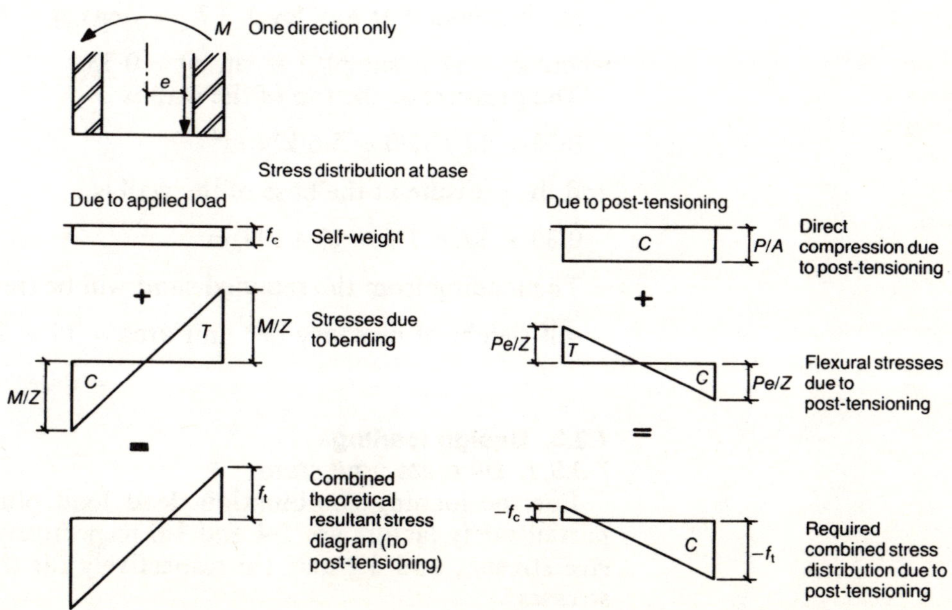

This is the minimum value of P_t necessary to produce the required post-tensioning stress. The corresponding value of its eccentricity e may now be calculated by substituting P_t into one of the original equations. Thus

$$(P_t/A) + (P_t e/Z) = +f_t$$

$$P_t e/Z = f_t - P_t/A$$

$$e = \left(f_t - \frac{P_t}{A} \right) \frac{Z}{P_t}$$

$$e = \left(\frac{f_t}{P_t} - \frac{1}{A} \right) Z$$

This provides the eccentricity corresponding to the minimum value of P_t already calculated. The eccentricity may be found to be larger than can be accommodated within the trial section selected. In such a case, the maximum value of e which can be accommodated should be inserted into the expression for e and a revised value of P_t obtained. This revised value of P_t will be larger than that originally calculated.

When the required post-tensioning force and its eccentricity have been calculated, the maximum combined compressive stresses (axial plus flexural) should be checked. The section must be checked to ensure that no flexural tensile stresses are developed, before there are losses of the post-tensioning force, and also for overall stability of the wall when subject to the post-tensioning force alone.

The moments and stresses in the reinforced concrete foundation should also be considered. However, this aspect of the design is outside the scope of this book.

7.3.2. Characteristic loadings

The retained material is assumed to comprise fine dry sand with an angle of internal friction of 30° and a density of 17 kN/m^3. Adequate land drainage is assumed to have been provided to ensure no build-up of water pressure. A surcharge above the retained sand, equivalent to 1·0 m depth of retained material, will also be applied.

Hence, from Rankine's formula

earth pressure at any level $= k_1 \times$ density $\times$ height

where $k_1 = (1 - \sin \phi)/(1 + \sin \phi) = 0.33$.

The pressure at the top of the wall is

$$0.33 \times 17 \times 1.0 = 5.6 \text{ kN/m}^2$$

and the pressure at the base of the wall is

$$0.33 \times 17 \times 3.1 = 17.4 \text{ kN/m}^2$$

The loading from the retained sand will be treated as a superimposed load

self-weight of masonry per unit area $= 19 \times 2.1/10^3$

$$= 0.04 \text{ N/mm}^2 \text{ at base level}$$

7.3.3. Design loadings
7.3.3.1. Ultimate limit state

For the loading combination dead load plus superimposed loading the partial safety factors are 1·4 and 1·6 respectively for the combined compressive stresses, and 0·9 and 1·6 respectively for the theoretical flexural tensile stresses.

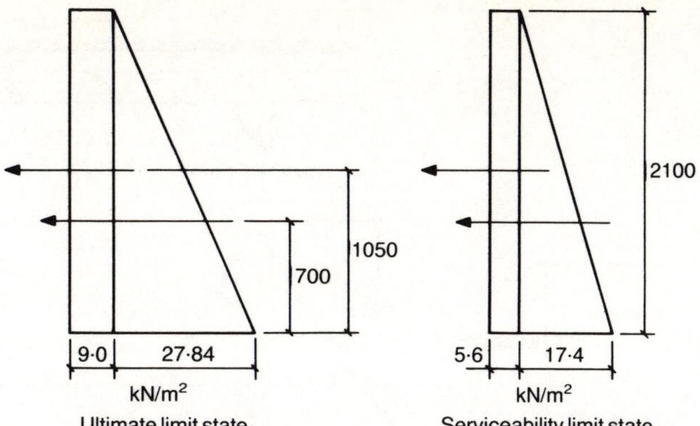

Fig. 7.22. Design wall:
design loading diagrams

The design superimposed load at the top of the wall is

$$1·6 \times 5·6 = 9·0 \text{ kN/m}^2$$

and that at the base of the wall is

$$1·6 \times 17·4 = 27·84 \text{ kN/m}^2$$

7.3.3.2. Serviceability limit state

The design superimposed load at the top of the wall is

$$1·0 \times 5·6 = 5·6 \text{ kN/m}^2$$

and that at the base of the wall is

$$1·0 \times 17·4 = 17·4 \text{ kN/m}^2$$

The design loading diagrams are shown in Fig. 7.22.

The design shear force at the base level (i.e. the ultimate limit state) is

$$(9 \times 2·1) + (27·84 \times 2·1/2) = 48·132 \text{ kN/m length of wall.}$$

7.3.4. Design bending moments at base level

7.3.4.1. Ultimate limit state

The ultimate limit state is calculated as follows

$$M_b = (9 \times 2·1 \times 1·05) + [27·84 \times (2·1/2) \times 0·7]$$

$$= 19·845 + 20·462$$

$$= 40·31 \text{ kNm}$$

7.3.4.2. Serviceability limit state

The serviceability limit state is calculated as follows

$$M_b = (5·6 \times 2·1 \times 1·05) + [17·4 \times (2·1/2) \times 0·7]$$

$$= 12·348 + 12·789$$

$$= 25·137 \text{ kNm}$$

7.3.5. Trial section

There are three dimensional variables which require consideration in order to arrive at the trial section for fuller analysis

- the overall depth of the wall D (see section 7.3.5.1)
- the thickness of the wall flanges T and the spacing of the cross-ribs B_r (see section 7.3.5.2)
- the thickness of the cross-ribs t_r (see section 7.3.5.3).

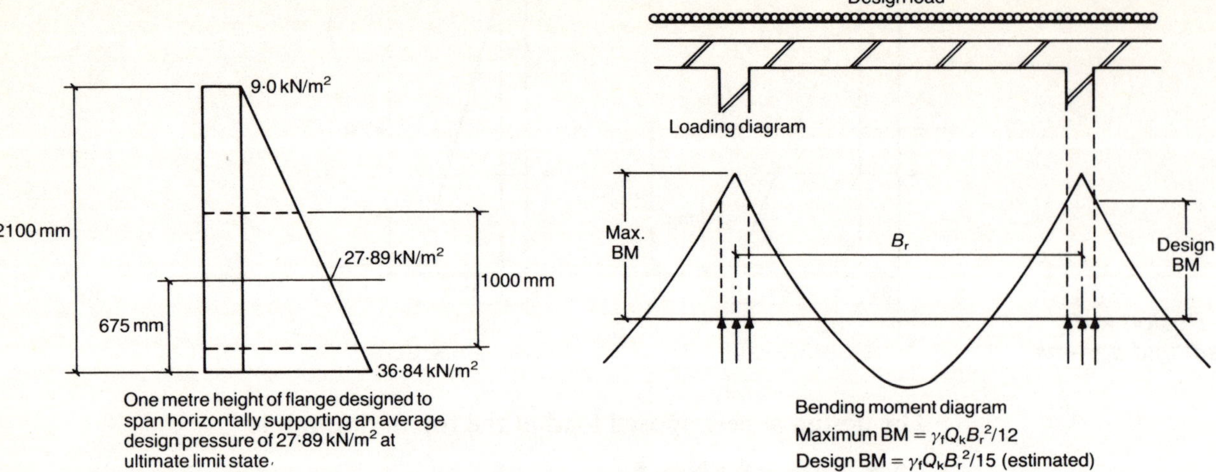

Fig. 7.23 (left). Design loading diagrams: earth pressure

Fig. 7.24 (right). Earth pressure for flange design bending moment for flange design

7.3.5.1. Overall depth

There is limited guidance which can be given for a reasonable assessment of the overall depth. Experience and familiarity with the design process will indicate to the designer the benefits of a deeper wall section, which benefits must be balanced against space requirements, quantity of walling materials and the effect on the size of the post-tensioning rods and the magnitude of the post-tensioning force. A greater depth of section will also assist in resisting shear forces, which can often be critical and are considered in the third of the three dimensional variables. For this example, the overall depth D is taken as 558 mm.

7.3.5.2. Flange thickness T and cross-rib spacing B_r

The wall flange, on the earth face, is required to support the earth and transfer its pressure to the cross-ribs by spanning horizontally between the cross-ribs. It is considered unreasonable that the maximum pressure at the base of the wall should be taken as the load to be supported on the horizontal span, as there is likely to be considerable resistance provided by the foundation and the flange will tend to cantilever for a certain height rather than span horizontally. Fig. 7.23 shows the design pressure diagram with loading assessed by the Authors as being that appropriate to the horizontal span of the flange.

The bending moment diagram for the design of the flange is shown in Fig. 7.24 in which the design bending moment intensity has been estimated as $\gamma_f Q_k B_r^2/15$ and occurs, as shown, at the intersection of the flange with the face of the cross-rib

$$\text{maximum } M_A = \gamma_f Q_k B_r^2/12$$

$$\text{design } M_a = \gamma_f Q_k B_r^2/15 \text{ (estimated)}$$

$$= 27\cdot89 B_r^2/15$$

$$= 1\cdot86 B_r^2$$

$$\text{design } MR = f_{kx} Z/\gamma_{mm}$$

$$= 2\cdot0 \times 1\cdot0 \times 0\cdot1025^2 \times 10^3/2\cdot3 \times 6$$

$$= 1\cdot523 \text{ kNm}$$

$$\text{design } MR \geqslant \text{design } M_A$$

Therefore

$$1 \cdot 523 = 1 \cdot 86 B_r^2$$

Hence

$$B_r = \sqrt{(1 \cdot 523 / 1 \cdot 86)}$$

$$= 0 \cdot 905$$

$$= \text{maximum span of flange}$$

However, the horizontal shear resistance is generally the more significant factor in the assessment of the size and centres of the cross-ribs in a diaphragm retaining wall design. Hence, to suit an acceptable bonding arrangement, the cross-ribs will be spaced for trial purposes at 675 mm centres, which is obviously within the capacity of the span of the flange as just designed.

7.3.5.3. Cross-rib thickness

Cross-rib thickness t_r may be assessed from a consideration of the horizontal shear force from the retained material which has previously been calculated as 48·132 kN/m at base level. The shear force per cross-rib is

$$V = 48 \cdot 132 \times 0 \cdot 675$$

$$= 32 \cdot 5 \text{ kN}$$

The maximum horizontal shear stress occurs on the centroid of the overall section and may be derived from the formula

$$V_h = V A_1 \bar{y} / I_{na} t_r$$

For this example (see Figs 7.25 and 7.26) $V = 32 \cdot 50$ kN per cross-rib, $A_1 = 0 \cdot 107$ m^2, $\bar{y} = 0 \cdot 178$ m, $I_{na} = 8 \cdot 07 \times 10^{-3}$ m^4, $Z = 0 \cdot 029$ m$^3 = 0 \cdot 043$ m^3/m, $A = 0 \cdot 214$ m$^2 = 0 \cdot 317$ m^2/m and $t_r = 0 \cdot 215$ m (assessed for trial purposes). Hence

$$V_h = \frac{32 \cdot 50 \times 0 \cdot 107 \times 0 \cdot 178}{8 \cdot 07 \times 10^{-3} \times 0 \cdot 215}$$

$$= 0 \cdot 357 \text{ N/mm}^2$$

shear resistance $= f_v / \gamma_{mv}$

where $f_v = 0 \cdot 35 + 0 \cdot 6 g_B$ for this example (clause 19.1.3.3 of BS 5628: Part 2[1]).

The unknown factor at this stage of the design process is the value of g_A, which is the summation of the self-weight of the masonry plus the post-

Fig. 7.25 (left). Properties relating to horizontal shear calculation

Fig. 7.26 (right). Trial section: dimensions and properties; $Z = 0 \cdot 029$ m^3 $= 0 \cdot 043$ m^3/m, $A = 0 \cdot 214$ m$^2 = 0 \cdot 317$ m^2/m

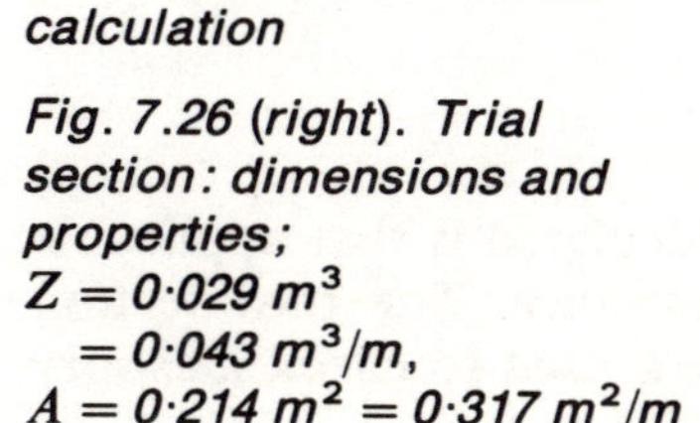
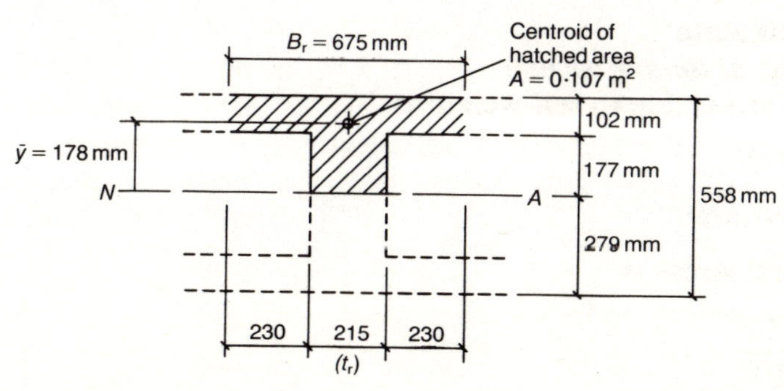

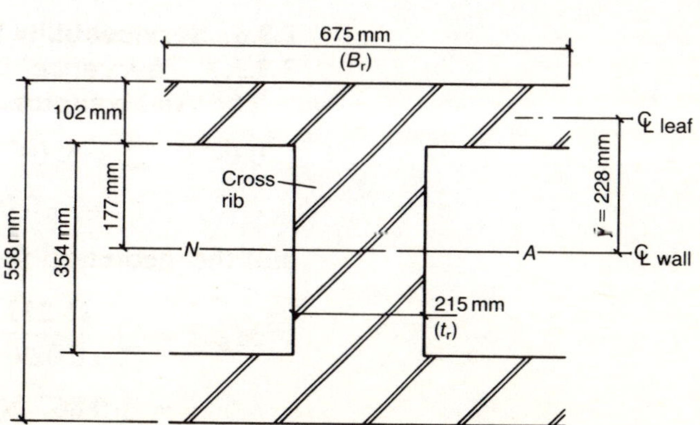

tensioning force. However, a minimum post-tensioning force required to provide the horizontal shear resistance may be calculated and later checked against the minimum post-tensioning force applied to eliminate the theoretical flexural tensile stress due to bending. The larger of the two forces calculated should then be used in the design

$$V_h = \frac{0.35}{\gamma_{mv}} + \frac{0.6 g_B}{\gamma_{mv}}$$

Therefore

$$0.357 = \frac{0.35}{2.0} + \frac{0.6 g_B}{2.0}$$

Therefore

$$g_B = 0.607 \text{ N/mm}^2$$

but

$$g_B = g_d + \text{design post-tensioning stress}$$
$$g_d = 0.9 G_k / A$$
$$= 0.9 \times 0.04$$
$$= 0.036 \text{ N/mm}^2$$

Hence

$$g_B = 0.036 + \text{design post-tensioning stress}$$
$$= 0.607 - 0.036$$
$$= 0.571 \text{ N/mm}^2$$

The design post-tensioning stress varies across the section owing to its eccentricity. However, the maximum value of horizontal shear stress occurs where the design post-tensioning stress has its average value. Hence, the design post-tensioning force (see Fig. 7.26) is

$$0.571 A = 0.571 \times 0.214 \times 10^3$$
$$= 122.2 \text{ kN per cross-rib}$$
$$= 181 \text{ kN/m}$$

At this stage the design post-tensioning force calculated is that applicable only to the development of horizontal shear resistance. The trial section derived is shown in Fig. 7.26 and this section is now used to check masonry stresses and to design post-tensioning rods.

7.3.6. Serviceability limit state
7.3.6.1. Theoretical flexural tensile stress

The axial compressive stress due to self-weight is

$$0.04 \times \gamma_{mm} = 0.04 \times 1.0$$
$$= +0.04 \text{ N/mm}^2$$

and the theoretical flexural stress is

$$M_b / Z = \pm \frac{25.237 \times 10^6}{0.029 \times 10^9}$$
$$= \pm 0.867 \text{ N/mm}^2$$

Therefore the net theoretical flexural tensile stress is

$$f_t = +0{\cdot}04 - 0{\cdot}867$$

$$= -0{\cdot}827 \text{ N/mm}^2$$

and

$$f_c = +0{\cdot}04 \text{ N/mm}^2$$

Hence an eccentric post-tensioning force to produce stresses of $f_c = -0{\cdot}04$ N/mm² and $f_t = +0{\cdot}827$ N/mm² is required.

7.3.6.2. Minimum required post-tensioning force: bending stresses

The minimum post-tensioning force required to eliminate the theoretical flexural tensile stress is

$$P = (f_t - f_c)A/2$$

$$= (0{\cdot}827 - 0{\cdot}04)0{\cdot}317 \times 10^6/2 \times 10^3$$

$$= 124{\cdot}74 \text{ kN/m}$$

This is less than that required to ensure adequate shear resistance (181 kN/m) as calculated in section 7.3.5.3. The higher value of the two is adopted for subsequent calculations.

It is first necessary to calculate the eccentricity at which the design post-tensioning force of 181 kN/m must be placed. This is given by the equation

$$e = \left(\frac{f_t}{P} - \frac{1}{A}\right)Z$$

$$= \left(\frac{0{\cdot}827}{181 \times 10^3} - \frac{1}{0{\cdot}317 \times 10^6}\right)0{\cdot}043 \times 10^9$$

$$= 61 \text{ mm}$$

(see Fig. 7.27).

The maximum practical eccentricity which can be provided in a wall section 558 mm deep is

$$e_{max} = \frac{558}{2} - 102 - \frac{\text{rod diameter}}{2}$$

$$= 177 - \frac{\text{rod diameter}}{2}$$

Assume that the maximum practical eccentricity is 160 mm. The required eccentricity is within this limit, i.e. by placing the calculated force 181 kN/m at an eccentricity of 61 mm, both the bending and shear stresses are limited to acceptable values.

If the calculation had indicated that an eccentricity greater than 160 mm were required, then it would have been necessary to use an increased post-tensioning force at a reduced eccentricity. The new force required would be determined by substituting the maximum practical value of e in the formula

$$P = \frac{f_t}{(1/A) + (e/Z)}$$

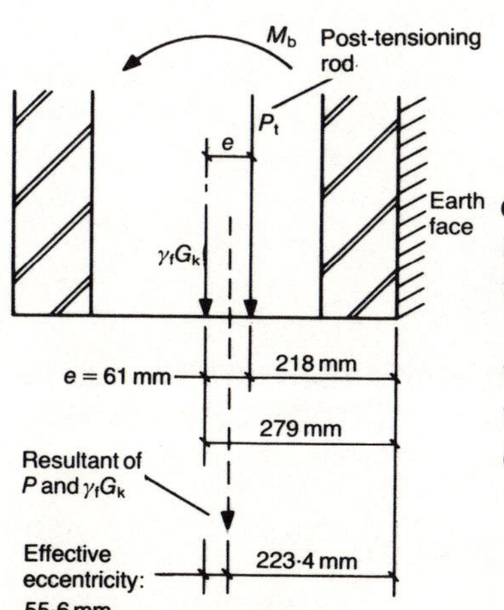

Fig. 7.27. Calculated eccentricity of post-tensioning force

7.3.7. Ultimate limit state: cracked section

As for the example in section 7.3.6.2, the design post-tensioning force calculated is used to check the moment of resistance of the cracked section to

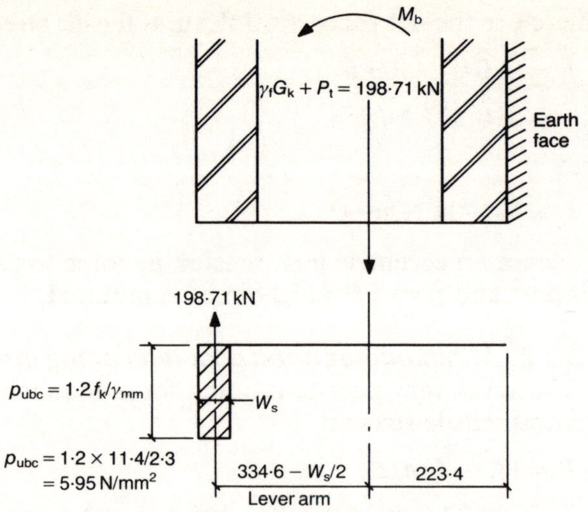

178   Design of reinforced and prestressed masonry

Fig. 7.28. Cracked section: ultimate resistance diagram

ensure that it exceeds the design moment for the ultimate limit state condition calculated earlier as 40·31 kNm

$$MR_s = (P \times \text{lever arm}) + (\text{design dead load} \times \text{lever arm})$$

but

$$P + \text{design dead load} = 181 + (1{\cdot}4 \times 0{\cdot}317 \times 19 \times 2{\cdot}1)$$
$$= 181 + 17{\cdot}71$$
$$= 198{\cdot}71 \text{ kN/m}$$

and the centroid of these loads is

$$\frac{(181 \times 0{\cdot}218) + (17{\cdot}71 \times 0{\cdot}279)}{181 + 17{\cdot}71} = 223{\cdot}4 \text{ mm from earth face}$$

(see Fig. 7.28).

$$MR_s = 198{\cdot}71 \times \text{lever arm}$$
$$= 198{\cdot}71 \times (334.6 - W_s/2) \times 10^{-3}$$

but

$$W_s = 198{\cdot}71 \times 10^3/5{\cdot}95 \times 1000$$
$$= 33{\cdot}4 \text{ mm}$$

Therefore

$$MR_s = 198{\cdot}71 \times (334{\cdot}6 - 33{\cdot}4/2) \times 10^{-3}$$
$$= 63{\cdot}17 \text{ kNm}$$

which exceeds the design moment of 40·31 kNm and is therefore acceptable for this loading condition.

7.3.8. Characteristic post-tensioning force

The characteristic post-tensioning force P_k is given by

$$P_k = P/\gamma_f$$
$$= 181/0{\cdot}9$$
$$= 201 \text{ kN/m}$$

This force is now used to check the design compressive stresses in the wall and to design the size and torque required for the post-tensioning rods, with due allowance being made for the anticipated 20% losses.

7.3.9. Compressive stresses in wall
7.3.9.1. Capacity reduction factors

The overall stability of the wall section and the local stability of the flanges (leaves) are checked under combined axial and flexural loading.

(a) *Overall stability*

$$h_{ef} = 2h$$

$$= 2 \times 2 \cdot 1$$

$$= 4 \cdot 2 \text{ m}$$

$$t_{ef} = \text{actual thickness}$$

$$= 0 \cdot 558 \text{ m}$$

$$SR = h_{ef}/t_{ef}$$

$$= 4 \cdot 2/0 \cdot 558$$

$$= 7 \cdot 53$$

The eccentricity of the resultant load is calculated by

effective eccentricity $= 55 \cdot 6$ mm (see Fig. 7.27)

$$\frac{\text{effective eccentricity}}{t_{ef}} = \frac{55 \cdot 6}{55 \cdot 8}$$

$$= 0 \cdot 1$$

Hence for $SR = 7 \cdot 53$ and $e_x = 0 \cdot 1t$

$$\beta = 0 \cdot 88$$

(b) *Local stability.* The flange is restrained against buckling by the cross-ribs which may be considered to provide enhanced resistance to lateral movement. Hence

$$SR = 0 \cdot 75 \times 675/102$$

$$= 5$$

The combined axial, flexural and post-tensioning stresses may be considered to be applied to the flange with zero eccentricity for the purpose of calculating β. Hence for $SR = 5$ and $e_x = 0$

$$\beta = 1 \cdot 00$$

7.3.9.2. Check combined compressive stresses

Consider dead loading plus post-tensioning force before losses. In anticipation of 20% loss of prestress, the characteristic post-tensioning force P should initially be increased by 20%. The post-tensioning force before losses is

$$201/0 \cdot 8 = 251 \text{ kN/m}$$

and the design post-tensioning force before losses is

$$\gamma_f \times 251 = 1 \cdot 4 \times 251$$

$$= 351 \cdot 4 \text{ kN/m}$$

The maximum design dead load is

$$1 \cdot 4 G_k = 17 \cdot 71 \text{ kN/m}$$

the maximum flexural compressive stress due to the post-tensioning force P before losses is

$$\frac{P}{A} + \frac{Pe}{Z} = \frac{351 \cdot 4 \times 10^3}{0 \cdot 317 \times 10^6} + \frac{351 \cdot 4 \times 10^3 \times 61}{0 \cdot 043 \times 10^9}$$

$$= 1 \cdot 108 + 0 \cdot 498$$

$$= 1 \cdot 606 \text{ N/mm}^2$$

and the minimum flexural compressive stress due to the post-tensioning force P before losses is

$$1 \cdot 108 - 0 \cdot 498 = 0 \cdot 61 \text{ N/mm}^2$$

The axial stress due to maximum dead load is

$$\frac{17 \cdot 71 \times 10^3}{0 \cdot 317 \times 10^6} = 0 \cdot 056 \text{ N/mm}^2$$

the maximum combined compressive stress is

$$0 \cdot 056 + 1 \cdot 606 = 1 \cdot 662 \text{ N/mm}^2$$

and the minimum combined compressive stress is

$$0 \cdot 056 + 0 \cdot 61 = 0 \cdot 666 \text{ N/mm}^2$$

7.3.9.3. Design strength before losses

The design strength before losses is

$$1 \cdot 2(1 \cdot 2 \beta f_{ki}/\gamma_{mm}) = 1 \cdot 2(1 \cdot 2 \times 1 \cdot 0 \times 11 \cdot 4/2 \cdot 3)$$

$$= 7 \cdot 14 \text{ N/mm}^2$$

where, for this example, it has been assumed that the post-tensioning force is not to be applied until the masonry has achieved its full design strength, i.e. $f_{ki} = f_k$. The capacity reduction factor β for this design check relates to the local stability of the flange.

The overall stability of the wall, with its relevant β value, is checked in section 7.3.10.

The design strength far exceeds the maximum combined compressive stress. Hence the wall is acceptable for this loading condition. The reserve of strength available indicates, by inspection, that the loading condition after losses will also be acceptable.

7.3.10. Overall stability of wall

Consider dead loading plus post-tensioning force after losses.

The design axial load is

$$\gamma_f G_k + \gamma_f P_k = (1 \cdot 4 \times 12 \cdot 65) + (1 \cdot 4 \times 201)$$

$$= 17 \cdot 71 + 281 \cdot 4$$

$$= 299 \cdot 11 \text{ kN/m}$$

the design axial stress is

$$\frac{299 \cdot 11 \times 10^3}{0 \cdot 317 \times 10^6} = 0 \cdot 944 \text{ N/mm}^2$$

and the design strength of the wall is

$$\beta f_k / \gamma_{mm} = 0{\cdot}88 \times 11{\cdot}4/2{\cdot}3$$

$$= 4{\cdot}36 \ \text{N/mm}^2$$

The design strength exceeds the design stress. Hence the wall is acceptable for this loading condition.

Now consider a wall subject to dead load plus superimposed loading plus post-tensioning force, after losses.

The axial stress due to G_k is $0{\cdot}056$ N/mm². The maximum flexural stress due to P_k is

$$\frac{\gamma_f P_k}{A} + \frac{\gamma_f P_k e}{Z} = \frac{1{\cdot}4 \times 201 \times 10^3}{0{\cdot}317 \times 10^6} + \frac{1{\cdot}4 \times 201 \times 10^3 \times 61}{0{\cdot}043 \times 10^9}$$

$$= 0{\cdot}888 + 0{\cdot}399$$

$$= +1{\cdot}287 \ \text{N/mm}^2$$

the minimum flexural stress due to P_k is

$$0{\cdot}888 - 0{\cdot}399 = +0{\cdot}489 \ \text{N/mm}^2$$

and the flexural stress due to applied M_b is

$$\pm \frac{M_b}{Z} = \pm \frac{40{\cdot}31 \times 10^6}{0{\cdot}043 \times 10^9}$$

$$= \pm 0{\cdot}937 \ \text{N/mm}^2$$

Hence the maximum combined compressive stress is

$$0{\cdot}056 + 0{\cdot}489 + 0{\cdot}937 = 1{\cdot}482 \ \text{N/mm}^2$$

and the minimum combined compressive stress is

$$0{\cdot}056 + 1{\cdot}287 - 0{\cdot}937 = 0{\cdot}406 \ \text{N/mm}^2$$

The stress diagrams for this working condition are shown in Fig. 7.29. The design strength of the wall is

$$1{\cdot}2\beta f_k / \gamma_{mm} = 1{\cdot}2 \times 1{\cdot}0 \times 11{\cdot}4/2{\cdot}3 = 5{\cdot}95 \ \text{N/mm}^2$$

Fig. 7.29. Stress diagram: overall stability of wall

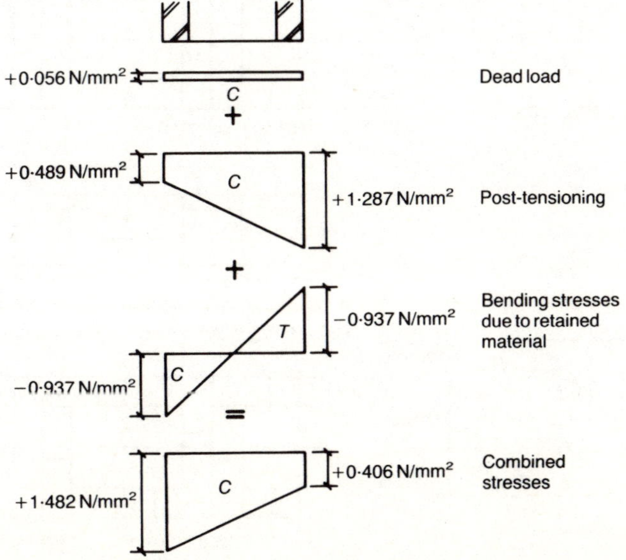

The design strength exceeds the combined design stress. Hence the wall is acceptable for this loading condition.

Check the minimum combined stress in which γ_f for dead loading and the post-tensioning force is 0·9. The axial stress due to G_k is

$$\frac{0 \cdot 9 \times 12 \cdot 65 \times 10^3}{0 \cdot 317 \times 10^6} = +0 \cdot 036 \text{ N/mm}^2$$

The maximum flexural stress due to P_k is

$$\frac{0 \cdot 9 \times 201 \times 10^3}{0 \cdot 317 \times 10^6} + \frac{0 \cdot 9 \times 201 \times 10^3 \times 61}{0 \cdot 043 \times 10^9} = 0 \cdot 571 + 0 \cdot 257$$

$$= +0 \cdot 828 \text{ N/mm}^2$$

the minimum flexural stress due to P_k is $0 \cdot 571 - 0 \cdot 257 = +0 \cdot 314$ N/mm² and the flexural stress due to applied M_b is $\pm 0 \cdot 937$ N/mm². Hence the maximum combined compressive stress is

$$0 \cdot 036 + 0 \cdot 314 + 0 \cdot 937 = +1 \cdot 287 \text{ N/mm}^2$$

and the minimum combined compressive stress is

$$0 \cdot 036 + 0 \cdot 828 - 0 \cdot 937 = -0 \cdot 073 \text{ N/mm}^2$$

From this it can be seen that virtually no tensile stresses are developed.

7.3.11. Check shear between leaf and cross-rib

A further check should be carried out on the shear stresses in the vertical plane at the junction of the cross-ribs and the leaves as shown in Fig. 7.30.

Shear stress V_h is

$$\frac{V A_2 \bar{y}}{I_{na} t_r} = \frac{32 \cdot 5 \times 10^3 \times 102 \times 675 \times 228}{8 \cdot 07 \times 10^9 \times 215}$$

$$= 0 \cdot 294 \text{ N/mm}^2$$

The cross-ribs are assumed to be half-bonded into the leaf at alternate courses, as shown in the bonding diagram Fig. 7.31. For shear failure to

Fig. 7.30 (below). Properties relating to vertical shear calculation

Fig. 7.31 (right). Bonding of cross ribs

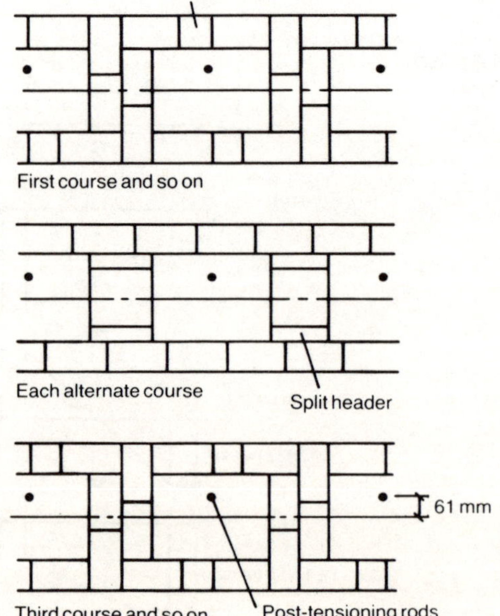

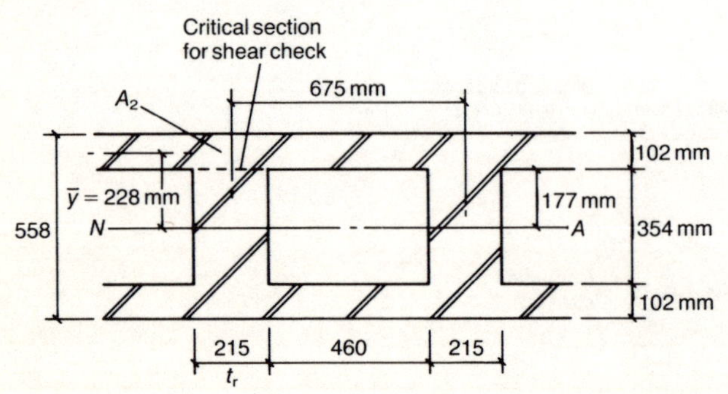

occur at the junction of the cross-rib with the leaf, the bonded bricks would need to snap. BS 5628[1] does not give values for such a shear failure, although research work is being carried out to establish such values.

By inspection of the design vertical shear stress calculated and by comparison with the compressive strength of the bricks being used (35 N/mm), the stress appears to be within the likely shear strength of the bricks. The horizontal shear stress was shown to be acceptable in section 7.3.5.3.

7.3.12. Design of post-tensioning rods

For post-tensioning rods the characteristic post-tensioning force required before losses is 251·0 kN. The characteristic post-tensioning force required per cell is

$$251 \times 0 \cdot 675 = 169 \cdot 4 \text{ kN}$$

The design strength of the rods is

$$0 \cdot 7 f_y / \gamma_{ms} = 0 \cdot 7 \times 460 / 1 \cdot 15$$
$$= 280 \text{ N/mm}$$

and the area of rod required per cell is

$$169 \cdot 4 \times 10^3 / 280 = 605 \text{ mm}^2$$

Two 20 mm dia. high-yield bars per cell are used and the remainder of the design of the rods should follow the same principles as were used for the examples in sections 7.3.10 and 7.3.11 using a concrete capping beam to distribute the post-tensioning force.

7.4. Post-tensioned beams

Preliminary designs for post-tensioned, prefabricated beams have been carried out and costed. Although there appears to be no problem in design or manufacture, such beams do not appear to compete with steel or precast concrete beams for speed and simplicity of fabrication and erection or total cost. Should aesthetics or other considerations outweigh costs, then such beams could be used and their design method would follow the basic principles established in chapter 6.

8

Post-tensioned masonry detailing

8.1. Introduction

As mentioned in chapter 5, detailing is the main tool used in the translation of design into construction. Certain details can be sensitive to the results of faulty translation. The detailer must translate the design in such a way as to minimise the effects of errors in construction and interpretation, i.e. to minimise the sensitivity of his details. The upper and lower anchorages of the post-tension bars are typical locations where the results of poor detailing could destroy the intended behaviour of the structural element.

This is not to say that this form of construction should be avoided; on the contrary, the design, detailing and construction are relatively simple compared with many other forms of construction. However, the general simplicity of the system tends to concentrate the critical details into particular locations, such as the anchorages.

The detailer's aim must be to keep the details simple and to translate clearly the designer's intentions into practical construction details which can easily be achieved and which will perform satisfactorily for the required design life. To do this the detailer must fully understand the principles involved, the effect on construction and performance of possible site variations to his details and the method of minimising such effects.

In addition to the basic aim of keeping the details simple and clear, other main objectives must be given equal importance. It is necessary to ensure that

(a)   the correct post-tensioning force is applied in the right location and distributed over sufficient cross-sectional area
(b)   the correct masonry strength is used
(c)   adequate anchorage at both ends of the post-tension rods is provided
(d)   the required force is maintained for the design life of the structure, i.e. losses and rod protection are catered for
(e)   simplicity of construction is achieved
(f)   adequate cover is provided to the rod.

To ensure that these objectives are attained requires a good knowledge of construction methods and difficulties as well as a full understanding of the designer's requirements and intentions. The generally adopted procedure for post-tensioning masonry is to anchor a post-tension rod at one end, provide the necessary durability protection and then to construct the required masonry section around it. The section constructed must contain a sufficient gap between the rod and the masonry to prevent any danger of interference with the required action of the rod both during and after the application of the post-tensioning force. A spreader plate is positioned over the other end of the rod and bearing on to the masonry or capping beam section to act as the second anchorage for the rod. The rod is then stressed either by jacking or by the use of a torque spanner screwing a nut on to the threaded end of the rod and tightening it to bear down on to the spreader plate. Fig. 6.3 shows a typical detail of the post-tension rod.

The basic system is simple and straightforward. However, certain aspects are more critical than others and these are now dealt with.

8.2. Buildability

The process of the construction of masonry (i.e. the laying of one unit on top of another using a wet bedding in between the units with no shuttering) makes the support of flimsy or thin strands of reinforcement or tendons very difficult.

For this reason, rods rather than strands are preferred for vertical post-tensioned members, i.e. for buildability, high tensile steel rods are generally chosen. The details provided in this book therefore indicate a similar choice. In terms of the general more minor decision-making related to buildability, notes are given as appropriate to each detail.

8.2.1. Horizontal members

In horizontal members the choice between details is more dependent on individual situations and preference. Strands can be more easily supported than in a vertical situation and bond ánchorage can be provided which can be more easily hidden behind the general face without changing from normal bonded facework.

Rods may also be used in horizontal members. However, if mechanical anchorages are used they may be difficult to incorporate without spoiling the facework (see section 8.10.1).

8.3. Rod and strand

While most post-tensioned masonry incorporates prestressing rods, wire strands can also be used. Both rods and wire strands should comply with the requirements of either BS 4486[8] or BS 5896.[9] The use of mild steel rods, however, should in most situations be avoided owing to the excessive loss of prestress resulting from creep in the steel.

As discussed in section 8.2, for most masonry elements post-tensioned high tensile steel rods are more practical than wire strands because they are more rigid and need less temporary support.

8.4. Location of post-tensioned rod or strand

A number of important design aims and requirements govern the location of the post-tensioned rod within the masonry section. These will vary to some extent, depending on the application, but some of the main aims and requirements which have to be balanced by the designer are

(a) to provide the force in the optimum location to maximise the resistance to the applied loading
(b) to fit the rod within the practical physical dimensions within a suitable bonding arrangement
(c) to provide a suitable force at the design location; if any adjustments to the designer's aims are made necessary by item (b), then the design must be adjusted to correct the required force for the revised location and to provide a suitable revised rod and prestress
(d) to provide adequate cover to the rod for protection against fire and/or corrosion and/or bond, depending on the location and method of post-tensioning used.

Aim (a) is determined by the magnitude and direction of the applied loading. Fig. 8.1 shows two laterally loaded walls: one resisting earth pressure in one direction only and the other resisting wind loading, which can occur in either direction. For the wind loading condition the pressure and

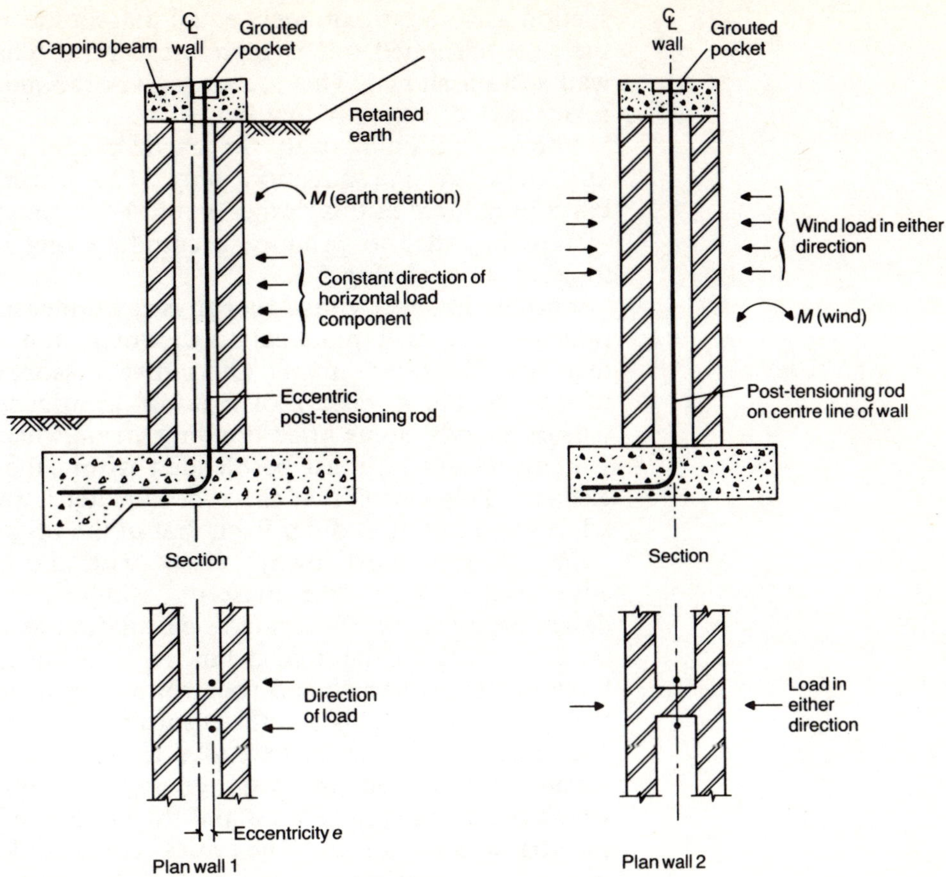

Fig. 8.1. Variable position of post-tensioning rods

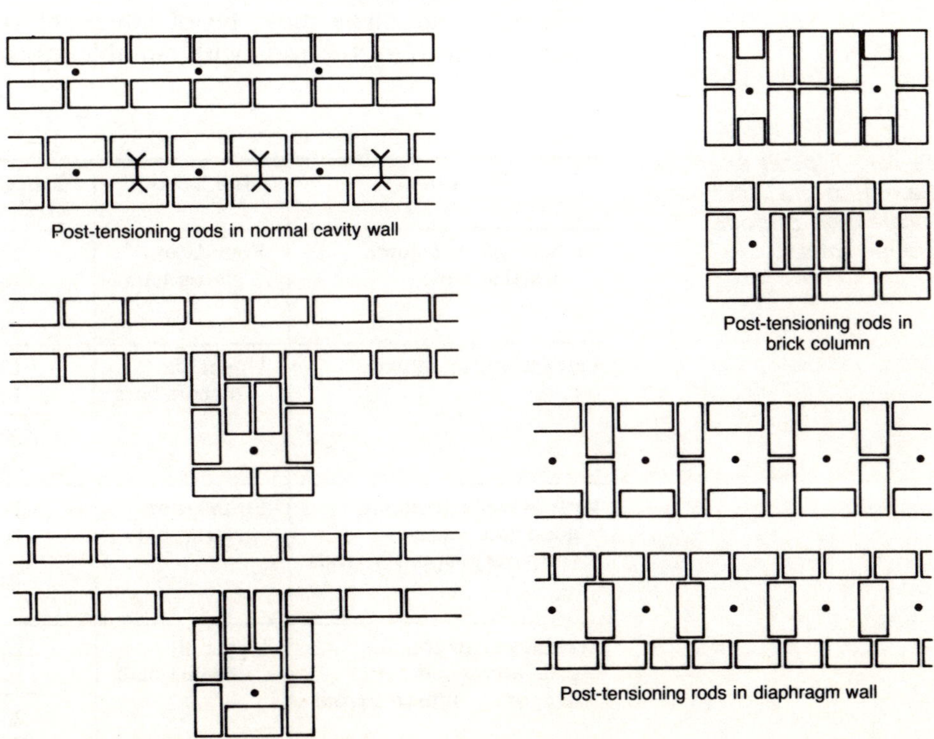

Fig. 8.2. Typical bonding patterns for post-tensioned walls

suction are of approximately equal magnitude and the optimum location for the post-tensioned rod is therefore central (concentric) to the symmetrical wall section shown. This location gives the maximum balance of resistance under each direction of loading.

For the retaining wall the earth pressure and surcharge loading are applicable in one direction only. The optimum location for the post-tensioning force in this design would be eccentric to the centroid of the wall section in order to produce a counteracting moment against the applied force, as shown in Fig. 8.1.

Fitting the rod within the physical dimensions of the section, as mentioned in (b), is a practical requirement and varies for different types of masonry. There are many alternative masonry bonded sections available which provide a variety of suitable locations to accommodate the post-tensioned rods. Some brick bonding arrangements are shown in Fig. 8.2.

Requirement (c) must be blended within the practical dimensions of the element. This means that the rod diameter, stress and location may require adjustment and thus differ from that of the original design requirement.

The detailer must always check with the designer before making any adjustments to suit the masonry bonding, to ensure that the necessary design adjustments, including re-calculation as necessary, are made, i.e. item (a) is satisfied on the final detail. In addition any notes for topping-up the force in the post-tensioned rods should be included on the detail and/or in the specification. This should ensure that the prestressing force is achieved and maintained without excessive losses. Cover requirements for the post-tensioned rod (item (d)) vary greatly depending on the condition under which the detail is to be used and the method of providing the anchorage for the post-tensioned force. The critical cover could be that required for protection against corrosion, bond or fire. In some cases the cover requirement dictates the size of the pocket. These requirements are dealt with in more detail in chapter 5. (Tables 5.1 and 5.2 list the minimum cover required for fire and durability protection.)

Cavities and ducts must be of adequate size to allow for unrestricted accommodation of the rods with suitable allowances for construction toler-

Table 8.1. Typical examples of maximum practical limits for projection of post-tensioned rods

Condition	Rod location	Diameter of rod, mm	Recommended maximum projection length, m
Vertical wall or column, normal access	Foundation starter bars	12 16	1·5
		20	2·0
Vertical wall or column	Upper lifts of main bars	12 16	2·0
		20 25	3·0
Vertical wall or column, good access and temporary support to rods	Foundation starter bars	12 16	2·5
		20 25	3·5
Vertical wall or column, good access and temporary support to rods	Upper lifts of main bars	12 16	4·0
		20 25	4·5

ances without critical design implications; the details must not be structurally sensitive to tolerance errors.

8.5. Curtailment and/or extension of bars

In translating the designer's requirements into practical details, it is essential that the detailer considers the construction methods and problems and blends the construction requirements in with those of the designer. One of the major considerations in post-tensioned masonry is the length of protruding rod from any section of construction. From the designer's point of view, jointing of the rods is often unnecessary in meeting his requirements, but it is essential if the design standards are to be met in the quality of the workmanship on site. Such practical requirements are also important in the cost of the work since the simpler the construction, the more economic the job will be. Table 8.1 is a guide for the detailer and indicates examples of general maximum practical limits for the projection of vertical post-tensioning rods for various construction conditions.

The values are a guide only because different conditions, contractors and design requirements can influence the practicality and need to vary from these values. The detailer should discuss the practicality of his details with the contractor, both before and after the work is carried out. From discussion both the detailer and the contractor can gain valuable experience of their combined requirements.

Extension of the rods in locations where joints are required can be achieved using a threaded ferrule as shown in Fig. 8.3, but the engineer must ensure that adequate anchorage length is provided within the ferrule for transfer of the applied load through the joint.

8.6. Applying the prestress

Newly constructed masonry needs time to mature and set before external forces over and above its own weight are applied. From a construction point of view the earlier the load can be applied, then the quicker the construction will be, but from a design point of view, the longer the masonry's mortar is allowed to cure, the stronger the section will be. Obviously again there is a practical period which can satisfy both requirements. Table 8.2 gives guidance on the time which should be allowed for different wall types. This guidance should not be treated as rigid, since curing under favourable conditions for a small post-tensioning force is less critical than curing under relatively poor conditions for a large force, i.e. a small prestress on well-cured mortar

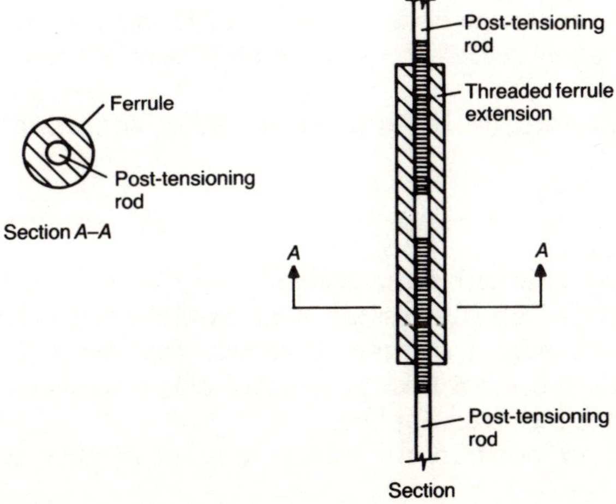

Fig. 8.3. Threaded ferrule rod extension detail

Table 8.2. Curing periods for mortars (i) and (ii)

Curing conditions	Curing period, days	Post-tensioning
Winter	14 days + number of days below specified minimum temperature	Post-tensioning after curing, and restress (top up) not less than 1 day later
Summer	14 days	As for winter

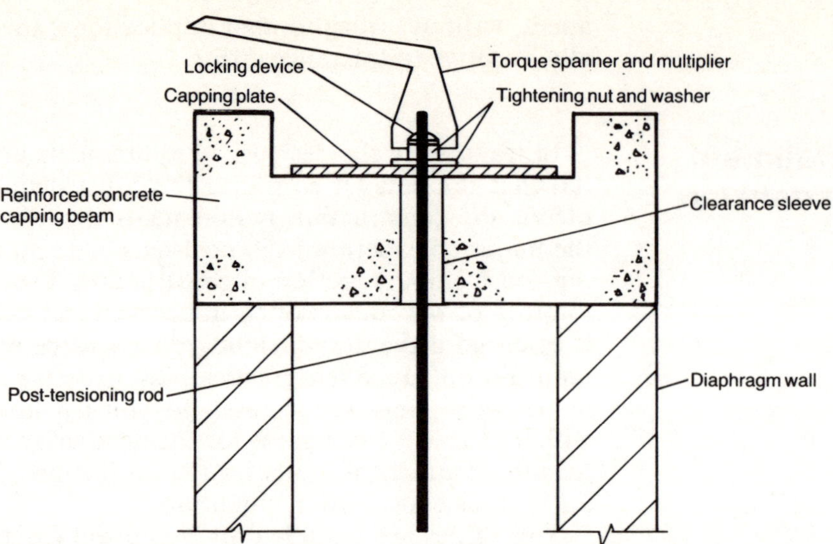

Fig. 8.4. Torque wrench stressing

as opposed to a large prestress on immature conditions. Adjustments to this advice can and should be made at the designer's discretion. For example, wherever possible the topping up should be carried out near to 28 days after construction.

The general system adopted for applying low post-tension force is that of tightening a nut against a steel spreader plate using a torque spanner and multiplier (see Fig. 8.4). For large forces hydraulic jacks are used as in prestressed concrete.

To ensure the correct application of the force to the rod when a torque spanner is used, certain critical conditions must be met.

- The type and pitch of the thread must comply with the design condition.
- The threads must be clean, lightly lubricated and the nut must be free to move.
- The clearance of all voids, holes or ducts must allow freedom of the rod to be torqued up without contact or influence on the strain being produced.

To ensure that the nut passes freely over the threads it should be specified that before applying the force the nut should be run up and down the threads, at least once, to a level below which it will finally rest, and the threads and nuts should be lightly lubricated. This will need to be done before construction of the upper anchorage because the lower threads would be below the level of the anchor plate before prestressing and therefore inaccessible at that stage.

In order to maintain a suitable force in the rod certain other conditions must be met.

(a) The shrinkage or expansion of the masonry must be compatible with the assumed design condition.
(b) The steel used must be that assumed in the design.
(c) The rod force must be topped up (restressed) in accordance with the design requirements after a period of time to reduce the losses. In addition, grouting of voids must not take place until all prestressing, including topping up forces, has been applied.
(d) The nut must be locked in position to prevent any possibility of slackening.

(*e*) Protection against corrosion suitable for the design life of the section must be provided.

(*f*) Any damp-proof courses must not compress under load to an extent that will affect either the initial prestress or its losses.

The specification on the details should therefore include

(*a*) the masonry strength, type and material

(*b*) the size and type of steel and rod

(*c*) the torque or jacking required and the sequence relating to any grouting

(*d*) the force topping-up requirements

(*e*) the method of locking the nut

(*f*) the system and details of corrosion protection

(*g*) limitations on any damp-proof course selected

(*h*) notes on running up and down of the nut prior to prestress.

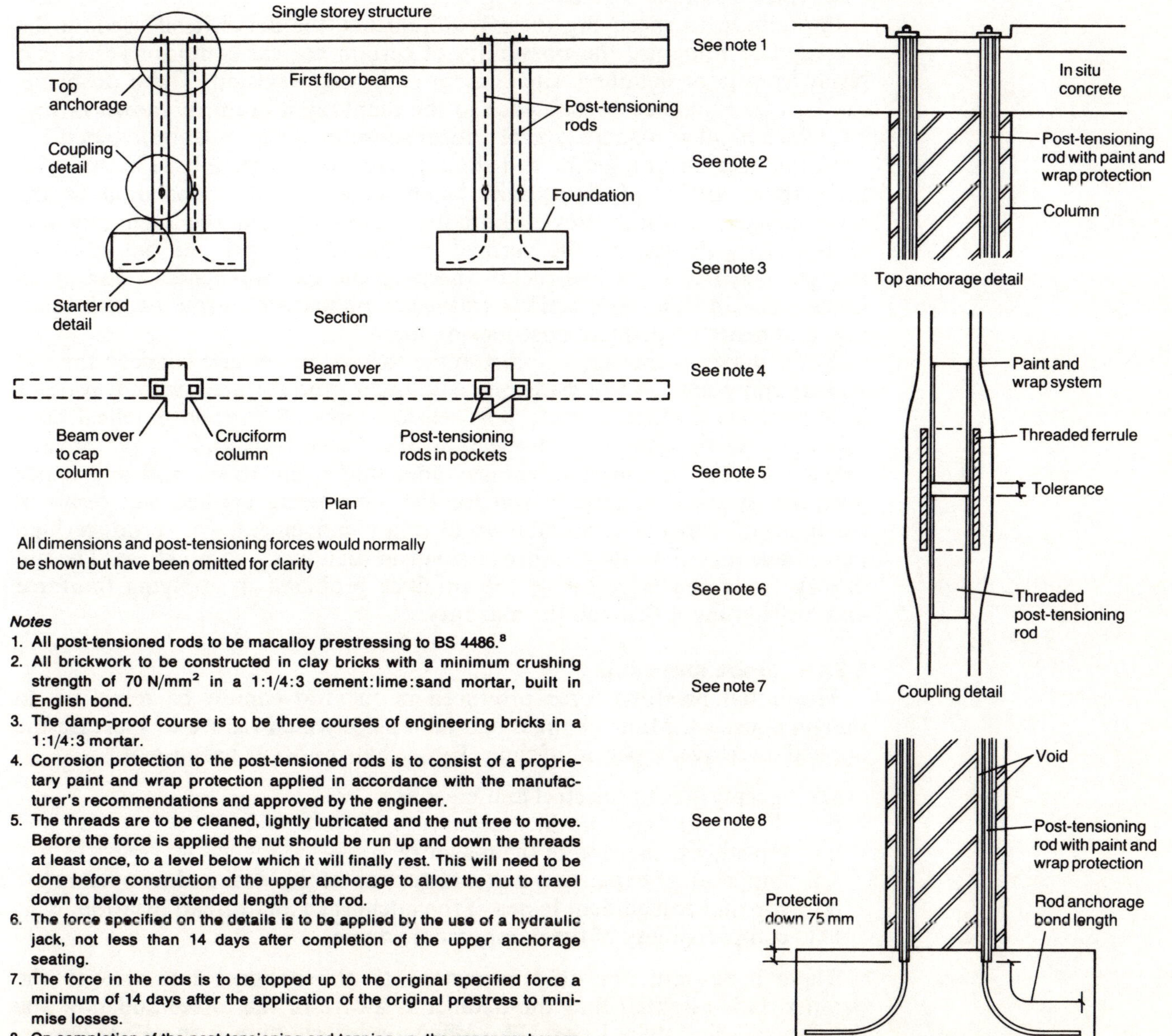

Fig. 8.5. Typical post-tensioning detail and specification: podium construction

All dimensions and post-tensioning forces would normally be shown but have been omitted for clarity

Notes

1. All post-tensioned rods to be macalloy prestressing to BS 4486.[8]

2. All brickwork to be constructed in clay bricks with a minimum crushing strength of 70 N/mm^2 in a 1:1/4:3 cement:lime:sand mortar, built in English bond.

3. The damp-proof course is to be three courses of engineering bricks in a 1:1/4:3 mortar.

4. Corrosion protection to the post-tensioned rods is to consist of a proprietary paint and wrap protection applied in accordance with the manufacturer's recommendations and approved by the engineer.

5. The threads are to be cleaned, lightly lubricated and the nut free to move. Before the force is applied the nut should be run up and down the threads at least once, to a level below which it will finally rest. This will need to be done before construction of the upper anchorage to allow the nut to travel down to below the extended length of the rod.

6. The force specified on the details is to be applied by the use of a hydraulic jack, not less than 14 days after completion of the upper anchorage seating.

7. The force in the rods is to be topped up to the original specified force a minimum of 14 days after the application of the original prestress to minimise losses.

8. On completion of the post-tensioning and topping up, the upper anchorage is to be solidly grouted to protect the plates and to lock up the anchorage.

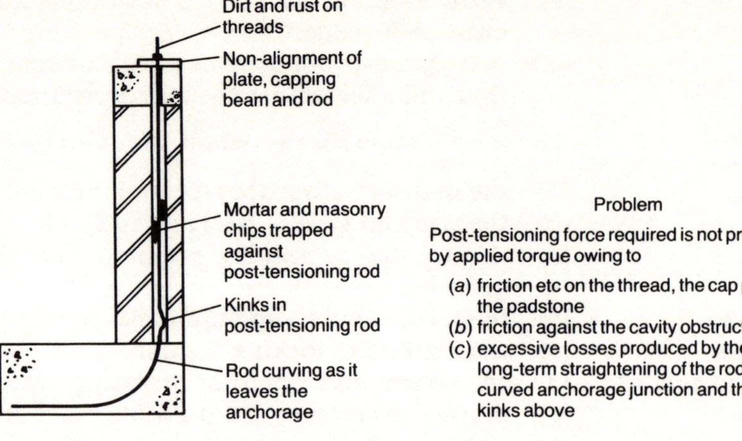

Fig. 8.6. Critical effects of poor detail and workmanship

A typical example is given in Fig. 8.5.

In addition to specifying these requirements, the detail itself needs to be prepared to minimise the possibility of certain critical conditions that can result from poor detailing, site control and/or supervision. These problems can best be explained by reference to the diaphragm example shown in Fig. 8.6, which highlights some typical critical conditions that must be avoided.

At the base anchorage the rod should leave the concrete base in line with the vertical portion of the rod and be embedded, or plate anchored, in line for a sufficient depth to prevent bursting of the surface of the anchorage due to rod straightening. In the vertical portion of the rod, the bar must be straight and free from obstruction because the bar will tend to straighten under load and the obstructions will cause premature torque or jack readings and hence inadequate post-tension force.

At the upper anchorage any dirt in the threads or contact between the rod and its end plates and/or padstone will again produce premature torque or jack readings and inadequate post-tension force. A further practical consideration to be given to the selection of suitable rods and torque or jack forces relates to the need to achieve adequate strain in the rod and hence adequate stress. Too large a rod for the force being applied will result in small strain, and the accumulation of minor movements can produce high percentage loss of strain and hence result in high losses of prestress. The use of large rods and large forces will produce problems in applying the force and distributing it through the masonry.

8.6.1. Losses and gains

The initial prestress force produced in the rod cannot be relied on to remain constant. Many changes can take place which reduce or increase the applied force over a period of time. These changes occur because of

- (a)   creep losses in the steel and masonry
- (b)   drying shrinkage in the masonry
- (c)   moisture expansion of the masonry
- (d)   temperature expansion and contraction of the steel and/or masonry
- (e)   slip and embedment losses of the anchorage and thread location
- (f)   compressibility of the damp-proof course.

These losses are not only important to the designer but also to the detailer. It is essential that the detailer is aware of the losses and how the detail may affect these losses. The following information is therefore important to the detailer. Creep losses in mild steel are far greater than for high tensile steel. Drying shrinkage of concrete and calcium silicate units is greater than that of clay masonry. Moisture expansion of clay masonry is usually greater than that for calcium silicate or concrete masonry. Locking of the upper nut by concreting in or using a locking device helps to reduce the losses due to slip but it should be noted that ordinary lock nuts can produce losses when the bottom nut is tightened against the upper and this loss would be over and above the normal 20% generally allowed for in design. Further discussion on damp-proof courses is included in section 8.9. The possible gain in prestress is usually ignored in design.

8.7. Cover and protection

Prevention of corrosion of the post-tension rod is one of the main considerations of the long-term performance of the system, particularly in exposed locations. The results of poor detailing of protection can result in cracking or spalling of the surrounding concrete or masonry, loss of cross-sectional area and hence loss of structural performance. There are many methods of preventing corrosion. The following list includes the main current systems adopted for rods in various locations

(*a*) treatment direct on to the surface of the rod to protect it from the local environment using a 'paint and wrap' process, i.e. a paint applied protection layer covered by a further protection layer using a tape wrapping

(*b*) similar direct protection layer using galvanising and a tape wrapping

(*c*) austenitic steels such as stainless steel

(*d*) dense grout cover to all steel after applying the force.

The adoption of any particular system should be related to that most suitable for the location and exposure of the element and its design life. The treatments listed can be used separately or in combination to give the necessary protection. However, the use of a grout cover which combines a requirement of bond should not be used in conjunction with a surface treatment which may slip. Examples of protection and comments on the details are shown in Figs 8.7–8.9.

Fig. 8.7. Typical rod protection detail: post-tensioned wall

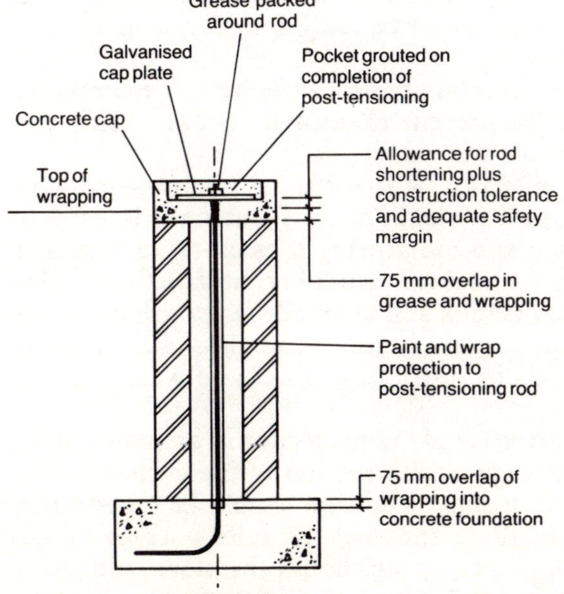

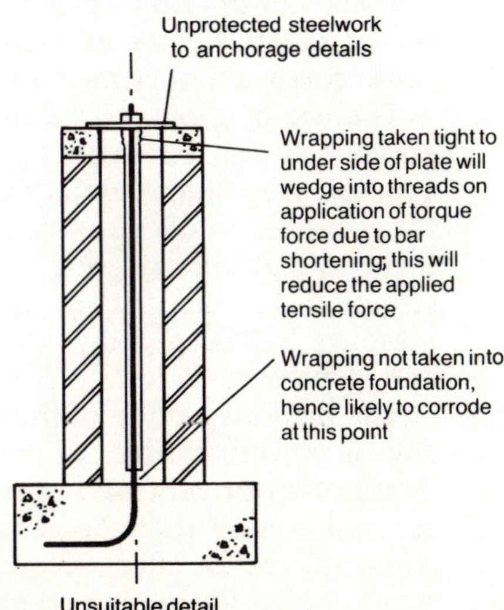

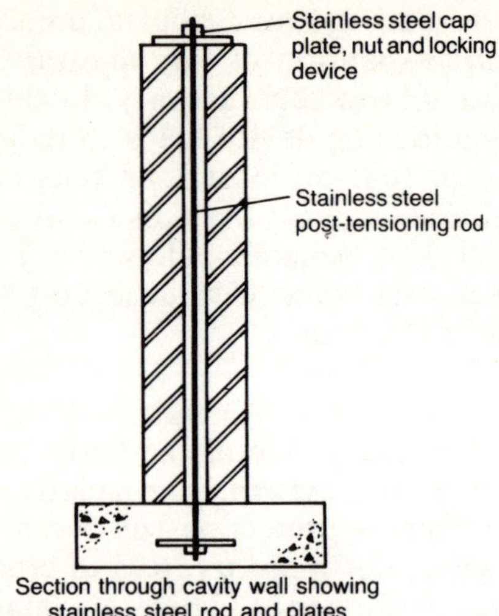

Fig. 8.8. Typical rod protection detail: cavity wall

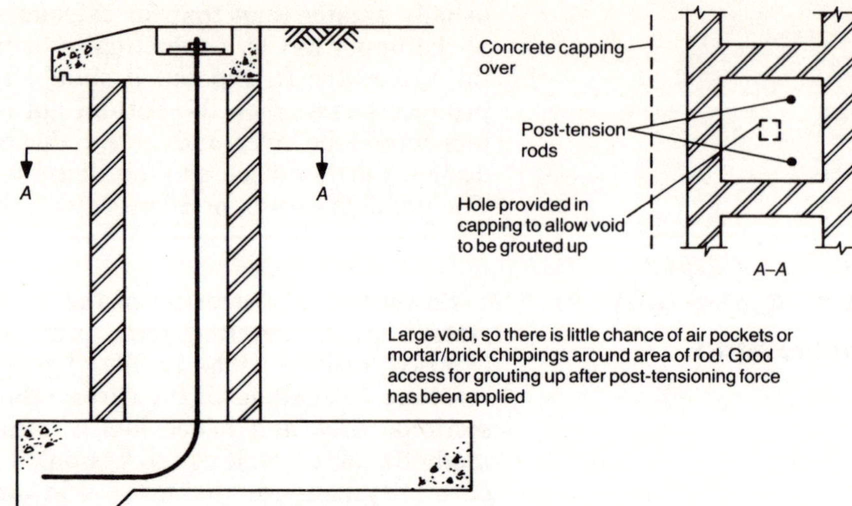

Fig. 8.9. Typical rod protection detail: diaphragm retaining wall

The choice of suitable protection depends on many considerations and for guidance certain points should be noted.

In the Authors' opinion, if galvanised steel is adopted for exposed situations it should be backed up by a coat of bitumen or semi-flexible moisture-resistant paint protection to cover the possible micro-cracking of the galvanising and thus extend the life of the element. This is particularly important for exposed and inaccessible locations.

In considering the cost of more expensive systems, such as the use of stainless steel, the effect on total cost should be considered. A high percentage increase on a small number of critical elements can be insignificant to the overall cost of the project, yet these can be the vital elements upon which the life expectancy of the project depends.

The use of a paint and wrap system (i.e. 'Denso' or similar, see Fig. 8.7) has been found to be both practical and effective in construction, supervision and performance. The protection can be easily inspected and checked prior to masonry construction and easily corrected if found to contain any workmanship defects.

When considering any system, or combination of systems, it is essential to ensure compatibility of materials to prevent electrolytic action which can cause corrosion of elements within the system.

The use of grout as the only protection is suitable for non-exposed situations but can be suspect for exposed locations. If grout only is used in exposed situations, careful detailing and monitoring is essential to minimise the possibility of air voids and/or mortar dropping obstructions. This involves particularly good practical details and close site supervision of construction. The fact that the grout protection cannot be applied until after the masonry has been constructed tends to restrict the quality of workmanship and inspection.

The final inspection of the effectiveness of the protection is extremely difficult, if not impossible, without damage or destruction of the elements. The depth of cover protection provided against corrosion should be based upon the assumption that the grout acts alone and that the brickwork or blockwork affords no effective protection. The cover should therefore be at least that adopted for reinforced masonry (see Tables 5.1 and 5.2).

The corrosion protection considerations do not relate to the rod alone; for example, the upper and lower anchorage positions are particularly vulnerable. Fig. 8.10 shows typical upper and lower anchorages for a vertical post-tensioned wall, with the lower anchorage cast into a concrete foundation.

The main factors relating to these details and their durability are those at the interfaces between separate elements.

For example, at base level the steel, as it enters the concrete anchorage, can be vulnerable to corrosion if attention to details and construction is not given to provide corrosion protection. The continuation of the rod protection into the foundation needs to be considered at the design stage when calculating the total length of anchorage required because the total length must allow for the unbonded length enclosed within the protective layer. At the upper anchorage the wrapping must not be allowed to interfere with the threads since this would reduce the force in the rod for a given prestress force. At the same time the rod must have corrosion protection and the threads need lubrication. Fig. 8.10 shows some methods of achieving these details.

8.8. Anchorages

8.8.1. Lower anchorage

The main structural requirement at the lower anchorage is to react the prestress force into the anchorage mass. The need to disperse the reaction

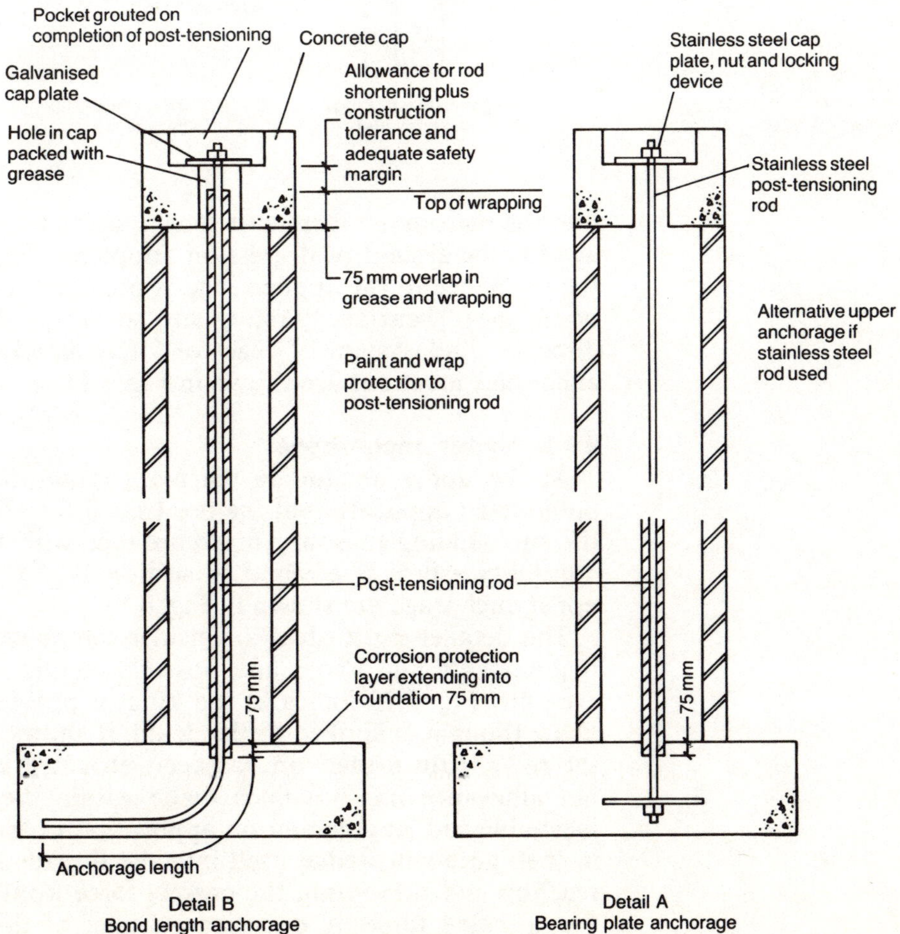

Fig. 8.10. Upper and lower anchorage details

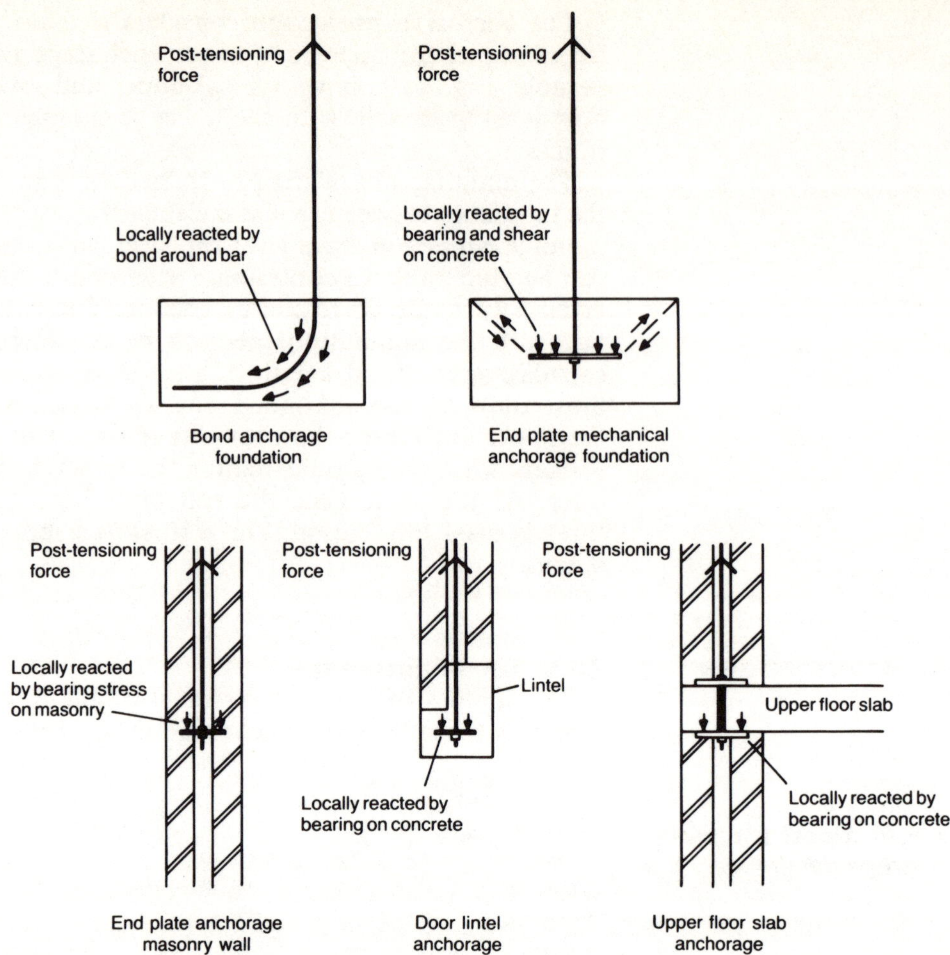

Fig. 8.11. Typical anchorage forces

into the masonry, concrete or steel governs the depth and size of the anchorage for the system of dispersion adopted. The anchorage must ensure that the resistance to the applied load is provided within the allowable stresses of bond, shear, bearing, bursting and so on for the materials being used. The direction and system of reactive forces involved for a number of various anchorage locations are shown in Fig. 8.11.

8.8.2. Upper anchorages

At the upper anchorage the main structural requirements are to allow unhindered application of the prestress force and to distribute this force into the surrounding masonry in accordance with the design requirements. The durability aspect is covered in section B.1. Typical details suitable for the upper anchorage are shown in Fig. 8.12.

The detailer must always consider the construction effects on the design requirements and avoid any possible conflict between the two needs. For example, Fig. 8.13 indicates the kind of problem which can result if inadequate thought is applied to the detail. It shows an upper anchorage in which the rod's main protection has been provided by a paint and wrap process, but allowance has not been made within the detail for the essential and inevitable rod lengthening on application of the force. The wrapping shown in the figure will wedge itself into the threads increasing the torque or jack reaction and preventing the correct force from being applied to the rod for the specified force. A detailing mistake of this kind could result in total failure of the element under the applied loads. The detail also shows inadequate durability protection at the interface of the rod with the capping plate.

8.9. Vertical and horizontal damp-proof membranes

Damp-proof membranes at right-angles to the direction of force or parallel to the direction of applied shear can be extremely critical to the performance of the post-tensioned element. They must not squeeze out or compress under the applied load nor slip under shear forces. These conditions must be met for the full range of temperatures to which the element will be subjected and without damage to the resistance of damp penetration.

For horizontal damp-proof membranes, consideration should be given to providing three courses of engineering brickwork or slate, particularly when the membrane will be subjected to the force directions indicated in Fig. 8.14.

For vertical damp-proof membranes located in joints designed as unbonded and where there are no vertical shears to be resisted, the choice of membrane is wide. In zones of vertical shear, however, the need to form a mechanical shear key makes the choice much more restricted. Metal shear keys designed to resist the applied shear as shown in Fig. 8.15 is the most common detail adopted.

This detail perforates the plane through which the damp-proof membrane will pass and hence an in situ membrane is required. This is usually achieved using a bituminous based brush-applied membrane to thoroughly filled and flushed up mortar joints, as shown in Fig. 8.16. Because of the obvious difficulties in construction and long-term performance, such jointing should, as far as possible, be avoided.

Fig. 8.12. Upper anchorage details

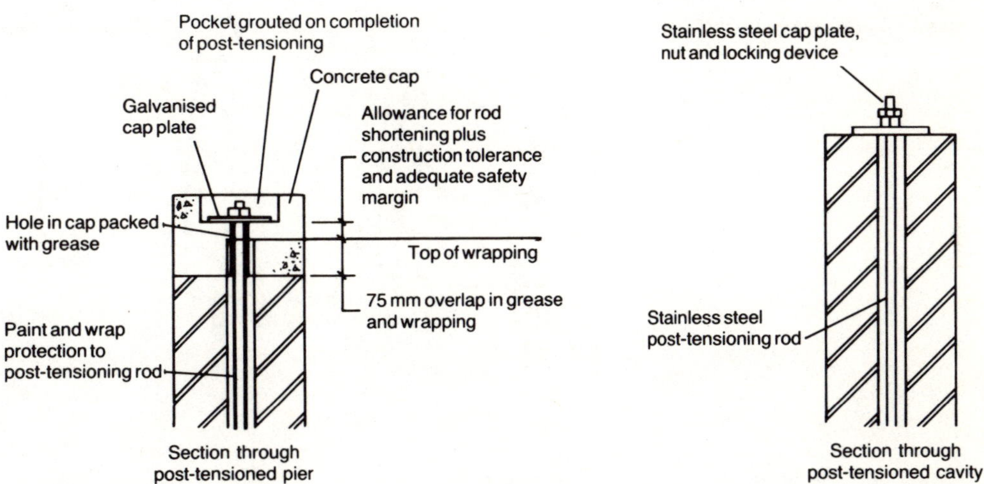

Fig. 8.13. Protective wrapping detail at upper anchorage

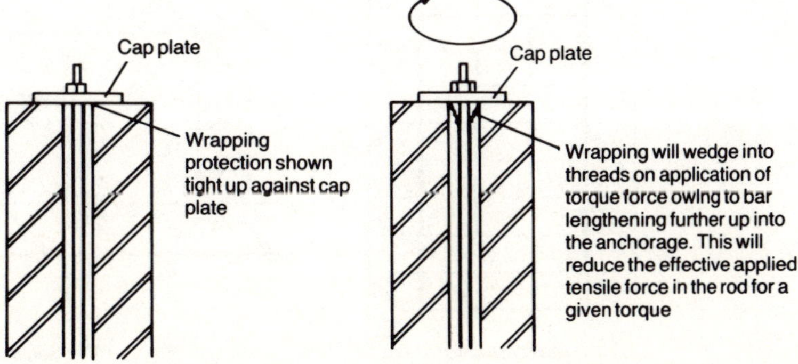

8.10. General arrangement details

As discussed in chapter 5, general arrangement drawings act as the basis of drawing information and provide an overall impression of the construction form, its mass and the materials being used. The drawing should provide

(*a*) scheme outlines including overall basic dimensions
(*b*) the strength of the masonry units, mortar and other material details

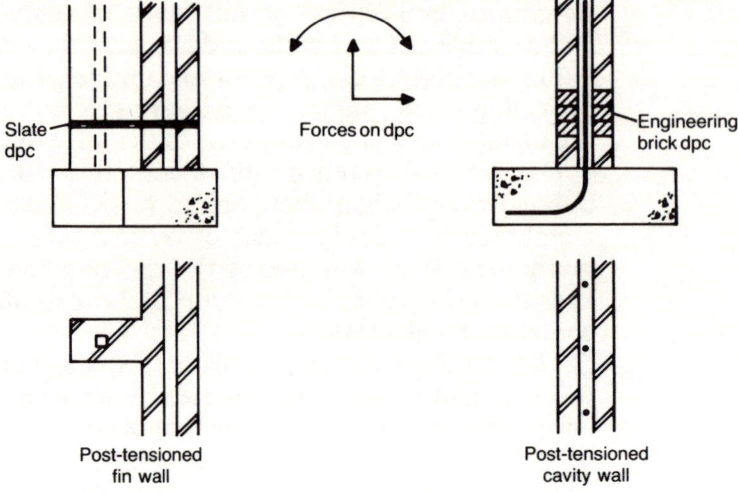

Fig. 8.14. Damp-proof membrane details

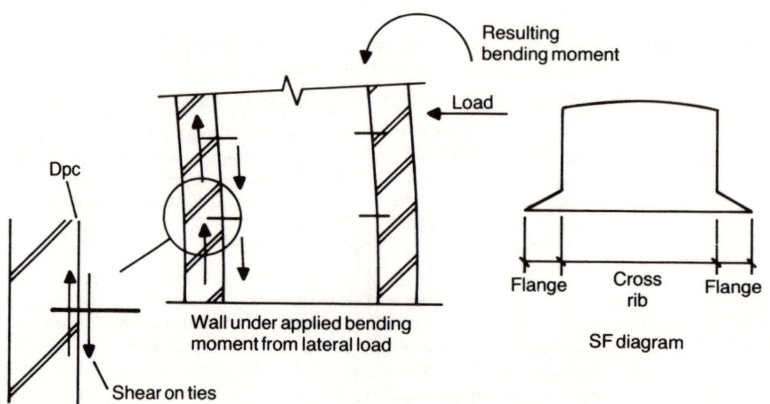

Fig. 8.15. Forces on metal shear keys in tied wall

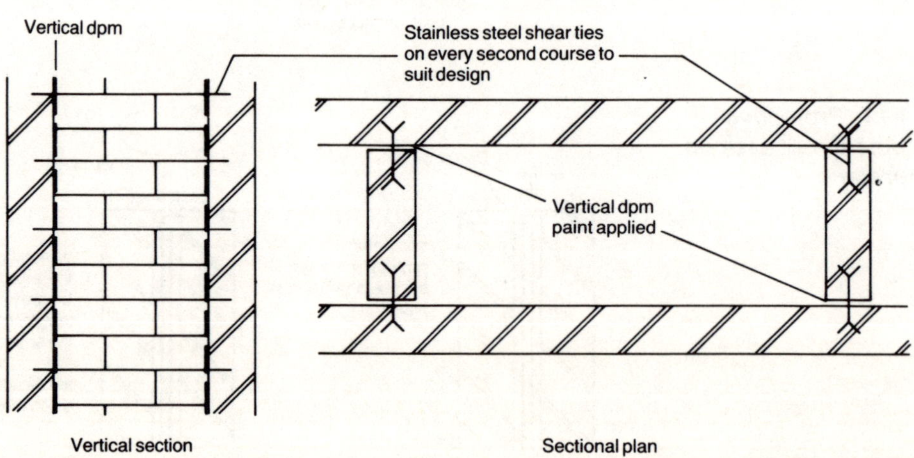

Fig. 8.16. Vertical damp-proof membrane in tied wall

Fig. 8.17. Post-tensioned channel piers to assembly hall

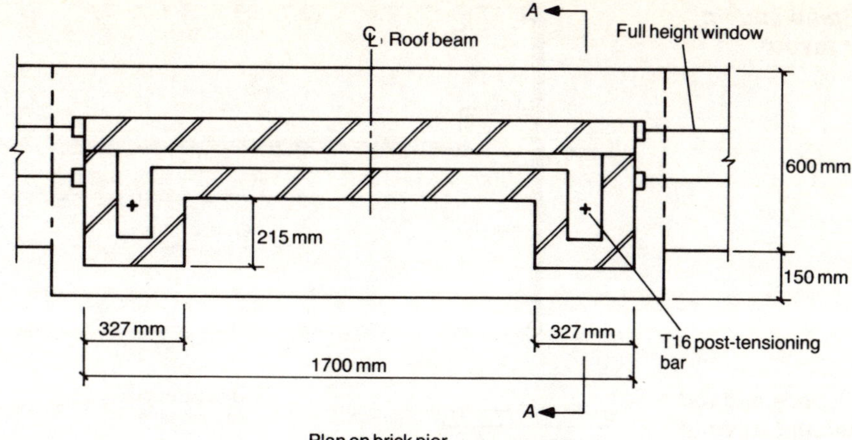

Fig. 8.18. Column and fin: typical floor layout

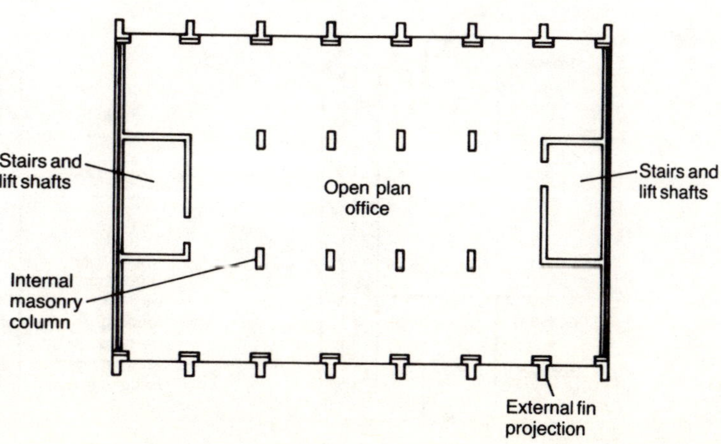

Fig. 8.19. Spine and fin: typical floor layout

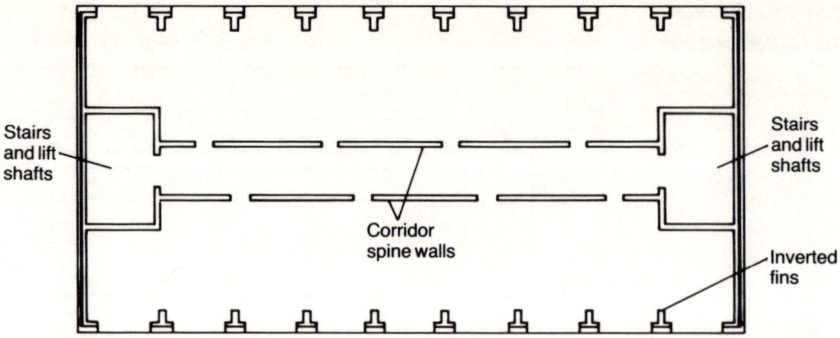

Fig. 8.20. Post-tensioned fins: internal and external

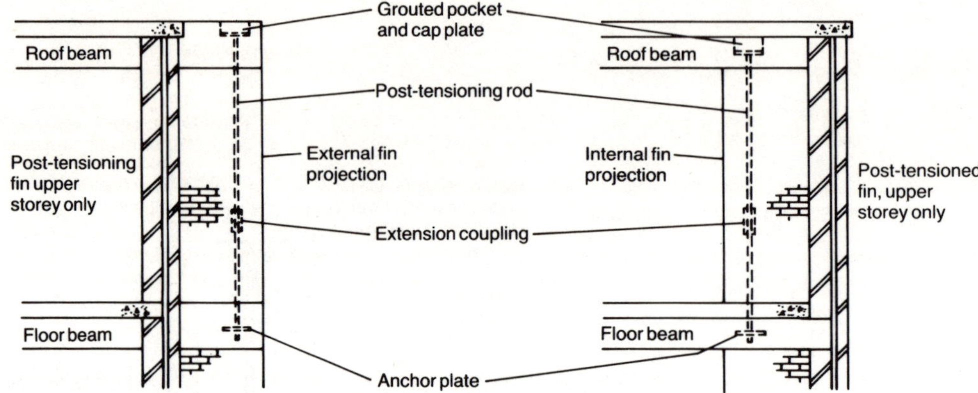

Fig. 8.21. Post-tensioned hollow fin: roof anchorage

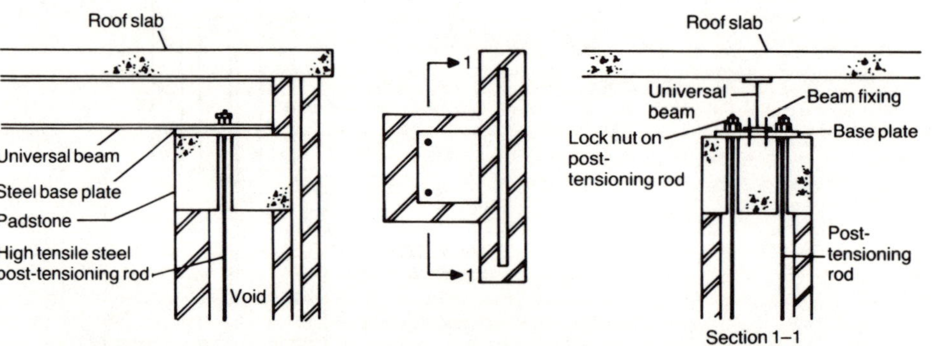

Details should also include information on such things as masonry strengths, torque to be applied, type and condition of thread, dimensions of cap plates and so on, durability specification for steel, rod clearances, torque sequence and topping-up, masonry strength at torquing, and permanent locking of torque nuts

Fig. 8.22. Post-tensioned hollow fin: upper floor anchorage

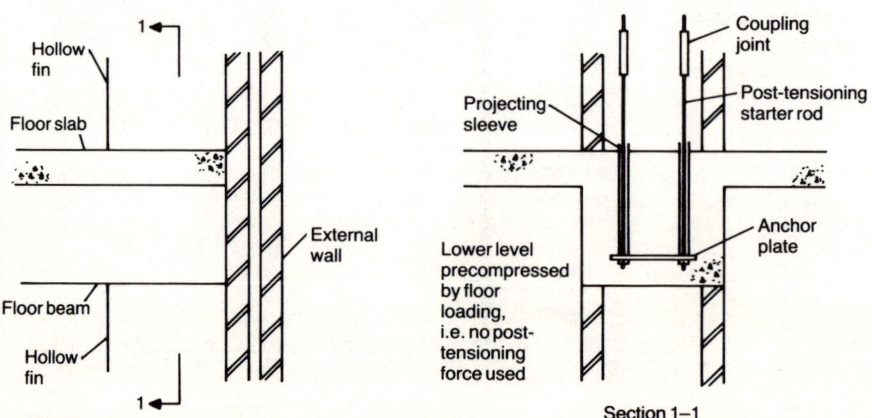

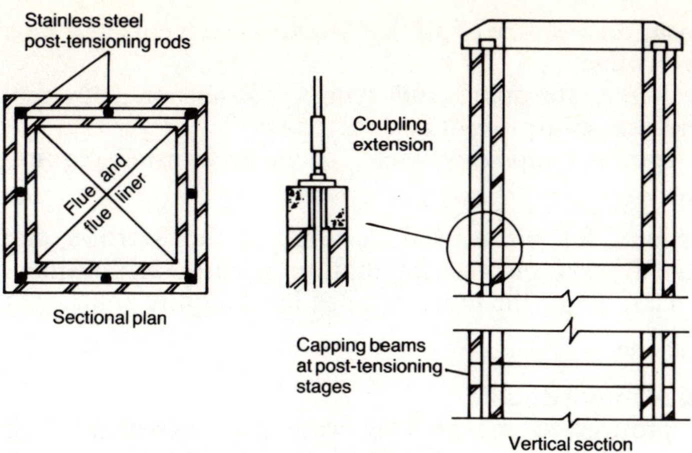

Fig. 8.23. Post-tensioned chimney

Fig. 8.24 (right). Post-tensioned fin retaining wall

Fig. 8.25. Post-tensioned box culvert

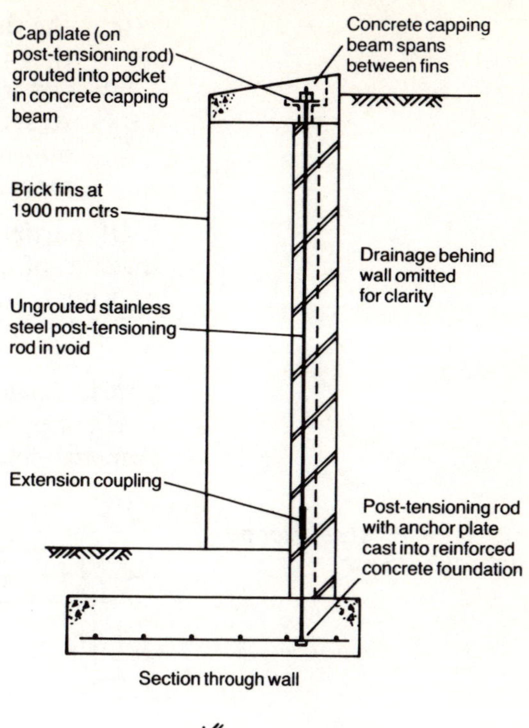

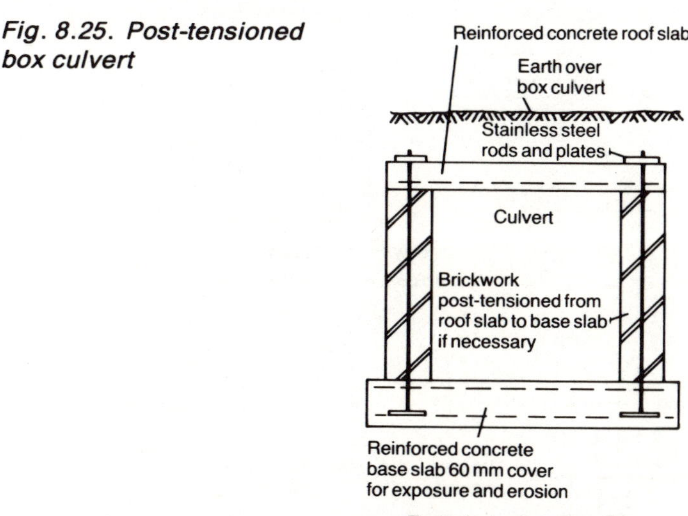

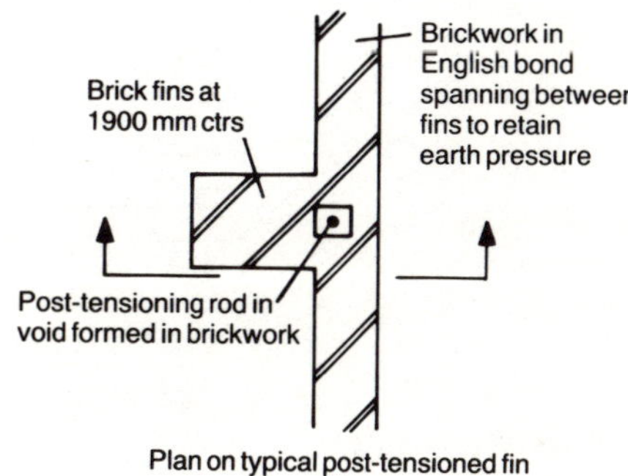

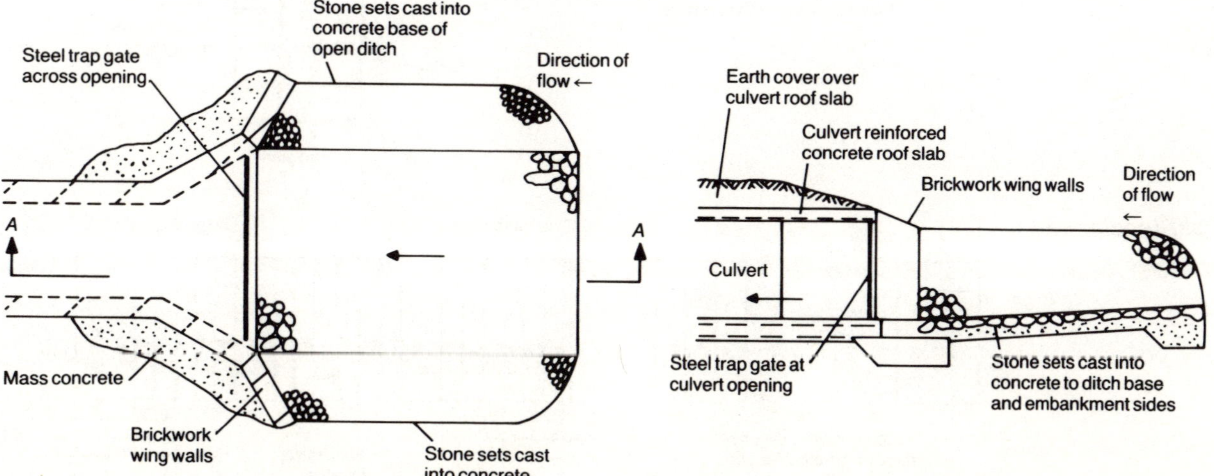

(c)   the thickness, location and size of all the structural elements and how they relate to each other

(d)   the location of joints, strapping and tying, post-tension rods and structurally significant damp-proof courses

(e)   information on special temporary works or other unusual requirements.

Of particular importance for prestressed masonry is the location and spacing of post-tensioned rods on the foundation general arrangement drawing together with reference to the more detailed local requirements.

Further information on general arrangements is given in reference 2.

8.10.1. Post-tensioned element details

The application of prestressed masonry is wide and increasing. The element details shown and discussed therefore mainly cover the more

Fig. 8.26. Post-tensioned flat soldier arch

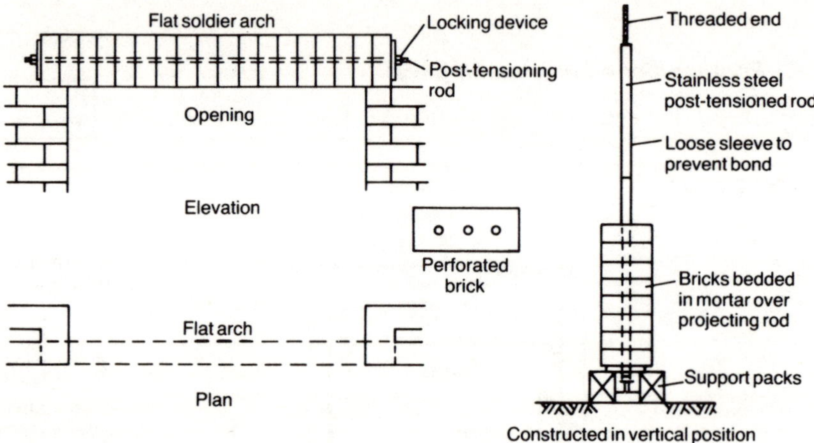

Fig. 8.27. Prestressed flat soldier arch

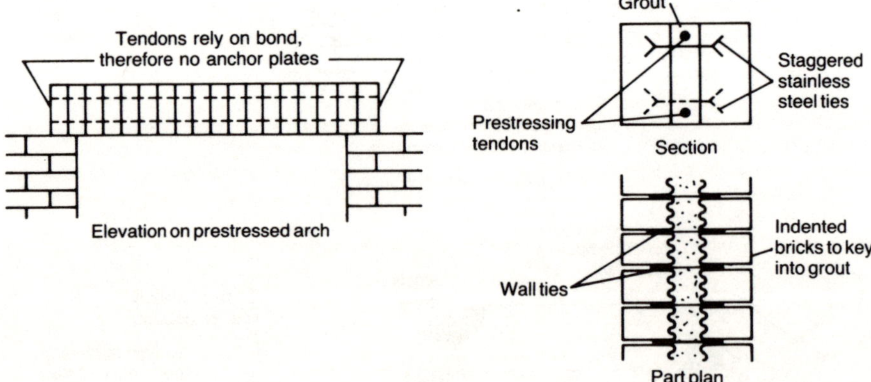

Fig. 8.28. Post-tensioned box beam

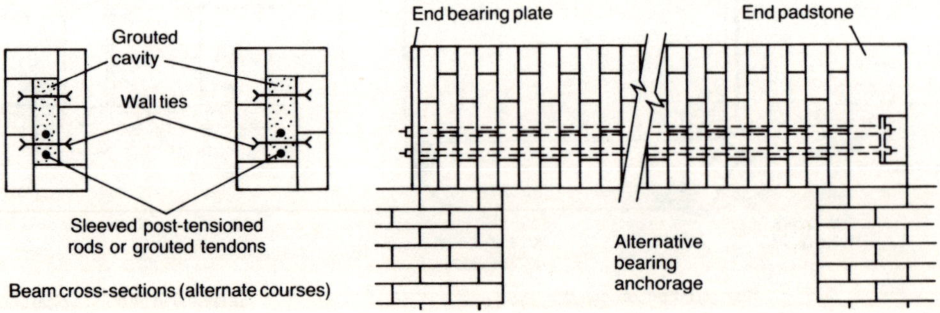

common elements with a limited number of the more unusual applications. The details, although limited, should provide basic guidance on the application and details of prestressed masonry. Figs 8.17–8.25 give a variety of notes for guidance but for clarity they do not indicate all the notation necessary on a completed working drawing. Complete working drawings must show clearly all the details necessary for construction, including anchorage, rod protection, masonry strength, capping plates, caps and/or beams, and

Fig. 8.29. Box beam: anchorage and bearing details

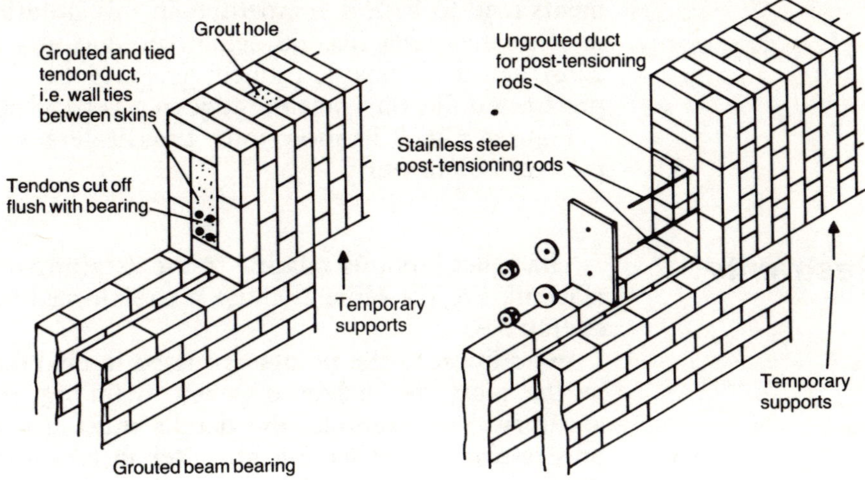

Fig. 8.30. Roof connection to post-tensioned wall

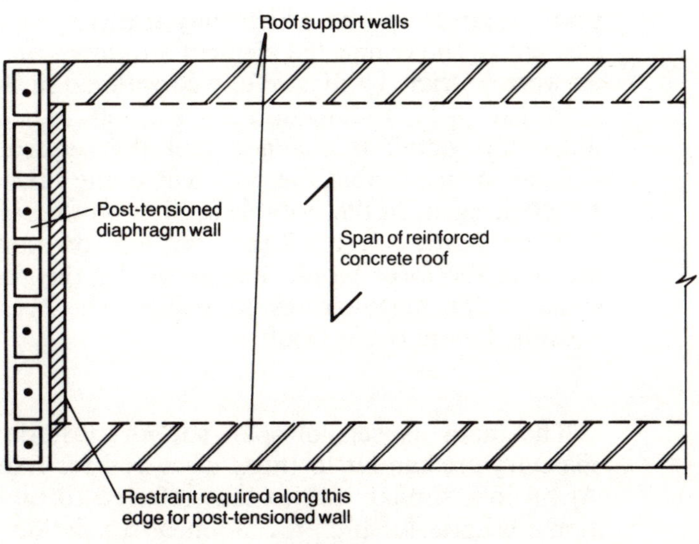

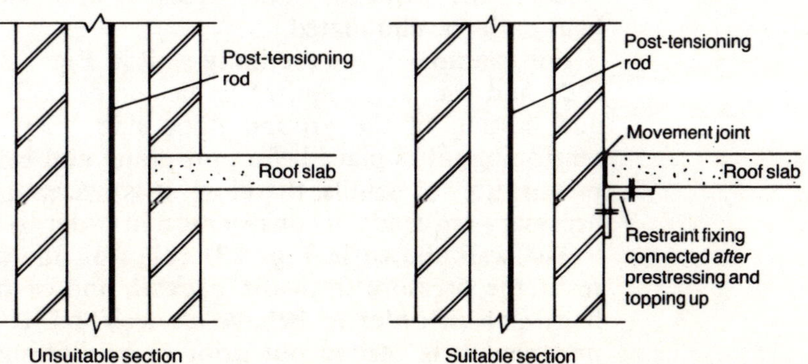

the sequence of construction and/or prestress force to be applied, including notes on the topping-up of prestress.

Figures 8.17–8.25 give some typical details and notes for various applications.

The use of prestressing for horizontal masonry elements, such as lintels and beams is not as popular as prestressing for vertical sections for a number of reasons. First, the advantages of post-tensioning in a vertical direction are not applicable to the horizontal direction, and masonry elements tend to be less competitive in this situation than elements constructed in other materials. Second, reinforced masonry is often cheaper and easier to construct than post-tensioned masonry for horizontal members, whereas prestressed masonry has the edge in a vertical situation.

Figures 8.26–8.30 show some typical details for horizontal elements using prestressed masonry.

8.11. Restraints

The general points relating to the detailing of restraints for post-tensioned masonry are the same as those for reinforced masonry, and are discussed in section 5.9.

In addition to the points discussed in chapter 5, however, further requirements must be achieved when detailing restraints for post-tensioned masonry. For example, the details must not interfere with or hinder the prestressing of the section nor alter in any way the stress path assumed in design.

For example, a restraint detail which is fixed in a post-tensioned element prior to stressing and which may transfer the load from the post-tensioned element to the connected element would be unsuitable unless the load transfer were restricted and carefully considered in the original design.

In Fig. 8.30, the section on the left indicates an unsuitable condition unless the detail is changed and the sequence of construction is clearly defined to ensure that the post-tensioning takes place before the roof is constructed, e.g. as in the suitable section on the right.

If the contractor fixed the roof and restraint prior to prestressing, then much of the force would be carried by the roof slab, and the compressive stress in the masonry resulting from the prestress would be reduced and possible failure could result.

8.12. Span and support details

The main aspects of span support criteria for detailing post-tensioned masonry are similar to those discussed for reinforced masonry in chapter 5. Again in a similar way to that referred to under restraint details, the additional criteria for the prevention of restriction to the process of prestressing the rods must be achieved. It is therefore essential that adequate attention be given to the sequence of construction and that the possibility of alternative load paths be eliminated.

For example, if the retaining wall in Fig. 8.31 incorporates post-tensioned rods and the wall requires to be temporarily propped prior to backfilling and casting of the ground floor slab, then it is essential that the post-tensioning takes place before propping and before backfilling. As part of the preparation of details, therefore, it is necessary to note on the drawing any necessary sequence of construction in order to achieve the correct end result.

The wall shown in Fig. 8.31 relies on the prestressing forces in order to resist the pressure from the backfill, and in turn relies on the tying of the floor slab in order to behave as designed. It is therefore essential that the prestressing is carried out prior to backfilling and prior to any temporary

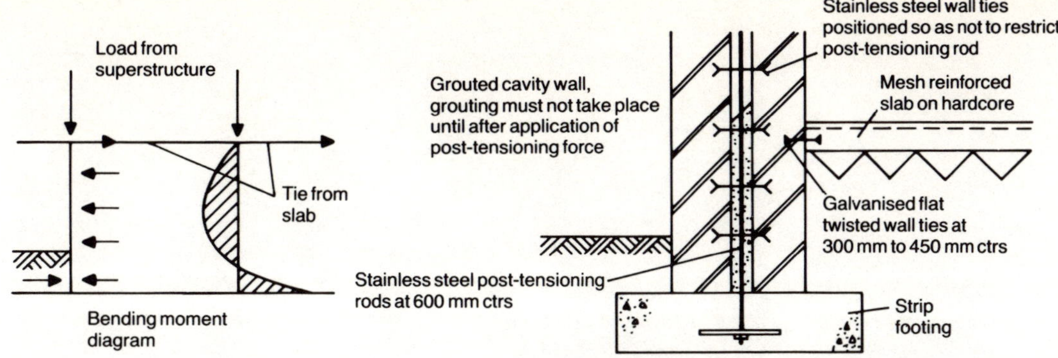

Fig. 8.31. Post-tensioned cavity retaining wall

propping. It is also essential that temporary propping be introduced after prestressing and before backfilling in order to make the wall stable for the temporary condition. Once the wall has been prestressed, it can be propped and backfilled up to the slab. Once the ground floor slab has been cast and cured adequately, the props can be removed.

9

Composite design

9.1. Composite action of load-bearing masonry panel walls supported on reinforced concrete beams

Composite action between masonry walls and reinforced concrete supporting beams is considerably under-exploited by many engineers. This is perhaps partly due to there being no design guidance given in current codes of practice. However, much research has been undertaken by various organisations and the Authors would recommend the adoption of the basic principles proposed by Wood and Simms.[20] This design method is considered to provide a sound and logical basis for the design of loadbearing masonry walls acting compositely with a reinforced concrete supporting beam.

The paper by Wood and Simms was related to the requirements of CP 111.[21] For this book it has been converted by the Authors to the requirements and philosophy of limit state design, and a number of other suggested adjustments have been made. The original paper is converted and updated and is discussed in the following pages. A worked example using this tentative design method is provided in section 9.6.

9.2. Introduction

In 1952 a design study[22] was published for brick walls supported by reinforced concrete beams subjected to gravity loading with an extra load on top of the wall. The design rules proposed were empirical and based on tests of typical house walls.

Because of arching action, the reaction between the wall and the beam was concentrated near the supports and this had the effect of considerably reducing the bending moment in the beam. Thus where there were no door or window openings near to the supports and provided that arching was (by inspection) possible over a door or window opening near to the centre of the span (or if there were no openings at all), then the equivalent bending moment, for design purposes, for a simply supported beam was reduced from $WL/8$, for a uniformly distributed load, to a value of $WL/100$, where W is the total load and L is the span.

There have been numerous applications in practice (this design procedure has been remarkably popular) and no case of failure has been reported so far as the Authors are aware.

However, there have been several cases where engineers have attempted to use these design rules for loadbearing walls, with much heavier loadings than house walls, again without any reported failures to date.

At the Building Research Establishment extra tests have been carried out to failure, which has shown that arching action is clearly taking place, leading eventually to crushing of the masonry near to the support.

The essential point is that if the loadbearing 'engineered' walls are heavily loaded, right up to their compressive limits, then there is little or no reserve of compressive strength left for the concentrated compressive stresses created in the masonry by composite action.

This chapter suggests a tentative design procedure, believed to be conservative, based on engineering judgment of the way composite action of heavily loaded walls would occur. The procedure has been developed in the light of extra tests carried out and taking due recognition of the increased safety factors used for masonry compared with reinforced concrete.

9.3. Proposed simple design method for heavily loaded walls acting compositely with reinforced concrete beams

In these conditions shear and bond will require analysis but the essential problem is to determine a sliding-scale equivalent bending moment.

The equivalent bending moment is $n_w L^2/Q_{ca}$ where n_w is the design load per unit length on the masonry (including self-weight), L is the span of the beam and Q_{ca} is a factor ranging from a maximum of 100 for full composite action with lightly loaded walls to a possible minimum of 8 where all composite action has disappeared and only uniform load can be allowed on the beam.

Moreover, in order to assess what is meant by lightly or heavily loaded walls it may be advisable to link this tentative design procedure with BS 5628[1] where possible.

It is proposed to use the distribution of interactive forces between wall and beam, as shown in Fig. 9.1. This simplified loading diagram will be conservative in that it probably overestimates the bending moment on the beam; however, it probably underestimates the actual peak stresses in the masonry locally by assuming complete stress redistribution. Nevertheless, equilibrium is preserved as there must be some local increase in the strength of the masonry in contact with the concrete.

On this basis there is an average stress concentration C_s in the masonry, given by

$$C_s = L/2x$$

where x is shown in Fig. 9.1. Therefore

$$x = L/2C_s$$

The moment anywhere near to the mid-span is

$$M = n_w L^2/Q_{ca}$$
$$= \frac{n_w L}{2} \times \frac{x}{2}$$
$$= n_w Lx/4$$

Fig. 9.1. Assumed equivalent beam loading at wall failure

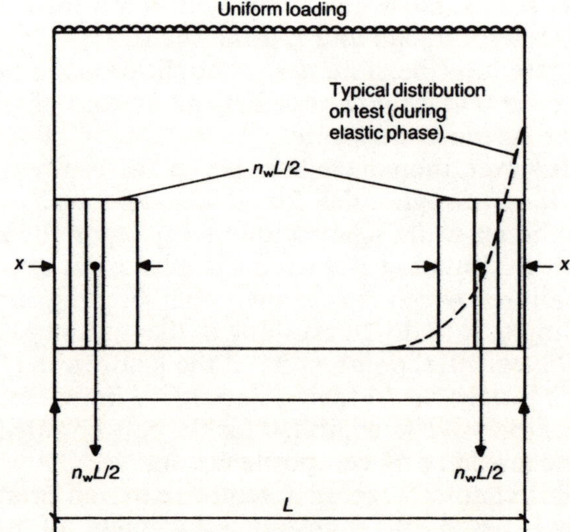

Bending moment factor, Q_{ca}	Stress extent, x/L	Stress concentration factor, C_s	Remarks
8	1/2	1·0	No composite action
12	1/3	1·5	Increase in composite action
24	1/6	3·0	
18	1/12	6·0	
100	1/25	12·5	Maximum composite action

This leads to

$$1/Q_{ca} = x/4L = 1/8C_s$$

or

$$C_s = Q_{ca}/8 \tag{9.1}$$

$$Q_{ca} = 8C_s \tag{9.2}$$

$$x/L = 4/Q_{ca} \tag{9.3}$$

where C_s is the stress concentration factor, Q_{ca} is the bending moment factor and x/L is the stress block extent.

This means that the bending moment factor, the stress block extent and the stress concentration factor have typical values such as are given in Table 9.1. It can be seen clearly that to make use of maximum composite action the wall must be considerably understressed as a whole.

BS 5628[1] suggests maximum allowable stresses for both uniformly distributed and concentrated loading. These are

$$f_a = f_k/\gamma_m$$

which is the basic allowable stress related to the strength of the brick or block and the mortar designation

$$f_{udl} = \beta f_k/\gamma_m$$

which is the allowable uniformly distributed compressive stress taking account of the slenderness of the wall, and f_c which is the allowable maximum stress resulting from a combination of uniformly distributed and concentrated loads.

BS 5628 also gives the following relationships

$$f_c = 1·5f_a$$

for local stress concentrations within 0·4 of the wall height measured from a horizontal support and taken as an average of the various bearing types given in BS 5628, and

$$f_{udl} = \beta f_a$$

$$= \beta f_k/\gamma_m$$

where β is the capacity reduction factor for wall slenderness.

Also, in practice, suppose the actual average wall stress f_w is less than the allowable f_{udl} by a reduction factor R (i.e. $f_w = Rf_{udl}$). Then

$$f_w = R\beta f_a$$

$$= R\beta f_k/\gamma_m$$

Taking advantage of the local allowable stress concentration factor, the stress concentration factor C_s is given by

$$C_s f_w < 1{\cdot}5 f_a$$

i.e.

$$C_s R\beta f_a < 1{\cdot}5 f_a$$

i.e.

$$C_s R\beta < 1{\cdot}5$$

and since, from equation (9.1)

$$C_s = Q_{ca}/8$$

then

$$R\beta < 12/Q_{ca} \qquad (9.4)$$

or

$$Q_{ca} < 12/R\beta \qquad (9.5)$$

where the design bending moment is $n_w L^2/Q_{ca}$.

Equation (9.5) is, therefore, effectively the basis of a design rule relating the required beam bending moment to the degree of loading on the wall.

Some typical examples are given in Table 9.2.

9.4. Possible modifications to the simple design method

At some risk of making the suggested design method less simple, the following modifications would refine it

- modification of the stress concentration factor of $1{\cdot}5$ so that it eventually becomes unity with the limiting case of completely distributed load on the beam (see section 9.4.1)
- allowance for increased allowable steel stress in the beam; this is also modified as the limiting case is approached (see section 9.4.2)
- taking account of the extra horizontal thrust H (see section 9.4.3 and Fig. 9.2).

Table 9.2. Composite action: horizontal thrust

Example	Type of wall	Required moment of resistance for beam design
(a)	Squat wall (slenderness ratio = 6), stressed to code limit: $\beta = 1{\cdot}0$, $R = 1{\cdot}0$	$n_w L^2/12$
(b)	Slender wall (slenderness ratio = 25), fully stressed; $\beta = 0{\cdot}5$, $R = 1{\cdot}0$	$n_w L^2/24$
(c)	Slender wall (slenderness ratio = 25), stressed to only one half the code limit; $\beta = 0{\cdot}5$, $R = 0{\cdot}5$	$n_w L^2/48$
(d)	Slender wall (slenderness ratio = 25), stressed to only a quarter of the code limit; $\beta = 0{\cdot}5$, $R = 0{\cdot}25$	$n_w L^2/96$

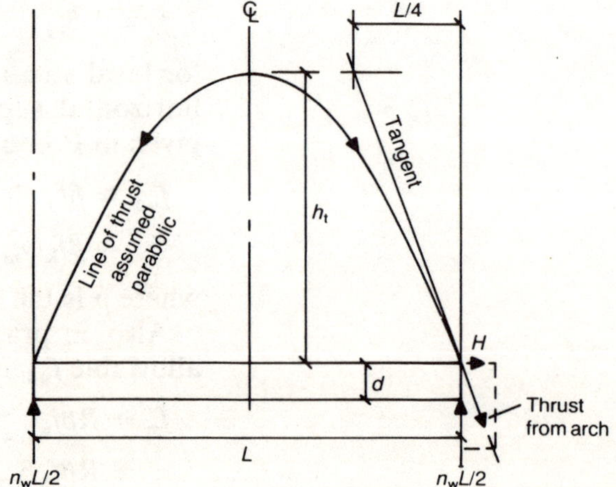

Fig. 9.2. Line of thrust and extra horizontal forces

9.4.1. Variable stress concentration factor F_{vs}

It would appear reasonable to allow the factor $1\cdot5$ when the maximum composite action is envisaged (i.e. design moment $= n_w L^2/100$) to be reduced linearly with respect to Q_{ca} towards the value $1\cdot0$ as the design moment approaches $n_w L^2/8$.

Putting

$$F_{vs} = aQ_{ca} + 6$$

gives

$$1\cdot5 = (a \times 100) + 6$$

$$1\cdot0 = (a \times 8) + 6$$

Hence

$$F_{vs} = (176 + Q_{ca})/184 \tag{9.6}$$

9.4.2. Variable steel stress

The present allowable steel stress for full composite design is 110 N/mm^2 (this was introduced by Wood[22] to control the extension of the tie, i.e. bottom steel supporting the arch). BS 8110[12] allows 250 N/mm^2 when there is no composite action.

Again putting

$$f_{sa} = (aQ_{ca}) + b$$

gives

$$110 = (a \times 100) + b$$

$$250 = (a \times 8) + b$$

Hence

$$f_{sa} = 254 - 0\cdot54Q_{ca} \tag{9.7}$$

which is the allowable steel stress corresponding to a given bending moment factor.

9.4.3. Horizontal thrust on beam

Providing there are no slip surfaces between the wall and the beam, such as damp-proof courses, then in full composite action the beam is acting mainly as a tie and in the elastic phase the equivalent bending moments, based on actual experiments, effectively correspond to the required stress in the ties. (This is the reason why the equivalent moment-arm (i.e. 2/3 of the depth of the wall) is more logical.) However, at ultimate load there is considerable redistribution of the stresses and the situations shown in Figs 9.1 and 9.2 should be combined.

If H is the horizontal thrust at the interface at the top of the beam, this is equivalent to a tension H at the beam centreline together with a negative bending moment of $-Hd/2$, where d is the depth of the beam.

As regards the negative bending moment, observing that the moment-arm is approximately equal to d (where d is the effective depth of the reinforced concrete beam), the induced compression in the bottom steel is

$$-\frac{Hd/2}{d} = -\frac{H}{2}$$

As regards the overall tension H on the centreline, and noting that the bending moment at the centre of the beam remains positive, then the tension

H would be redistributed between the compressive stress block in the beam and the bottom reinforcement, decreasing the compression in the former and increasing the tension in the latter. It would therefore be conservative to assume that all the tension should be taken on the bottom reinforcement. This means that the net required design tension in the reinforcement is $+H - H/2 = +H/2$.

Moreover, assuming a parabolic line of thrust with height h_t (see Fig. 9.2), the tangent of the inclination of thrust at the supports is h_t divided by $L/4$. Hence

$$H = \frac{n_w L}{2} \times \frac{L}{4h_t}$$

$$= \frac{n_w L^2}{8h_t}$$

Simplifying the complication of different heights, it is noted that the previous maximum allowed value of h_t is $0\cdot 7L$, whereas the minimum allowed value of h_t is $2/3(0\cdot 6L) = 0\cdot 4L$, because the smallest wall height is $0\cdot 6L$. The mean value of h_t is $0\cdot 55L$ which gives the probable value of

$$H = \frac{n_w L}{8} \times \frac{1}{0\cdot 55}$$

$$= \frac{n_w L}{4\cdot 4}$$

Moreover, the range of allowable beam depths varies from $L/20$ to $L/15$ and an average beam depth could be taken as $L/17\cdot 5$ approximately.

This means that the equivalent additional bending moment to produce a design tension of $H/2$ is approximately equal to

$$Hd/2 = n_w L^2/8\cdot 8 \times 17\cdot 5$$

$$= n_w L^2/154$$

However, the horizontal thrust must be assumed to reduce to zero when the bending moment is $n_w L^2/8$ (i.e. when the load is a uniformly distributed load) and to be fully operative when the bending moment is equal to $n_w L^2/100$.

It would appear reasonable to incorporate a reduction factor of $(Q_{ca} - 8)/92$ giving

$$\text{total effective bending moment} = \frac{n_w}{2} \times \frac{Lx}{2} + \frac{n_w L^2}{154}\left(\frac{Q_{ca} - 8}{92}\right)$$

so that

$$\frac{1}{Q_{ca}} = \frac{x}{4L} = \frac{1}{154}\left(\frac{Q_{ca} - 8}{92}\right) \tag{9.8}$$

Since it is already known that $x/4L = 1/8C_s$, it will be found that

$$C_s = \frac{19\cdot 25 Q_{ca}}{154 - Q_{ca}(Q_{ca} - 8)/92} \tag{9.9}$$

Following the previous argument involving the stress reduction factors in the wall, the corresponding equation is

$$C_s R\beta f_a < F_{vs} f_a$$

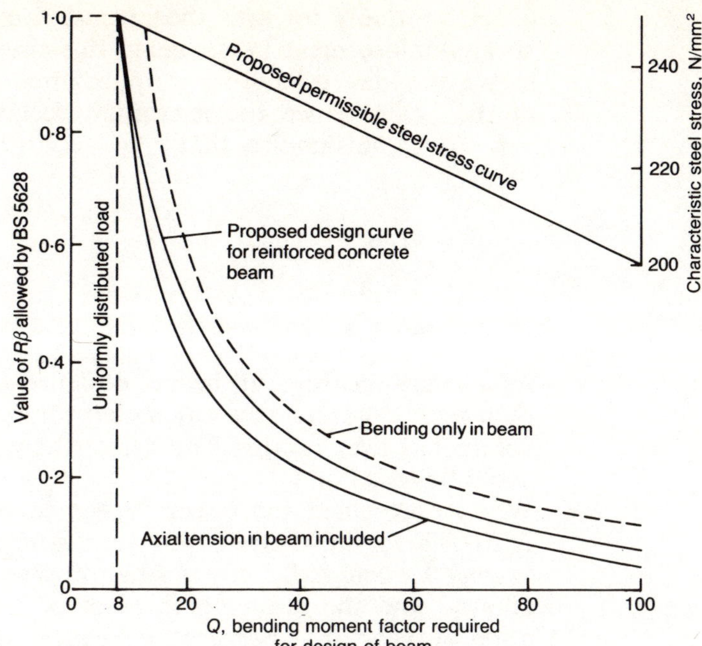

Fig. 9.3. Composite action design curves—the proposed permissible steel stress curve is applicable to both equations

which leads to

$$R\beta < \frac{176 + Q_{ca}}{184}\left(\frac{154 - Q_{ca}(Q_{ca} - 8)/92}{19\cdot25Q_{ca}}\right) \tag{9.10}$$

which should be compared with equation (9.4).

This is best plotted graphically, as shown in Fig. 9.3 for the curve including axial tension in the beam. This curve is more conservative than that for equation (9.4). It would be reasonable to regard equation (9.4) as giving an optimistic upper bound for this design bending moment factor and equation (9.10) as giving a conservative lower bound.

This new curve, including axial tension, would require the following bending moments for the cases (a)–(d) in Table 9.2.

- case (a) $\beta = 1\cdot0$: $R = 1\cdot0$ requires $n_w L^2/8$
- case (b) $\beta = 0\cdot5$: $R = 1\cdot0$ requires $n_w L^2/17$
- case (c) $\beta = 0\cdot5$: $R = 0\cdot5$ requires $n_w L^2/34$
- case (d) $\beta = 0\cdot5$: $R = 0\cdot25$ requires $n_w L^2/64$

It is suggested that the design of reinforced concrete beams be based on the proposed curve shown in Fig. 9.3 until further experimental data become available.

9.5. Summary of proposed design rules for load-bearing walls on beams

Providing there are no door or window openings near the supports which would destroy an arching system then the following points apply.

- The equivalent bending moment may be calculated conservatively using the proposed design curve shown in Fig. 9.3.
- The allowable steel stress may be obtained from the same figure and varies with the degree of composite action.
- No beam supporting a loadbearing wall is presumed to have a smaller bending moment than $n_w L^2/100$.
- The minimum height of the wall must be $0\cdot6L$.
- A property of composite action is that the compression in the arch

collects radially towards the nearest firm support. Hence, as regards shear reinforcement in the beam this should be theoretically unnecessary if the loaded length x is approximately no greater than the depth of the beam. Since the maximum recommended depth agrees with $d/L = 1/15$, this implies that

$$Q_{ca} \simeq 4L/x$$

$$= 4L/d$$

$$= 4 \times 15$$

$$= 60$$

As a safety measure, all beams designed for bending moments greater than $n_w L^2/60$ should be calculated for resistance to shear. (No shear failure has ever occurred on tests where maximum composite action could be relied on.)

- Wood[22] expected the beam depths to lie within the range $L/15$ to $L/20$. These limits are still recommended. When, however, with a heavily loaded wall, only a small degree of composite action can be allowed and the beam depth must be increased, either for consideration of shear or to prevent the beam from being over-reinforced, then the designer should use the required depth, even if this exceeds $L/15$. When considerable composite action is present (by inspection of Fig. 9.3, when the bending moment factor exceeds, say, 40) and only a small percentage of steel is required, if the designer has practical reasons for preferring a depth of beam greater than $L/15$, then the steel area should be calculated as though the beam still had a limiting depth of $L/15$. This is to avoid having too small an amount of steel acting as a tie.

- All the tests described by Wood[22] were made using a reinforced concrete beam in the composite wall–beam system, but the beneficial composite action observed could reasonably be expected with other types of supporting beam with the same bond conditions at the wall–beam interface, say for instance a concrete encased steel joist. However, where test data are not available for a beam type giving this equivalent interface bond, the designer may assume that the design of the beam can be obtained from the proposed curve (Fig. 9.3) provided that, for the time being, the bending moment has an upper limit of $n_w L^2/50$ and the horizontal shear resistance is checked using an appropriate frictional value of the beam–wall interface.

9.6. Design example

The following design example illustrates the application of this tentative design method.

The wall panel in Fig. 9.4 is supported by a reinforced concrete beam. There are no openings in positions that would prevent the panel from arching, so it is proposed to take advantage of the composite action between the panel and the beam.

The wall panel comprises clay bricks with a crushing strength of 20 N/mm² set in a designation (iii) mortar; the partial safety factor for materials γ_m is taken as 2·5 for this example. (Note the use of the plain masonry γ_m factor.)

The bending moment factor Q to be used in the design of the beam is calculated by both the simple and the modified methods.

First, check the ratio of height to span. It is recommended that this should

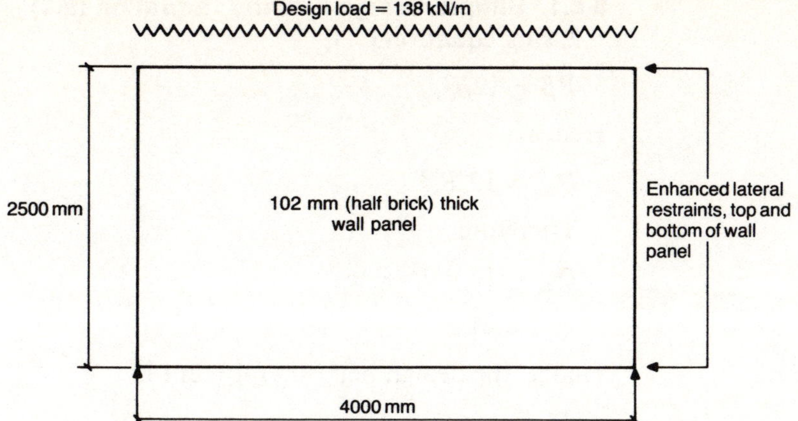

Fig. 9.4. Wall panel: composite action design: (a) bending only in beam; (b) axial tension in beam included

not be taken as greater than unity for composite action to be taken into account in the design

$$\text{height/span} = 2500/4000 = 0\cdot625$$

Composite action may therefore be considered. For a 102 mm thick wall with effective height of $0\cdot75 \times 2500$ and slenderness ratio of $0\cdot75 \times 2500/102 = 18$, from table 7 of BS 5628: Part 1[1] $\beta = 0\cdot77$.

The characteristic compressive strength of masonry f_k taken from table 2(a) of BS 5628: Part 1 is

$$f_k = 5\cdot8 \text{ N/mm}^2$$

Therefore, the basic allowable compressive stress f_a is

$$f_a = f_k/\gamma_m$$

$$= 5\cdot8/2\cdot5$$

$$= 2\cdot3 \text{ N/mm}^2$$

and the allowable (uniformly distributed) compressive stress f_{udl} is

$$f_{udl} = \beta f_a$$

$$= 0\cdot77 \times 2\cdot3$$

$$= 1\cdot8 \text{ N/mm}^2$$

The allowable concentrated stress f_c is given by

$$f_c = 1\cdot5 f_a$$

$$= 1\cdot5 \times 2\cdot3$$

$$= 3\cdot5 \text{ N/mm}^2$$

and the stress in the wall at the design load is

$$g_A = 138 \times 10^3/102 \times 10^3$$

$$= 1\cdot35 \text{ N/mm}^2$$

Therefore

$$R = g_A/f_{udl}$$

$$= 1\cdot35/1\cdot8$$

$$= 0\cdot75$$

9.6.1. Simple design method: equation (9.4)

Using equation (9.4)

$$R\beta = 12/Q_{ca}$$

Hence

$$Q_{ca} = 12/R\beta$$

Therefore

$$Q_{ca} = 12/0{\cdot}75 \times 0{\cdot}77$$

$$= 21$$

Hence, the design bending moment is

$$M = n_w L^2/21$$

The steel stress is

$$f_{sa} = 254 - 0{\cdot}54 Q_{ca} \tag{9.7}$$

$$= 254 - (0{\cdot}54 \times 21)$$

$$= 243 \text{ N/mm}^2$$

This is a working stress.

9.6.2. Modified design method: equation (9.7) and Fig. 9.3

Using equation (9.7)

$$R\beta = 0{\cdot}75 \times 0{\cdot}77$$

$$= 0{\cdot}58$$

The corresponding value of Q_{ca} taken from Fig. 9.3 is 17. Hence, the design bending moment is

$$M = n_w L^2/17$$

The steel stress is

$$f_{sa} = 254 - (0{\cdot}54 Q_{ca}) \tag{9.7}$$

$$= 254 - (0{\cdot}54 \times 17)$$

$$= 245 \text{ N/mm}^2$$

9.6.3. Design of reinforced concrete

For both the simple and the modified design methods, the design of the reinforced concrete supporting beam should be carried out on the basis of CP 114[23] or BS 8110[12] using the working steel stresses calculated in sections 9.6.1 and 9.6.2. The compressive stresses in the concrete do not require modification.

The designer should also check the design stresses in the supporting beam during the period of construction before the masonry is capable of acting compositely due to the mortar not having achieved its design strength, i.e. while the masonry is still wet.

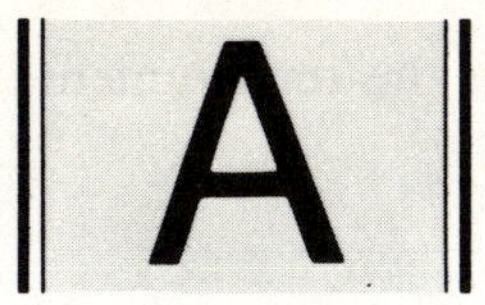

Materials, components and workmanship

A.1. Introduction

As a general guide (and basis for specification) materials and components should comply with the relevant British Standards and codes where appropriate and covered.

Where new materials not covered by standards are used, they should be tested by a reputable testing authority, have an agreement certificate and provide guarantees of long-term performance.

Where British Standards do not apply (e.g. in overseas work), other local, recognised standards may be used. These should be checked with British Standards and any differences allowed for in design, detailing, specification and construction.

A.2. Materials

A.2.1. Masonry units

Units should comply with British Standards as follows

(a) calcium silicate bricks: BS 187[24]
(b) clay bricks and blocks: BS 3921[25]
(c) precast concrete bricks and blocks: BS 6073: Part 1[7]
(d) reconstructed stone masonry units: BS 6477[26]
(e) natural stone blocks: BS 5390[27]

A.2.2. Selection of masonry units

The selection of units is based mainly on strength and cost. The other serviceability considerations, including durability, fire resistance, thermal and acoustic properties, should follow the recommendations given in BS 5628: Part 3.[1]

Masonry units from demolished or deteriorating structures may be re-used provided that they have been thoroughly cleaned, have been inspected for deterioration and conform to the recommendations in British Standards and codes for similar new materials. The high labour cost of cleaning and inspection generally makes this an uneconomic exercise, except for some examples of excellent natural stone or rebuilding and conserving fine old brick buildings.

A.2.3. Steel

A.2.3.1. Reinforcing steel

Reinforcing steel (including bed joint and nominal reinforcement) should comply with British Standards as follows

(a) hot rolled steel bars: BS 4449[10]
(b) cold worked steel bars: BS 4461[11]
(c) hard drawn low carbon steel wire: BS 4482[28]
(d) steel fabric: BS 4483[29]
(e) austenitic stainless steel (bar, wire or fabric): BS 970: Part 1,[30] grades 304S15, 316S31 or 316S33, excluding free machining specifications.

A.2.3.2. Prestressing steel

Prestressing wire, strands and bars should comply with the requirements of BS 4486[8] or BS 5896.[9]

A.2.3.3. Corrosion protection of reinforcing steel

To increase corrosion protection reinforcement may be

(a) galvanised, after manufacture, in accordance with the recommendation of BS 729;[31] the Authors advise that such galvanising can be liable to micro-cracking in prestressing tendons

(b) clad with a layer of nominated thickness not less than 1 mm of austenitic stainless steel

(c) provided with cover as recommended in section 5.7.2.

A.2.3.4. Corrosion protection of prestressing

The corrosion protection of prestressing is dealt with in section 8.7.

A.2.4. Cement

Cements should comply with the requirements of BS 12,[33] BS 146[33] or BS 4027.[34]

Masonry cements and high alumina cement should not be used.

Plasticisers for concrete infill, grouts and mortars should only be used at the engineer's discretion and with his written approval. The manufacturer's written, unconditional guarantee of long-term durability should be obtained before permission is granted. The contractor should also give a written, unconditional guarantee that he will follow the manufacturer's instructions on quality, quantities, mixing and mixing times. If the use of plasticisers is permitted they should comply with the recommendations given in section A.9.

A.2.5. Aggregates

Aggregates for mortar should follow the recommendation given in BS 5628: Part 3.[1] Sand grading should be checked for workability and its effect on mortar strength.

Aggregates for concrete infill (grout) should follow the recommendations given in BS 8110: Part 1.[12]

A.2.6. Mortars

Mixing of mortars and their use should be in accordance with the recommendations of BS 5628: Part 3[1] or, where appropriate, BS 5390.[27]

The proportions and mean compressive strength of mortars should be as given in Table A.1. Weigh batching of mortar is not yet common on construction sites in the UK. Where such methods are considered appropriate the engineer should carry out site testing.

Testing of volume batched mortar, should be in accordance with rec-

Mortar designation	Type of mortar (proportions by volume)		Mean compressive strength at 28 days from preliminary (laboratory) tests, N/mm^2
	Cement: lime: sand	Cement: sand with plasticiser	
(i)	1:0 to $\frac{1}{4}$:3	—	16·0
(ii)	1:$\frac{1}{2}$:4 to $4\frac{1}{2}$	1:3 to 4	6·5

Table A.1. Recommendations for mortar

ommendations in appendix A.1 of BS 5628:[1] Part 1, for both trial mixes and the site control of mortar.

Ready-mixed lime and sand mortars should comply with the requirements of BS 4721[35] and the addition of cement should be done only on site.

Wet ready-mixed lime and sand mortars with retarded cement should be used only at the discretion and written permission of the engineer after written guarantees have been obtained from both the supplier and the contractor. Such mortars should comply with BS 4721.[35]

If colouring agents in mortars are requested by the architect, and permitted by the engineer, they should comply with the requirements of BS 1014,[36] not exceed 10% by mass of the cement in the mortar, be evenly distributed in the mortar and, above all, should not reduce the long-term strength and durability of the mortars.

If the use of carbon black in mortars is considered it should not exceed 3% by mass of the cement. In the Authors' opinion, carbon black should not be used when the flexural tensile strength of masonry is exploited in design.

A.2.7. Lime

The addition of lime to the mortar improves its workability but since lime mortars gain strength slowly (by loss of water and by carbonation) the proportions of the mix should be controlled.

Non-hydraulic (calcium) lime, semi-hydraulic (calcium) limes and magnesium limes should conform to the requirements of BS 890.[37]

A.2.8. Concrete infill

For reinforced and prestressed masonry the concrete infill should be either specified or designed to comply with minimum grade 25 of BS 5328[38] or it should be a specified mix by volume of 1:(0–1/4):3:2 cement, lime, dry sand and 10 mm (maximum sized) aggregate, the maximum size of the aggregate not exceeding the reinforcement cover less 5 mm.

For pretensioned masonry, where the infill is to act structurally, the minimum grade should be 40 (see BS 5328[28]).

For workability considerations, the slump should be between 75 and 150 mm depending on the size, shape of void to be filled and congestion of reinforcement or tendons. Where plasticisers are permitted by the engineer it is advisable to carry out site tests to determine the allowable slump. Similarly, when filling small sections, site tests should be conducted on slump and height of pour to ensure that complete and satisfactory compaction and filling results. Where narrow ducts cannot be avoided the concrete infill may be substituted for neat sand–cement grout of minimum cube strength at seven days of 17 N/mm^2. The sand for grout should pass a 1·18 mm sieve.

Site testing is advisable where there is lack of experience of interaction of the concrete infill, or grout, with the masonry. Porous units may suck too much water from the mix and some dense units may have lower bond with the infill or grout. (Units of high suction rates tend to need wetting before being laid.) Such testing is helpful in specifying the slump to be used, checking any difficulties in compaction, complete filling of voids and so on. Test specimens can be checked for strength and broken up to check for bond and filling.

All sampling and testing of fresh and hardened infill concrete should be carried out in accordance with the recommendations of BS 1881.[39] Prescribed mixes should be assessed and judged on the basis of the specified proportions of the mix and required workability, unless the engineer has specified otherwise. Designed mixes should be assessed on their seven-day cube strength.

Table A.2. BS references
for admixtures

Type of admixture	Standard to be complied with
Concrete admixtures	
• Accelerating admixtures, retarding admixtures and water-reducing admixtures	BS 5075: Part 1[43]
• Air-entraining admixtures	BS 5075: Part 2[43]
Mortar plasticisers	BS 4887[44]

A.2.9. Admixtures

Where admixtures are permitted, their effect on the durability and strength of the concrete or mortar combination with other ingredients to form deleterious compounds and risk of corrosion to the steel should be checked.

The use of more than one type of admixture in a mix should not be permitted until the long-term compatibility, interaction and effect on strength of the admixtures has been checked.

If the use of admixtures is permitted, it is vital to ensure that there is strict adherence to the manufacturer's instructions on quality, mixing times, curing, and so on.

Admixtures should comply with the relevant British Standards listed in Table A.2.

Calcium chloride should not be used as an additive because there can be a long-term corrosive action on the wall ties, reinforcement and prestressing steel.

A.2.10. Chloride

The corrosive effect of chloride is well documented. Chlorides should not be added to the mix. Calcium chloride ions occur naturally in some materials. The percentage content of these ions should be ascertained and should not exceed the following limits.

(a)   For admixtures, the chloride ion content should not exceed 2% by mass of the admixture or 0·03% by mass of the cement.

(b)   For sands, the chloride ion content should not exceed 0·15% by mass of the dry sand.

(c)   For mixes, the chloride ion content of aggregates, lime, water, sand, cement or other sources should not exceed the limits given in Table A.3.

Table A.3. Chloride content
of mixes

Type of use of concrete or mortar	Maximum total chloride content expressed as percentage of chloride ion by mass of cement
Prestressed concrete; heat-cured concrete containing embedded metal	0·1
Concrete or mortar made with cement comply with BS 4027[34]	0·2
Concrete or mortar containing embedded metal and made with cement complying with BS 12[32] or BS 146[33]	0·4

Fig. A.1. Wall ties in grouted cavity walls

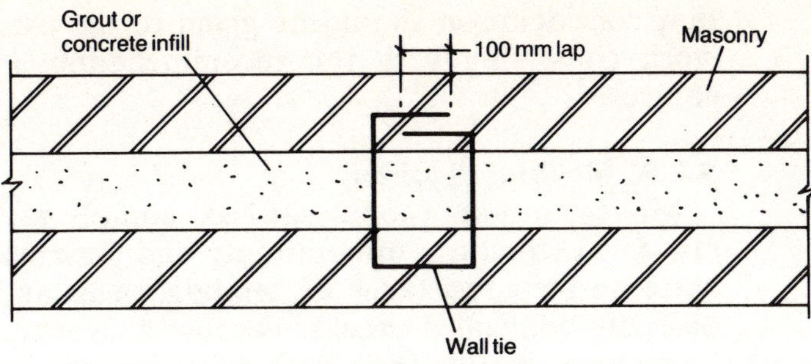

Fig. A.2. Typical spacing of wall ties

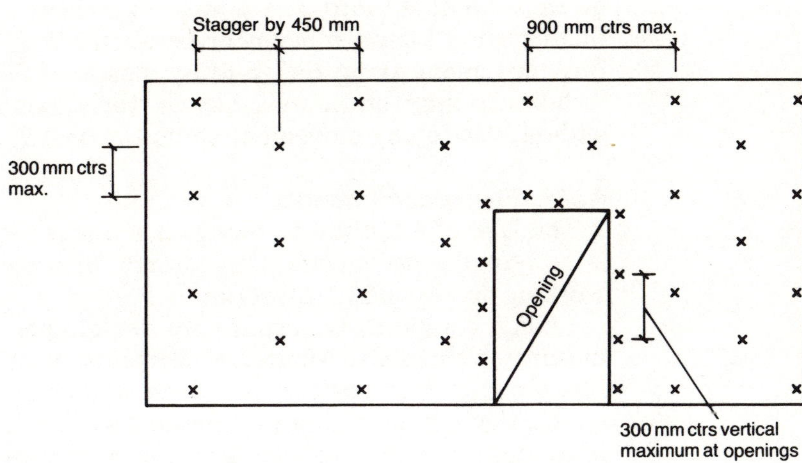

A.3. Components

A.3.1. Wall ties

For low-lift grouted cavity construction, wall ties should be vertical-twist type complying with BS 1243[40] and the spacing should follow the recommendations given in BS 5628: Part 1.[1]

For high-lift grouted cavity construction, wall ties should be of 6 mm minimum diameter, galvanised low carbon steel, resin-coated galvanised mild steel or austenitic stainless steel. They should be bent as links, as shown in Fig. A.1, to ensure adequate tie action.

The ties must be of sufficient strength, number and spacing to resist the bursting forces occurring during placing and compaction of the grout or concrete infill. The bursting force should be resisted by the ties alone and no reliance should be placed on the masonry. Ties should not be placed at greater spacing than that shown in Fig. A.2. The cover should be as that given for carbon steel reinforcement (see section B.2).

The resistance to corrosion of the wall ties should be at least equal to that of the reinforcement or prestressing tendon; where necessary the minimum mass of zinc coating on galvanised steel wires should be 940 g/m^2. Where the ties are of different metal from the reinforcement or prestressing, the two metals should not come into contact, otherwise electrolytic action, leading to corrosion, could take place.

A.3.2. Damp-proof courses

Damp-proof courses should comply with current British Standards.

Designers may need to be reminded that damp-proof courses form structural discontinuities and that their resistance to compression (and therefore contraction), tension, shear, bending stress and sliding under vertical and lateral loading must be checked. Where only unreliable or inadequate information is available from the manufacturer, or where the resistance is too low, consideration should be given to the use of class A engineering brick-work complying with the recommendations of BS 743[41] as damp-proof courses.

A.3.3. Movement joints

Movement joints should at least comply with BS 5628.[1] Movement joints also form structural discontinuity and designers are advised to place no reliance on their resistance to bending, shear and so on. Movement joints are basically controlled cracks, because masonry can expand and/or contract. Some recommendations for spacing and positioning of movement joints are given in BS 5628: Part 3. Attempts to transfer shear forces across such joints or to restrict lateral movement between adjacent panels (while still permitting movement along the wall) by means of debonded dowel bars or other techniques may be inadvisable, as such bars can be difficult to position without destroying movement along the wall.

A.3.4. Fixing components

The lack of attention to fixings is a major cause of structural distress and it is essential to ensure that design assumptions are translated to the working details and construction.

Fixing components, usually of metal, are used to connect structural masonry elements to structural elements in other materials. Typical examples are steel roof and concrete floor slab connections, restraining and tieing straps, shear transfer and horizontal restraint fixings. There is a wide variety of numerous types available and many of these are not covered by British Standards. The designer should consider the strength, durability, suitability, cost, test data, compatibility with adjoining materials and so on before specifying such fixings.

Further information on fixings is given in reference 2.

A.3.5. Copings

Copings, sills and similar provisions to shed rainwater from masonry should be designed and fixed in accordance with the recommendations of BS 5628: Part 3.[1]

A.4. Workmanship

All construction should conform to the special category of construction as specified in BS 5628: Part 3.[1]

A.4.1. Materials

All materials should comply with the recommendations given above in section A.2.

Storage and handling of masonry units and the storage, mixing and handling of mortars should comply with the recommendations given in BS 5628: Part 3.

The storage and mixing of materials for concrete (and concrete infill) and the storage, handling and fixing of both reinforcement and prestressing tendons should comply with the recommendations given in BS 8110: Part 1.[12]

A.4.2. Reinforced construction: general

The specification for level, line and plumb of masonry and its protection in adverse weather and temporary stoppage of work, and so on should comply with the recommendations of BS 5628: Part 3.[1]

The maximum height of masonry laid per 24 hours should not exceed

1·5 m. The masonry should rise uniformly and the practice of building corners and other intersections higher than the remainder of the wall should be discouraged.

Details for infill concrete are given in section A.2.8 and for reinforcement in section A.2.3. The fixing in accordance with the details, the provision and maintenance of cover, cleaning reinforcement of loose rust and similar considerations should comply with normal good practice in reinforced concrete construction.

Detailing is important because the brick or block layers may not be familiar with fixing reinforcement or with reinforcing details. It is therefore essential that the working drawings are easily understood and it must be appreciated that the normal conventions of reinforced concrete detailing may need to be amended.

A.4.3. Grouted cavity construction

In wall construction the normal precautions in good masonry construction should be adhered to, e.g. keeping the cavity clear of mortar droppings, proper embedment, and positioning and quality of wall ties.

In reinforcement fixing the normal precautions in good practice should be followed, e.g. correct positioning, tieing, maintenance of cover, and adequate lapping. Reinforcement should be fixed sufficiently in advance of masonry construction so as not to cause delays.

Low-lift wall construction is most commonly used for relatively narrow cavities of 50–75 mm wide. The concrete infill or grout should be placed/poured as the wall rises (making sure that no overflow stains or splashes face work) and compacted immediately. The level of the pour should be kept about 50 mm below the level of the masonry, kept clean of mortar droppings between lifts and no lift should exceed 400 mm. To prevent bursting of freshly built masonry (which has had insufficient time to gain strength to resist the lateral pressure from the head of the infill) it is advisable to limit lifts to about 1 m/day. Any masonry which has bulged, opened up, burst or in any way distorted should be condemned, demolished and rebuilt.

In high-lift wall construction all these recommendations apply with the following variations.

(a)   Lifts may increase but not exceed 3 m.
(b)   Infill should not take place earlier than three days after wall construction; they should be placed in two lifts and each lift recompacted after any initial settlement caused by water absorption of the masonry units and before the initial set.
(c)   Clean-out holes, at the base of each lift on one face (at least 150 mm × 200 mm at 500 mm centres), should be provided to enable cleaning of the cavity to be carried out. After cavity cleaning the clean-out holes should be blocked and the infill units propped before the next lift is cast.
(d)   Wall ties as described in section A.3.1 for high-lift grouted cavity construction must be provided.

A.4.4. Reinforced hollow blockwork

In wall and beam (e.g. lintel) construction, normal good masonry practice should be followed, e.g. laying on a full bed of mortar, cleaning mortar debris from core before placing infill, building true to line, level and plumb. Normally the reinforcement is fixed after completion of the masonry construction.

The recommendations given in section A.3.1 for low-lift grouted cavity construction apply also to blockwork. As hollow blockwork acts as a permanent shutter the height of lift pour can be doubled to 800–900 mm.

The recommendations given in section A.3.1 for high-lift brickwork cavity construction apply also to blockwork. The following additions or alterations should be considered.

(*a*) When only some of the cores are to be filled (to resist accidental damage, to increase blast-resistance and so on) a clean-out hole of block size (and not less than 100 mm × 100 mm) should be provided. When many or all of the cores are to be filled, instead of numerous (and awkward to construct) clean-out holes being left, the base course may be built of bricks spaced to suit block sizes.

(*b*) High-lift grouting should, in general, not be used for walls less than 200 mm wide.

A.4.5. Quetta bond, rat-trap and similar bonded walls

The voids formed by the bonding around reinforcement should be filled with mortar as the work proceeds. Where the cavities are larger, as in a 260 mm diaphragm wall, they may be filled with infill concrete using the low-lift or high-lift techniques described in section A.3.1. If secondary reinforcement, in the bed joints, is considered necessary or desirable it should be fixed (and given adequate cover) as the work proceeds.

A.4.6. Pocket-type walls

Pocket-type walls are generally used as earth retaining walls so that reinforcement cover needs to be increased. The main reinforcement is better fixed before masonry construction, and any bed joint reinforcement to assist in tieing the masonry to the reinforced concrete column and aiding composite action should be fixed as masonry construction proceeds. The formwork to the backface of the pocket (column) should follow normal reinforced concrete construction in that it should be properly propped to withstand the forces caused by concrete placing and compaction. It should also be adequately sealed to prevent loss of grout from the infill concrete. The height of pour lift should follow normal reinforced concrete column construction practice and should not take place until the masonry has gained a specified strength.

A.4.7. Columns

The main reinforcement in brick columns should be fixed before masonry construction starts and the links may be fixed as the work proceeds. The voids for the reinforcement should be filled with mortar as the work proceeds.

For hollow block columns, with the exception of starter bars, the main reinforcement can be fixed after the masonry construction is complete. A clean-out hole must be left open at the base of the column so that the mortar droppings can be cleared out and the reinforcement cage fixed to the starter bars. Normal concrete practice for high-lift pours should be adopted after the masonry construction has reached a specified strength or earlier, provided that propping and other precautions are taken to prevent damage.

A.4.8. Beams

For reinforced brick beams, the soffit shutter should remain in position and propped for approximately double the time of reinforced concrete construction or until the engineer is satisfied that the brickwork has acquired sufficient strength. When the void is U shaped and the brickwork acts as a permanent shutter, the reinforcement cage may be positioned (maintaining proper cover) and concreted in when the brickwork has given sufficient strength to resist the bursting force of placing and compacting the concrete. Where normal bonding is used, the bottom reinforcement and links may be fixed after the soffit brickwork has been laid and the top reinforcement may be fixed as the work proceeds.

In hollow blockwork, channel section blocks are commonly used and after laying and cleaning out of the void the reinforcement cage may be fixed. When the beam is to act compositely with blockwork over the positioning of links taller than the U blocks, it should be positioned to line up with the perpends of the blockwork over. This produces a better keying together and reduces the possibility of horizontal shear cracking between the beam and blockwork over.

A.4.9. Prestressed construction: general

In general the recommendations given in section A.4.2 and in the subsequent appropriate sections should be adopted with the following variations.

(*a*) The positioning, tensioning and so on should be carried out in accordance with the recommendations given in BS 8100.[42]

(*b*) The transfer stress should not exceed the specified strength of the masonry.

(*c*) The tendons must be positioned correctly, kept clear of obstructions and protected from corrosion, the threads must be kept clean and lightly oiled, and the recommendations given in the sections on prestressing detailing adhered to.

A.4.10. Filling of cavities, voids and ducts

The quantity of concrete infill in cavities, grout in prestressing ducts, mortar in voids and so on should be measured and checked to ensure complete filling. Vent holes should be used in vertical elements to assist in this operation and the holes should be plugged as the infill rises to their level. The importance of complete filling and thorough compaction should be conveyed to the site labour force.

A.4.11. Provision for chases, ducts, openings and fixings

Under no account should reinforced or prestressed masonry be cut, drilled, sawn, chased or in any other way disturbed after construction without the written approval of the engineer. Such actions can be highly dangerous structurally. Openings, holes, chases, bolt fixings and the like should be preplanned and built in as the work proceeds.

A.4.12. Jointing and pointing

No raking out of joints should be permitted without the written approval of the engineer. Any request for such action should either be considered and designed for in the planning stage or permitted after construction only if the reduced effective cross-sectional area of the walls or other element is permissible.

Durability, buildability and other serviceability requirements

B.1. Durability

After the post Second World War rash of building failures there was natural concern with new materials and methods. Masonry is a most durable material and its correct application to modern building can produce durable structures. The methods of design and detailing for the durability of plain masonry structures are well documented, e.g. see references 1, 2 and 13. Experience has shown that, with care, reinforced masonry is durable, and in the Authors' experience, post-tensioned masonry (after 20 years) has proved even more durable than other masonry of the same age. The major, and understandable, concern among designers regarding reinforced and pre-stressed masonry is the corrosion of the steel. This concern arises from the fact that both bricks and blocks are porous and so moisture will reach the steel. However, the cover provided by bricks and blocks is disregarded and the steel is either encased in concrete infill, grout or mortar with depth of cover at least as good as that for reinforced and prestressed concrete, or is itself corrosion-resistant.

Good detailing, specification, supervision and construction will, as in other construction, solve this problem.

B.2. Corrosion resistance of metal

The required resistance will depend on both the type of steel and the degree of exposure to corrosion. These in turn affect the depth of cover. The main types of steel used, in ascending resistance to corrosion, are

- low carbon steel
- high yield galvanised steel, with or without a resin coating
- austenitic stainless steel.

B.2.1. Classification of exposure situations

Exposure is due partly to weather (the combination of wind and rain) and partly to the protection and exposure situation of the positions of the structural element in the structure. BS 5628: Part 3[1] tabulates the weather classification as shown in Table B.1.

BS 5628: Part 2 classifies four exposure situations

- E1, where internal work and inner skin of ungrouted external cavity walls and behind surfaces are protected by an impervious coating that can readily be inspected, or external parts are built where the exposure category given in table 10 of BS 5628: Part 3 is 'sheltered' or 'very sheltered'
- E2, where buried masonry and masonry are continually submerged in fresh water, or external parts are built where the exposure category given in table 10 of BS 5628: Part 3 is 'sheltered/moderate' or 'moderate/severe'
- E3, where masonry is exposed to freezing while wet, subjected to heavy condensation or exposed by cycles of wetting by fresh water and

drying out, or where external parts are built where the exposure category given in table 10 of BS 5628: Part 3 is 'severe' or 'very severe'

- E4, where masonry is exposed to salt or moorland water, corrosive fumes, abrasion or salt used for de-icing.

B.2.1.1. Exposure situations requiring special attention

Special consideration should be given to any feature that is likely to be subjected to more severe exposure than the remainder of the building or structure. In particular, parapets, sills, chimneys (particularly those exposed to corrosive gases) and the details around openings in external walls should be examined. Normally such situations should be considered equivalent to exposure situation E3.

Table B.1. Classification of exposure to local wind-driven rain

Exposure category	Local spell index calculated as described in DD 93	Exposure category in CP 121:[45] Part 1*
	L/m² per spell	
Very severe	98 and over	Severe
Severe	68 to 123	
Moderate/severe	46 to 85	Moderate
Sheltered/moderate	29 to 58	
Sheltered	19 to 37	Sheltered
Very sheltered	24 or less	

* CP 121: Part 1 defined three exposure categories, namely severe, moderate and sheltered, corresponding to values of Lacy's annual mean driving rain index: >7 m²/s, 3 m²/s to 7 m²/s and <3 m²/s, respectively (see BRE *Driving rain index*.[46] Developments since the publication of that code, such as the introduction of insulation into cavity walls and the advent of improved meteorological data, have made it necessary to increase the number of exposure categories.

Table B.2. Selection of reinforcement for durability

Exposure situation	Minimum level of protection for reinforcement, excluding cover	
	Located in bed joints or special clay units	Located in grouted cavity or Quetta bond construction
E1	Carbon steel galvanised following the procedure given in BS 729,[31] minimum mass of zinc coating 940 g/m² *	Carbon steel
E2	Carbon steel galvanised following the procedure given in BS 729;[31] minimum mass of zinc coating 940 g/m²	Carbon steel or, where mortar is used to fill the voids, carbon steel galvanised following the procedure given in BS 729[31] to give a minimum mass of zinc coating of 940 g/m²
E3	Austenitic stainless steel or carbon steel coated with at least 1 mm of stainless steel	Carbon steel galvanised following the procedure given in BS 729;[31] minimum mass of zinc coating 940 g/m²
E4	Austenitic stainless steel or carbon steel coated with at least 1 mm of stainless steel	Austenitic stainless steel or carbon steel coated with at least 1 mm of stainless steel

* In internal masonry other than the inner leaves of external cavity walls carbon steel reinforcement may be used.

B.2.1.2. Effect of different masonry units

The protection against corrosion provided by brickwork tends to be improved if high strength, low water absorption bricks are used in strong mortar. Where bricks that have a greater water absorption than 10% or concrete blocks having a net density less than 1500 kg/m³, measured as described in BS 6073:[7] Part 2, are used, the steel recommended for the next most severe exposure situation or, where appropriate, stainless steel, should be used, unless protection to the reinforcement is to be provided by concrete cover (see Table B.2).

B.2.2. Cover

The higher the corrosion resistance of the steel, the less are the cover requirements for durability. In the ultimate austenitic stainless steel requires no cover. (Such steel does need embedment in mortar or concrete to develop bond stress.)

The depth of cover also depends on the location with respect to bed joints, cavities, hollow blockwork and pockets.

For bed joints, all reinforcement should have 15 mm absolute minimum depth of mortar cover to the exposed face, or faces, of the masonry. For grouted cavities and voids, the following points should be noted.

(a) Carbon steel cover requirements, depending on concrete grade are given in Table B.3.

(b) Galvanised steel reinforcement requires 20 mm or bar diameter, whichever is the greater depth of mortar or concrete cover.

(c) Stainless steel reinforcement does not require cover for durability but should have 15 mm cover for bond.

Figure B.1 shows minimum depth of concrete cover for carbon steel reinforcement. Consideration must be given to the tolerances of the

Table B.3. Minimum concrete cover for carbon steel reinforcement, mm

Exposure situations	Concrete grade in BS 5328[38]			
	25	30	35	40
	Minimum cement content, kg/m³			
	250	300	350	350
E1	20	20	20	20
E2	—	30	30	25
E3	—	40	35	30
E4	—	—	—	60

Fig. B.1. Minimum cover in hollow blockwork and pocket walls

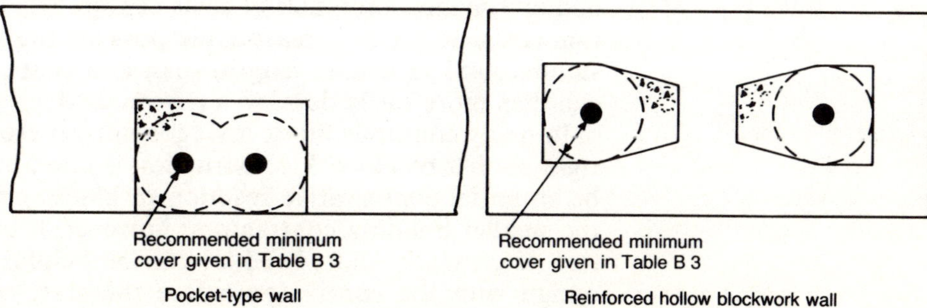

masonry, reinforcement, aggregate and shuttering. The ends of all bars should have the same cover as carbon steel for the appropriate exposure condition.

B.2.3. Corrosion protection of prestressing tendons

Where rods or tendons are encased in concrete infill, grout or mortar, the depth of cover should be the same as for reinforcement. It is important that the anchorages, particularly when exposed to the weather, should have adequate and full protection.

When placed in open ducts they should be of austenitic stainless steel, galvanised using a minimum mass of zinc coating of 940 g/m^2, or painted with bituminous or other water-resistant paint and wrapped in proprietary waterproof tape. (Similar treatment should be given to carbon steel reinforcement placed in open ducts.)

B.3. Buildability

Buildability is defined and described in reference 47.

If the details lack good buildability then it is equally possible that the structure will lack good durability and economy.

The detailer should avoid as far as possible the need for special-shaped bricks and blocks, cutting bricks and blocks, the use of difficult bonding (such as Quetta bond), a mixture of units of differing strength, type and material and similar causes of site delays and problems.

Reinforcement details should be particularly clear and simple because the fixing may be done by bricklayers inexperienced in reinforcement fixing or details. The reinforcement should be easy to fix in adequately sized voids and all reinforcement dimensions should relate to masonry dimensions.

It will be apparent from study of the worked examples that the comments in section 1.8 (on the decision to reinforce or prestress brickwork and blockwork) are reasonable. For example, the output of bricklayers building a reinforced Quetta bond retaining wall will be lower than those building a post-tensioned diaphragm. A reinforced Quetta bond has been described by a builder as 'a three-dimensional jig-saw puzzle' whereas basically a post-tensioned diaphragm was described as 'only a wide cavity with some rods in'. Reinforced blockwork is regarded by many builders as almost as simple as substituting temporary timber formwork for permanent concrete blockwork.

Simplicity of construction generally equates with quality, speed and economy of construction. The designer and detailer should (particularly at preliminary design and sketch stage), as always, thoroughly examine the proposed design for methods of improving construction simplicity—without sacrificing structural safety or efficiency.

Standardisation is also important. It is important to minimise the number and types of difficult masonry units and the mixture of bricks and blocks. Each different type will require separate storage, transport and handling, and any savings on material costs can quickly be lost in increased construction costs. So far as is reasonably possible the structural elements should be standardised in height, length, span and shape. The brick and block layers can then more easily develop a rhythm and feel of the construction.

In many countries block laying is not yet established as a trade in its own right so that blockwork construction is often built by bricklayers. There can be an initial conservative reaction to blockwork both by the craftsman and the smaller building contractors; however, it has been found that this reaction is generally short-lived. It can be helpful if the designer discusses his design with the contractor before the start of construction and suggests, tactfully, that the contractor studies the blockwork publications of the Cement and Concrete Association and the blockmaker's literature.

The designer should be aware that large, solid and dense blocks are much heavier to handle than bricks—some are so heavy that they require two men to lift them in position. Until a small and simple clamp crane is developed to reduce hard and heavy work, thought should be given to the use of such blocks in difficult positions.

Blockwork, being usually more porous than brickwork, requires more stringent attention during construction than brickwork during wet weather. Discussion with the contractor on his proposals to protect the work during rainfall is useful.

Brickwork, because of the smaller size of unit, is more easily built in complex shapes (e.g. serpentine walls and tubular columns of small radius of curvature, and tapered fins) and where these are required for aesthetic reasons the extra cost may be acceptable.

Brickwork usually has a higher relative characteristic compressive strength than blockwork, and where it has not it can often be built in elements of higher section modulus than blockwork to compensate.

The Authors would advise those designers without wide experience in reinforced and prestressed masonry to prepare alternative preliminary designs in the various options, carry out rough cost exercises, investigate buildability and so on before deciding on structural form, technique and material. Each successive trial run on future projects becomes simpler and faster; it will also ultimately save design and drawing office costs and result in a more satisfactory structure for the client.

B.4. Fire-resistance

The recommendations for fire resistance of reinforced and prestressed concrete given in BS 8110[12] should be followed except that the masonry can be taken as providing fire-resistant cover.

Reinforced column design graphs

Figures C.1–C.20 which are shown on the following pages plot N/btf_k against M/bt^2f_k for different values of γ_{mm}. These reinforced column design graphs are derived from formulae given in chapter 3.

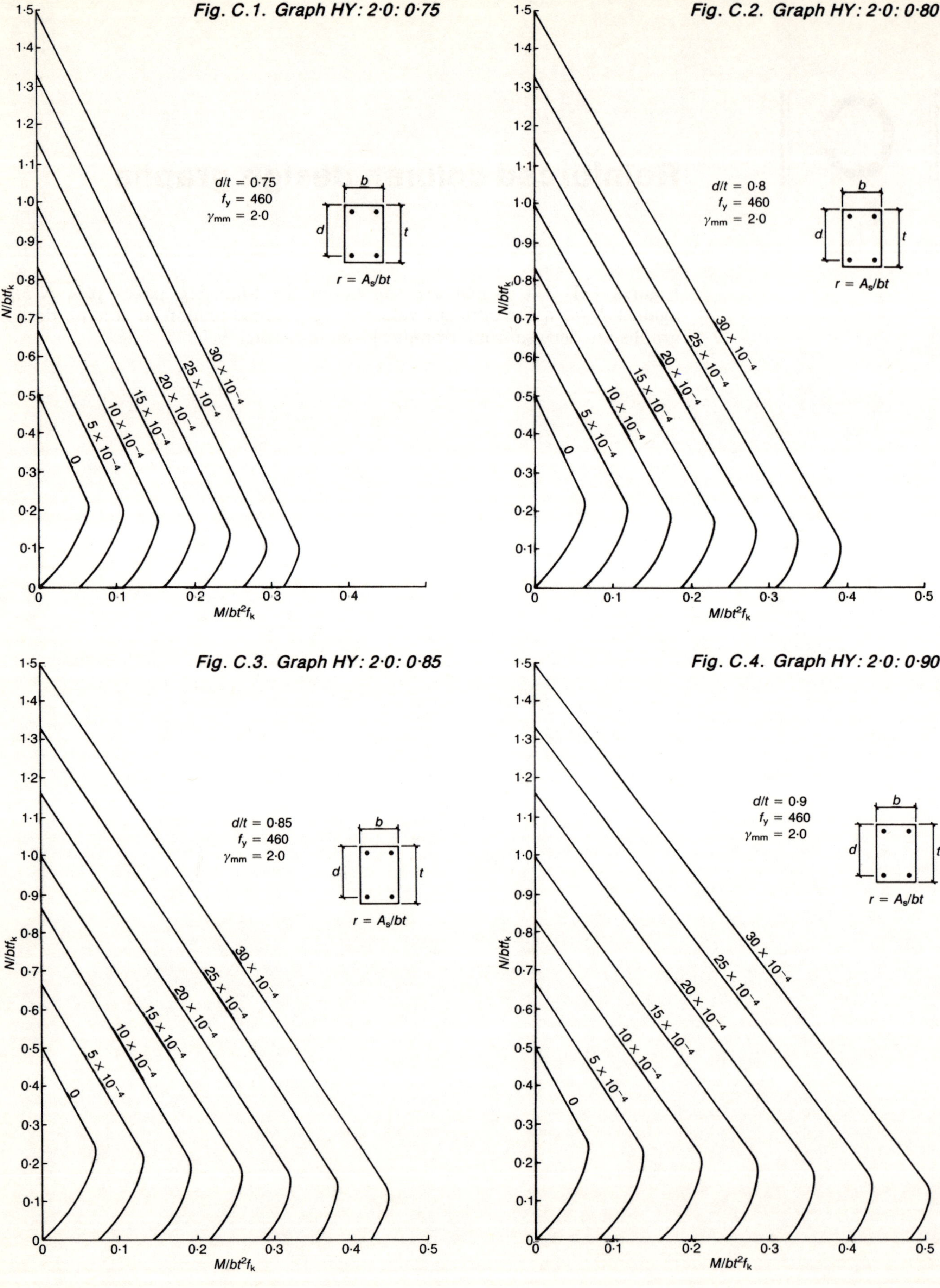

Fig. C.1. Graph HY: 2·0: 0·75

Fig. C.2. Graph HY: 2·0: 0·80

Fig. C.3. Graph HY: 2·0: 0·85

Fig. C.4. Graph HY: 2·0: 0·90

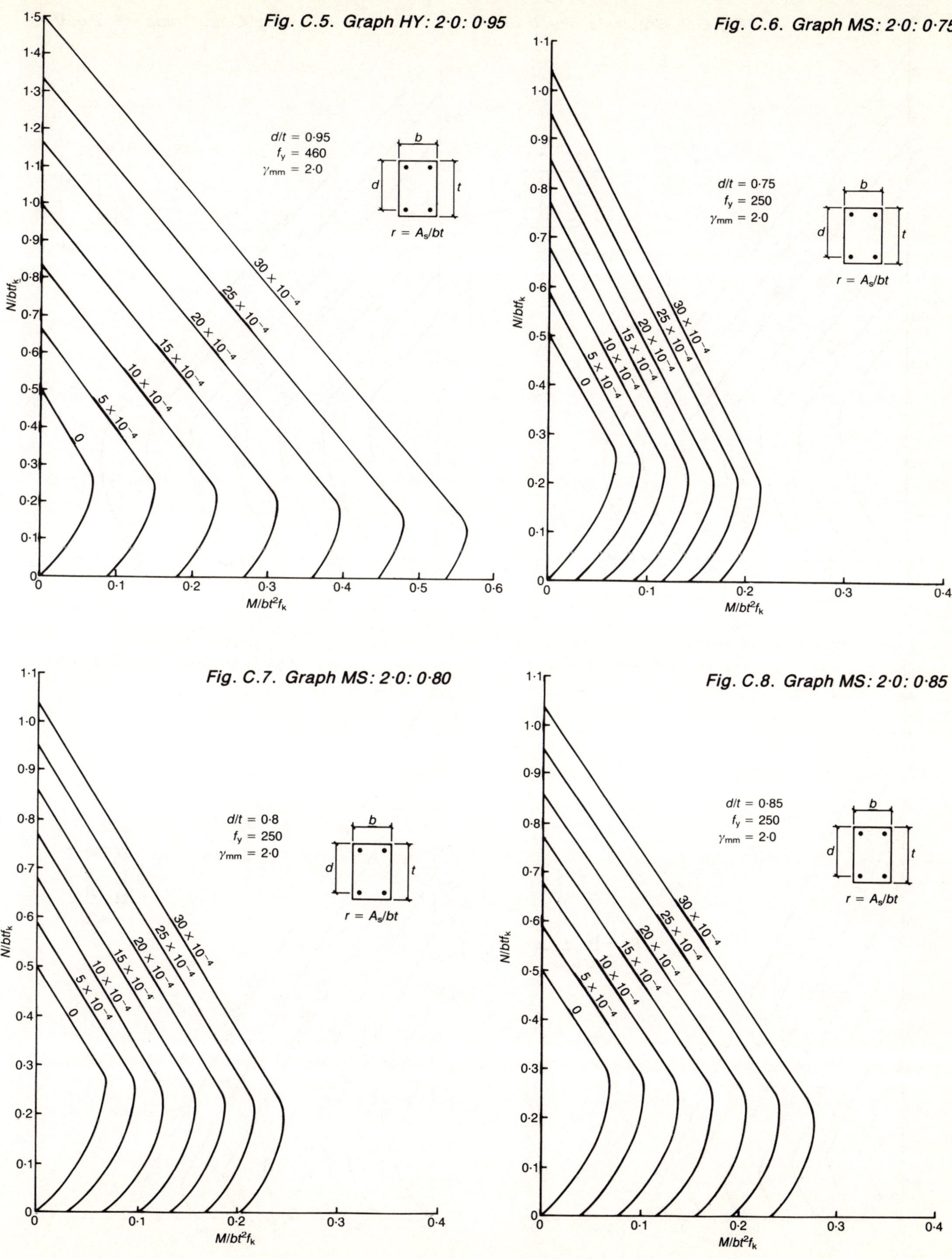

Fig. C.5. Graph HY: 2·0: 0·95
1·5
1·4
1·3
1·2
1·1
1·0
0·9
0·8
0·7
0·6
0·5
0·4
0·3
0·2
0·1
0
N/btf_k
M/bt²f_k
0
0·1
0·2
0·3
0·4
0·5
0·6
d/t = 0·95
f_y = 460
γ_mm = 2·0
b
d
t
r = A_s/bt
30 × 10⁻⁴
25 × 10⁻⁴
20 × 10⁻⁴
15 × 10⁻⁴
10 × 10⁻⁴
5 × 10⁻⁴
0

Fig. C.6. Graph MS: 2·0: 0·75
1·1
1·0
0·9
0·8
0·7
0·6
0·5
0·4
0·3
0·2
0·1
0
N/btf_k
M/bt²f_k
0
0·1
0·2
0·3
0·4
d/t = 0·75
f_y = 250
γ_mm = 2·0
b
d
t
r = A_s/bt
30 × 10⁻⁴
25 × 10⁻⁴
20 × 10⁻⁴
15 × 10⁻⁴
10 × 10⁻⁴
5 × 10⁻⁴
0

Fig. C.7. Graph MS: 2·0: 0·80
1·1
1·0
0·9
0·8
0·7
0·6
0·5
0·4
0·3
0·2
0·1
0
N/btf_k
M/bt²f_k
0
0·1
0·2
0·3
0·4
d/t = 0·8
f_y = 250
γ_mm = 2·0
b
d
t
r = A_s/bt
30 × 10⁻⁴
25 × 10⁻⁴
20 × 10⁻⁴
15 × 10⁻⁴
10 × 10⁻⁴
5 × 10⁻⁴
0

Fig. C.8. Graph MS: 2·0: 0·85
1·1
1·0
0·9
0·8
0·7
0·6
0·5
0·4
0·3
0·2
0·1
0
N/btf_k
M/bt²f_k
0
0·1
0·2
0·3
0·4
d/t = 0·85
f_y = 250
γ_mm = 2·0
b
d
t
r = A_s/bt
30 × 10⁻⁴
25 × 10⁻⁴
20 × 10⁻⁴
15 × 10⁻⁴
10 × 10⁻⁴
5 × 10⁻⁴
0

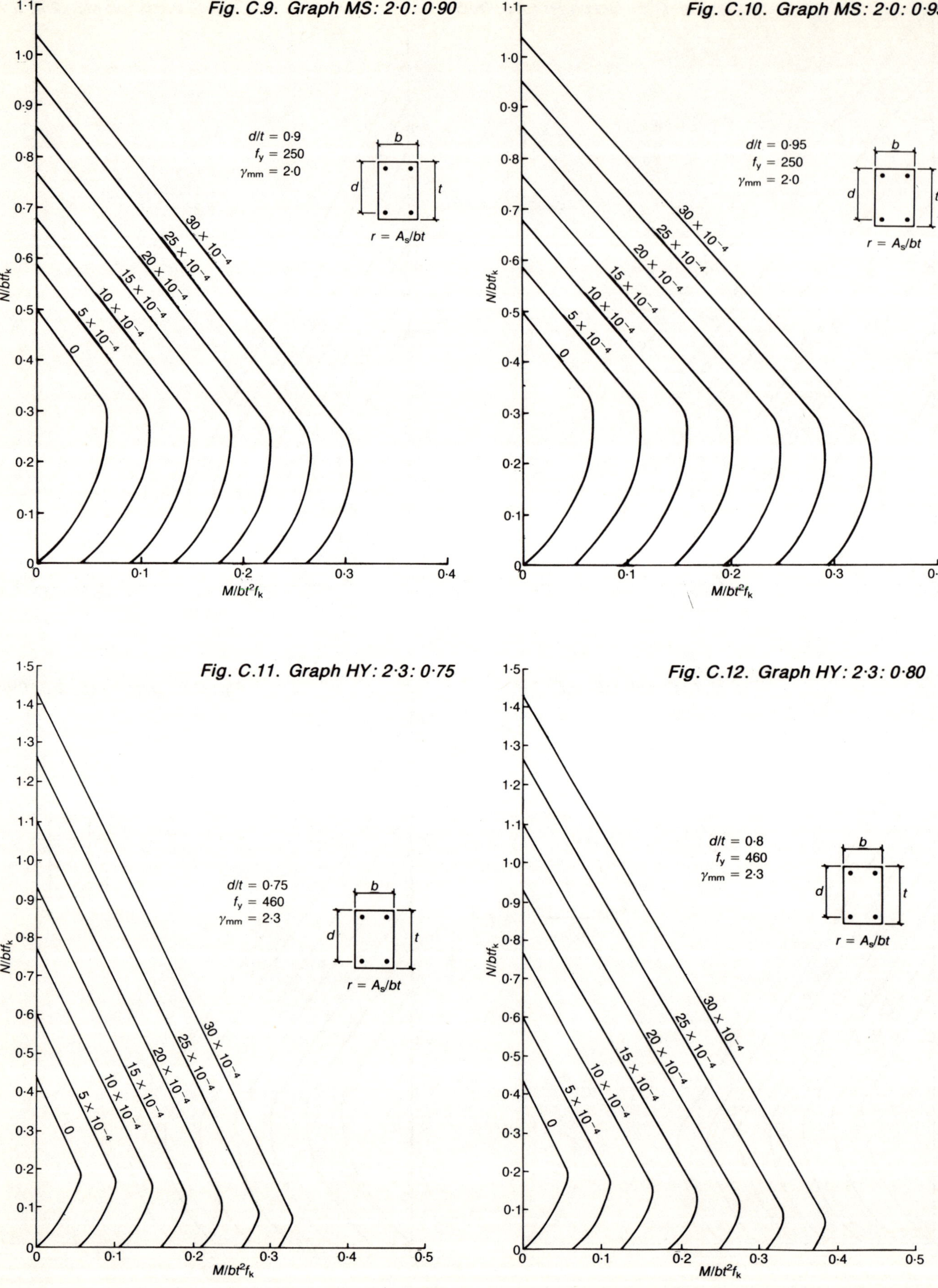

Fig. C.9. Graph MS: 2·0: 0·90
Fig. C.10. Graph MS: 2·0: 0·95
Fig. C.11. Graph HY: 2·3: 0·75
Fig. C.12. Graph HY: 2·3: 0·80
N/btf_k
M/bt^2f_k
$d/t = 0.9$
$f_y = 250$
$\gamma_{mm} = 2.0$
$d/t = 0.95$
$f_y = 250$
$\gamma_{mm} = 2.0$
$d/t = 0.75$
$f_y = 460$
$\gamma_{mm} = 2.3$
$d/t = 0.8$
$f_y = 460$
$\gamma_{mm} = 2.3$
$r = A_s/bt$
30×10^{-4}
25×10^{-4}
20×10^{-4}
15×10^{-4}
10×10^{-4}
5×10^{-4}
0
b
d
t

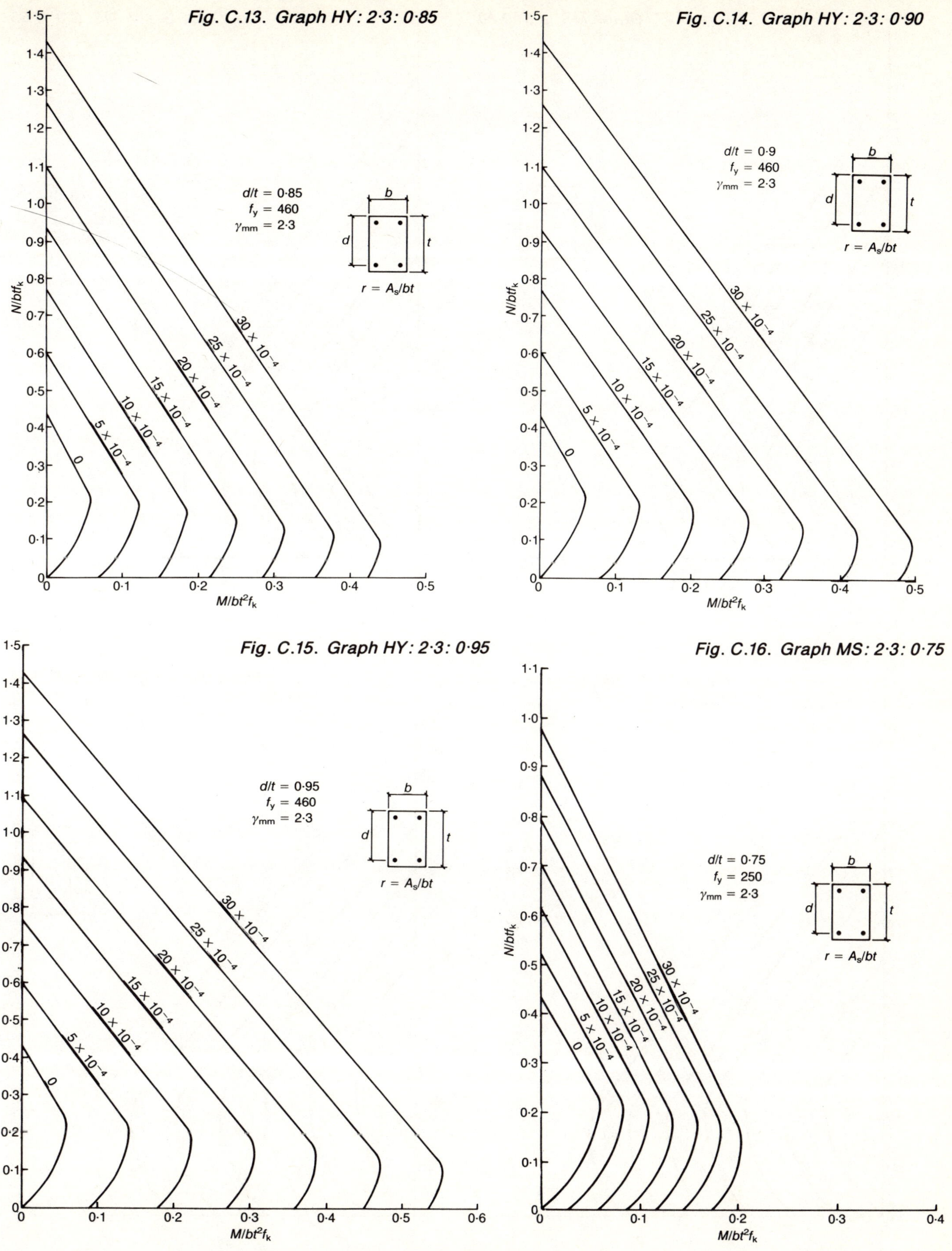

Fig. C.13. Graph HY: 2·3: 0·85
N/btf_k
1·5
1·4
1·3
1·2
1·1
1·0
0·9
0·8
0·7
0·6
0·5
0·4
0·3
0·2
0·1
0
d/t = 0·85
f_y = 460
γ_mm = 2·3
b
d
t
r = A_s/bt
30 × 10⁻⁴
25 × 10⁻⁴
20 × 10⁻⁴
15 × 10⁻⁴
10 × 10⁻⁴
5 × 10⁻⁴
0
0   0·1   0·2   0·3   0·4   0·5
M/bt²f_k

Fig. C.14. Graph HY: 2·3: 0·90
N/btf_k
1·5
1·4
1·3
1·2
1·1
1·0
0·9
0·8
0·7
0·6
0·5
0·4
0·3
0·2
0·1
0
d/t = 0·9
f_y = 460
γ_mm = 2·3
b
d
t
r = A_s/bt
30 × 10⁻⁴
25 × 10⁻⁴
20 × 10⁻⁴
15 × 10⁻⁴
10 × 10⁻⁴
5 × 10⁻⁴
0
0   0·1   0·2   0·3   0·4   0·5
M/bt²f_k

Fig. C.15. Graph HY: 2·3: 0·95
N/btf_k
1·5
1·4
1·3
1·2
1·1
1·0
0·9
0·8
0·7
0·6
0·5
0·4
0·3
0·2
0·1
0
d/t = 0·95
f_y = 460
γ_mm = 2·3
b
d
t
r = A_s/bt
30 × 10⁻⁴
25 × 10⁻⁴
20 × 10⁻⁴
15 × 10⁻⁴
10 × 10⁻⁴
5 × 10⁻⁴
0
0   0·1   0·2   0·3   0·4   0·5   0·6
M/bt²f_k

Fig. C.16. Graph MS: 2·3: 0·75
N/btf_k
1·1
1·0
0·9
0·8
0·7
0·6
0·5
0·4
0·3
0·2
0·1
0
d/t = 0·75
f_y = 250
γ_mm = 2·3
b
d
t
r = A_s/bt
30 × 10⁻⁴
25 × 10⁻⁴
20 × 10⁻⁴
15 × 10⁻⁴
10 × 10⁻⁴
5 × 10⁻⁴
0
0   0·1   0·2   0·3   0·4
M/bt²f_k

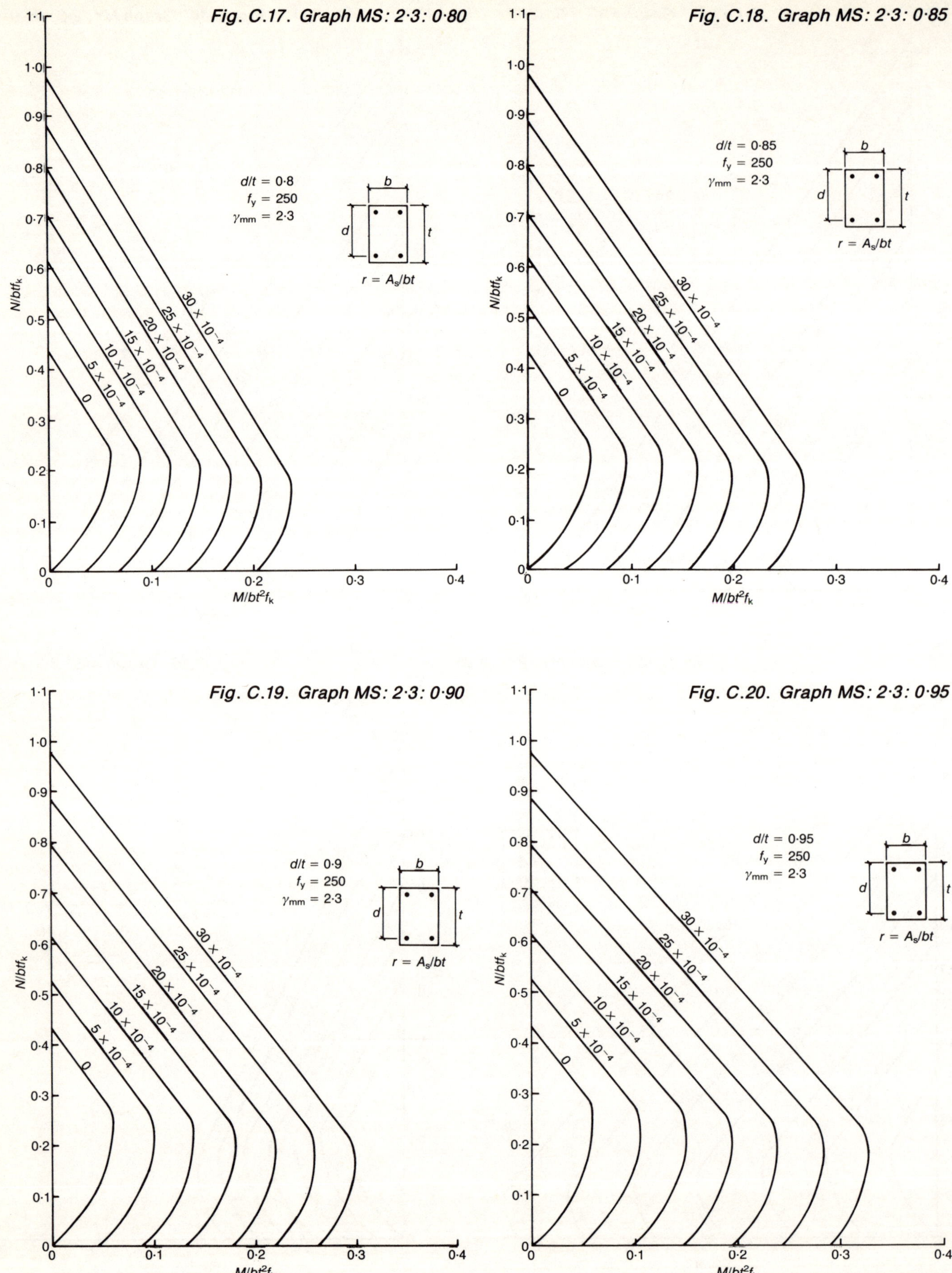

Fig. C.17. Graph MS: 2·3: 0·80

Fig. C.18. Graph MS: 2·3: 0·85

Fig. C.19. Graph MS: 2·3: 0·90

Fig. C.20. Graph MS: 2·3: 0·95

References

1. British Standards Institution. *Code of practice for the use of masonry.* BSI, London, BS 5628, Part 1, 1978 (1985); Part 2, 1985; Part 3, 1985.
2. Curtin W.G. *et al. Structural masonry detailing.* Granada, London, 1984.
3. Gage M.T. and Kirkbride T.V. *Design in blockwork.* Architectural Press, London, 1978.
4. British Standards Institution. *Loadings for buildings.* BSI, London, 1984, BS 6399.
5. British Standards Institution. *Code of basic data for the design of buildings.* Chapter V: Loading. BSI, London, Part 2, 1972.
6. British Standards Institution. *Code of practice for foundations.* BSI, London, 1986, BS 8004.
7. British Standards Institution. *Precast concrete masonry units.* BSI, London, 1981, BS 6073.
8. British Standards Institution. *Specification for hot rolled and hot rolled and processed high tensile alloy steel bars for the prestressing of concrete.* BSI, London, 1980, BS 4486.
9. British Standards Institution. *Specification for high tensile steel wire and strand for the prestressing of concrete.* BSI, London, 1980, BS 5896.
10. British Standards Institution. *Specification for hot rolled steel bars for the reinforcement of concrete.* BSI, London, 1978 (1984), BS 4449.
11. British Standards Institution. *Specification for cold worked steel bars for the reinforcement of concrete.* BSI, London, 1978 (1984), BS 4461.
12. British Standards Institution. *Structural use of concrete.* BSI, London, BS 8110, Part 1, 1985; Part 2, 1985; Part 3, 1985.
13. Curtin W.G. *et al. Structural masonry designers' manual,* 2nd edn. Blackwell Scientific, Oxford, 1987.
14. British Standards Institution. *Structural use of concrete.* BSI, London, 1970, CP 110 (replaced by BS 8110).
15. Powell B. and Hodgkinson H.R. The determination of stress/strain relationship of brickwork. *Proc. 4th Int. Brick Masonry Conf., Bruges,* 1976, pp 2.A.5–2.A.5-5.
16. Turnsek V. and Cacovic F. Some experimental results on the strength of brick masonry walls. *Proc. 2nd Int. Brick Masonry Conf., Stoke-on-Trent,* 1971, pp 149–156.
17. Lenczner D. Creep in brickwork. *Proc. 2nd Int. Brick Masonry Conf., Stoke-on-Trent,* 1971, pp 44–49.
18. Lenczner D. Creep and prestress losses in brick masonry. *Struct. Engr,* 1986, **64B**, 57–62.
19. Timoshenko S.P. *Strength of materials.* Van Nostrand, Wokingham, 1969.
20. Wood R.H. and Simms L.G. *A tentative design method for the composite action of heavily loaded brick panel walls supported on reinforced concrete beams.* Building Research Establishment, Garston, 1969, CP 26/69.
21. British Standards Institution. *Structural recommendation for loadbearing walls.* BSI, London, 1970, CP 111 (replaced by BS 5628, Parts 1 and 2).
22. Wood R.H. *Studies in composite construction,* part 1. The composite action of brick panel walls supported on reinforced concrete beams. HMSO, London, 1952. National Building Studies, Research paper 13.

23. British Standards Institution. *The structural use of reinforced concrete in buildings.* BSI, London, 1969, CP 114.
24. British Standards Institution. *Specification for calcium silicate (sandlime and flintlime) bricks.* BSI, London, 1978, BS 187.
25. British Standards Institution. *Specification for clay bricks.* BSI, London, 1985, BS 3921.
26. British Standards Institution. *Specification for water repellents for masonry surfaces.* BSI, London, 1984, BS 6477.
27. British Standards Institution. *Code of practice for stone masonry.* BSI, London, 1976 (1984), BS 5390.
28. British Standards Institution. *Specification for cold reduced steel wire for the reinforcement of concrete.* BSI, London, 1985, BS 4482.
29. British Standards Institution. *Specification for steel fabric for the reinforcement of concrete.* BSI, London, 1985, BS 4483.
30. British Standards Institution. *Specification for wrought steels for mechanical and allied engineering purposes.* BSI, London, BS 970, Part 1, 1983.
31. British Standards Institution. *Specification for hot dip galvanized coatings on iron and steel articles.* BSI, London, 1971 (1986), BS 729.
32. British Standards Institution. *Specification for ordinary and rapid-hardening Portland cement.* BSI, London, 1978, BS 12.
33. British Standards Institution. *Specification for Portland blastfurnace cement.* BSI, London, 1973, BS 146, Part 2.
34. British Standards Institution. *Specification for sulphate-resisting Portland cement.* BSI, London, 1980, BS 4027.
35. British Standards Institution. *Specification for ready mixed building mortars.* BSI, London, 1981 (1986), BS 4721.
36. British Standards Institution. *Specification for pigments for Portland cement and Portland cement products.* BSI, London, 1975 (1986), BS 1014.
37. British Standards Institution. *Specification for building limes.* BSI, London, 1972, BS 890.
38. British Standards Institution. *Methods for specifying concrete, including ready mixed concrete.* BSI, London, 1981, BS 5328.
39. British Standards Institution. *Testing concrete.* BSI, London, BS 1881.
40. British Standards Institution. *Specification for metal ties for cavity wall construction.* BSI, London, 1978, BS 1243.
41. British Standards Institution. *Specification for materials for damp-proof courses.* BSI, London, 1970, BS 743.
42. British Standards Institution. *Lattice towers and masts.* BSI, London, 1986, BS 8100.
43. British Standards Institution. *Concrete admixtures.* BSI, London, 1982, BS 5075.
44. British Standards Institution. *Mortar admixtures.* BSI, London, 1986, BS 4887.
45. British Standards Institution. *Code of practice for walling.* BSI, London, 1951, CP 121 (withdrawn).
46. Building Research Establishment. *Driving rain index.* BRE, Garston, 1976.
47. Construction Industry Research and Information Association. *Buildability—an assessment.* CIRIA, London, 1983, Report SP 26.

Index